THE ECONOMICS OF COMPETITION IN THE TELECOMMUNICATIONS INDUSTRY

CHARLES RIVER ASSOCIATES INCORPORATED

Charles River Associates, located in Boston, Massachusetts, has been providing research and consulting support to decision makers in industry and government since 1965. CRA's work covers a wide spectrum, including the following areas:

Antitrust policy
Commodity-market forecasting
Combined economic/engineering feasibility for new ventures
Communications
Consumer behavior
Economic development
Fuel industries, electric power, and energy economics
Industry regulation
International trade
Minerals, metals, and durable-goods industries
Regional economics
Science and technology policy
Transportation planning
Urban and intercity transportation economics

The Economics of Competition in the Telecommunications Industry

John R. Meyer
Robert W. Wilson
M. Alan Baughcum
Ellen Burton
Louis Caouette

A Charles River Associates Research Study

 Oelgeschlager, Gunn & Hain, Publishers, Inc.
Cambridge, Massachusetts

International Standard Book Number: 0-89946-056-9
Library of Congress Catalog Card Number: 80-17414

Printed in the United States of America

Library of Congress Cataloging in Publication Data
Main entry under title:

The Economics of competition in the telecommunications industry.

  Bibliography: p.
  1. Telecommunication—United States. 2. Telephone—
United States. I. Meyer, John Robert.
HE7775.E28 1980   384'.065'73   80-17414
ISBN 0-89946-056-9

Contents

List of Tables   vii
List of Figures   xiii
Preface   xv

Chapter 1   **Introduction: The Industry in Perspective**   1
            Antecedents   1
            The Regulatory Problem   3
            The Policy Issues   10
Chapter 2   **The Impact of Growth**   23
            Historical Competition and Growth in the
                Telephone Industry   23
            Potential Growth for the Telephone Industry   38
            Competition and Diversion   42
            Forecasting the Impact of Competition and
                Growth on AT&T's Local Service Rates and
                Financial Performance   49
            Summary and Conclusions   68
Chapter 3   **Pricing**   75
            Goals and Criteria   75
            The Present Structure of Telephone Tariffs   81
            Alternative Rate Structures   89
            Summary   101
Chapter 4   **Natural Monopoly and Economies of Scale**   111
            The Natural Monopoly Issue   112
            Econometric Evidence on Economies of Scale   125
            Engineering and Simulation Studies of Scale
                Economies in Long Distance
                Telecommunications   134
            The Impact of Competition on Future
                Telecommunications Costs in Comparison
                with the Direct Costs of Regulation   143
            Summary   146
Chapter 5   **Market Structure and Regulation**   155
            Market Structure and Technological Innovation   158
            Extending Service to Rural and Other Low-
                Density Areas   172

Regulation and the Creation of "Preferred"
Market Structures: Local Telephone Service as
a Case Example 184
Summary 193
Chapter 6 Public Policy 201
Regulation and the Conventional Wisdom 201
Entry 202
Rate Regulation 207
Predatory Pricing and Uneconomic Entry 210
Toward a More Rational Domestic
Telecommunications Policy: Summary 217
**Appendix A Distributional Impact of an Alternative
Rate Structure** 225
Background Data and Information 226
Overall Change in Family Expenditures for
Telephone Service 238
Welfare Change as Measured by Price Indices 238
**Appendix B Sources and Methods Used in Projecting
AT&T Financial Performance Under
Various Hypotheses** 245
Description of Historical Financial Ratios 245
Pro Forma Financial Statements, 1975–1985 253
Scenarios and Results for Projection Years 261
Comparative Results: Testing the Validity of
the Financial Ratios Used in the Projections 268
Appendix C Alternative Rate Structures 299
**Appendix D Econometric Studies of Economies of
Scale in Telephone Systems** 317

Index 331

About the Authors 343

List of Tables

1–1   Bell System Operating Revenues, 1950–1978   4

1–2   AT&T Estimates of the Costs of Providing Interstate
Private Line Service (High Capacity Transmission
Facilities), 1971–1991   11

1–3   AT&T Summary of Results by Service Categories,
1976   14

2–1   Key Periods and Dates in Telephone Industry History   24

2–2   Historical Patterns of Telephone Industry Growth   30

2–3   Revenues and Expenses of Bell Operating
Companies, 1895–1913   34

2–4   Bell System Growth Rates Before and After the 1968
Carterfone Decision   37

2–5   Recent U.S. Growth Experience, 1970–1976   39

2–6   Potential Growth in the Volume of Household
Telephone Usage, 1977–1990, Based on Usage by
Households in 1977   41

2–7   Potential Telecommunication Volume Growth for
Intercity Transmission in 1990   41

2–8   Current and Potential Revenues of AT&T
Competitors   47

2–9   Bell Revenues Assumed Vulnerable to Erosion by
Rate Reduction, 1975   48

2–10	Annual Bell System Growth Rates, 1973–1975	51
2–11	General Economic and Telecommunications Industry Growth Rates, 1973–1975	51
2–12	Experienced Growth Rates for Various Elements of AT&T, 1971–1977	52
2–13	Impact of Competition on AT&T's Financial Performance in 1985	54
2–14	Telephone Company and General Economic Rate Indexes for 1985 under Different Financial and Development Scenarios	59
3–1	Sample of Bell Local Exchange Rates	85
3–2	Telephone Revenues, 1975	93
3–3	Alternative Rate Structure Parameters and Expected Revenue for Local Telephone Service	95
3–4	Local Residential Demand	96
3–5	Alternative Rate Structure Parameters and Expected Revenue for Long Distance Telephone Service	97
3–6	Comparison of Initial and Alternative Prices and Volumes of Message Toll Minutes Assuming High Elasticity and Moderate Capacity	99
3–7	Total Revenue under Selected Rate Structures	100
3–8	Comparison of Selected Price Structures with Littlechild and Mitchell Studies	101
4–1	Point Estimates of Scale Elasticity in Telecommunications Systems from Econometric Studies	129
4–2	Statistical Significance of Mantell's Scale Elasticity Estimates	130
4–3	Relation Between Scale Elasticity and Average Cost Elasticity	136
4–4	Book Cost of Long Lines Plant and Equipment, 1975	139
4–5	AT&T Multiple Supplier Network Study: Cost of Providing Interstate Private Line Demand (Excluding OCC Investment Cost Prior to 1977)	141
4–6	Percent Increase in Telephone System Total Costs in 1985 under Various Circumstances	144
5–1	U.S. Telephone Industry Operating Revenues and Number of Telephones, 1976	177
5–2	Comparison of Operating Expenses per Telephone for Bell System and Independent Telephone Companies, 1976	178
5–3	Cross-Section Cost Estimates for Bell System and Independent Companies, 1976, Logarithm Forms	180

5–4	Distribution of Total Operating Expenses for Bell Companies as Compared to Independents, 1976	183
5–5	Distribution of Interstate Long Distance Billing	186
5–6	Proportion of Interstate Business MTS and WATS Revenues Generated by Largest Metropolitan Areas	187
5–7	Cable Television Coverage in the 25 Largest U.S. Cities	190
A–1	Telephone Service Rates and Revenues in 1975 and under Rate Structure A–5.	227
A–2	Telephone Rates and Revenues in 1975 and under Rate Structure A–5 Deflated to 1972–1973	230
A–3	Telephone and Consumption Expenditures by Income Class, 1972–1973	232
A–4	Average Residential Expenditures for Local Telephone Service by Income Class: Initially and under Rate Structure A–5.	234
A–5	Residential Expenditures for Toll Telephone Service by Income Class: Initially and under Rate Structure A–5.	236
A–6	The Change in Distributed Business Telephone Service Expenditures from the Existing Rate Structure to Rate Structure A–5.	239
A–7	Estimated Increase (+) or Decrease (−) in Total Annual Expenditures by a Family for Telephone Services under Rate Structure A–5, by Income Class	240
A–8	Index Numbers as Indicators of Welfare Change	241
B–1	Financial Ratios Used in AT&T Pro Forma Income Statements, 1971–1977	247
B–2	Financial Ratios Used in AT&T Pro Forma Balance Sheets, 1971–1977	249
B–3	Modifications to AT&T's Embedded Direct Cost (EDC) Study for 1975	254
B–4	Sample AT&T Pro Forma Income Statement, 1975–1985	255
B–5	Sample AT&T Pro Forma Balance Sheet, 1975–1985	258
B–6	Historical A Scenario AT&T Pro Forma Income Statement, 1975–1985	270
B–7	Historical A Scenario AT&T Pro Forma Balance Sheet, 1975–1985	271
B–8	Historical B Scenario AT&T Pro Forma Income Statement, 1975–1985	272

B–9   Historical B Scenario AT&T Pro Forma Balance
       Sheet, 1975–1985   273
B–10   Historical C Scenario AT&T Pro Forma Income
       Statement, 1975–1985   274
B–11   Historical C Scenario AT&T Pro Forma Balance
       Sheet, 1975–1985   275
B–12   No-Inflation Scenario AT&T Pro Forma Income
       Statement, 1975–1985   276
B–13   No-Inflation Scenario AT&T Pro Forma Balance
       Sheet, 1975–1985   277
B–14   Nominal A Scenario AT&T Pro Forma Income
       Statement, 1975–1985   278
B–15   Nominal A Scenario AT&T Pro Forma Balance
       Sheet, 1975–1985   279
B–16   Nominal B Scenario AT&T Pro Forma Income
       Statement, 1975–1985   280
B–17   Nominal B Scenario AT&T Pro Forma Balance
       Sheet, 1975–1985   281
B–18   Nominal C Scenario AT&T Pro Forma Income
       Statement, 1975–1985   282
B–19   Nominal C Scenario AT&T Pro Forma Balance
       Sheet, 1975–1985   283
B–20   Nominal D Scenario AT&T Pro Forma Income
       Statement, 1975–1985   284
B–21   Nominal D Scenario AT&T Pro Forma Balance
       Sheet, 1975–1985   285
B–22   Nominal E Scenario AT&T Pro Forma Income
       Statement, 1975–1985   286
B–23   Nominal E Scenario AT&T Pro Forma Balance
       Sheet, 1975–1985   287
B–24   Nominal E Scenario Deferred Credits Worksheet,
       1975–1985   288
B–25   Comparison of "Actual" and "Projected" AT&T
       Income Statement for 1976   289
B–26   Comparison of "Actual" and "Projected" AT&T
       Balance Sheet for 1976   290
B–27   Comparision of "Actual" and "Projected" AT&T
       Income Statement for 1977   292
B–28   Comparison of "Actual" and "Projected" AT&T
       Balance Sheet for 1977   293
B–29   Comparison of "Actual" and "Projected" AT&T
       Income Statement for 1978   295

B–30   Comparison of "Actual" and "Projected" AT&T
Balance Sheet for 1978   296

C–1   Estimates of Price Elasticities for Telephone Service
Categories   300

C–2   Telephone Statistics, 1975   301

C–3   Local Charges, Volume, and Revenue for Telephone
Services   303

C–4   Initial Prices and Volumes of Message Toll Minutes   305

C–5   Daytime Intrastate Toll Rates for Selected U.S. Cities
and States   306

C–6a   Estimated Prices and Volumes of Message Toll
Minutes Assuming Low Elasticity and Low
Capacity   307

C–6b   Estimated Prices and Volumes of Message Toll
Minutes Assuming Low Elasticity and Moderate
Capacity   308

C–6c   Estimates Prices and Volumes of Message Toll
Minutes Assuming Low Elasticity and Low
Capacity, and Maintaining Initial Revenue Level   309

C–6d   Estimated Prices and Volumes of Message Toll
Minutes Assuming High Elasticity and Low
Capacity   310

C–6e   Estimated Prices and Volumes of Message Toll
Minutes Assuming High Elasticity and Moderate
Capacity   311

C–7   Alternative Rate Structure Parameters and Expected
Revenue for Local and Long Distance Telephone
Service   312

C–8   Comparison of Price Structures with Littlechild and
Mitchell Studies   313

D–1   Summary of Econometric Estimates of Scale
Elasticities for Telephone Systems as a Whole,
Estimated from Aggregate Production or Cost
Functions   318

List of Figures

1–1   Telephone Service Rates versus Price and Wage
        Changes, 1940–1972   3
1–2   A Partitioning of 1975 Costs and Revenues of AT&T   5
2–1   Bell System Telephones, 1876-1976   32
2–2   Reduction in the Average Yearly "Bell" Exchange
        Revenue Derived from Certain Groups of Cities
        Arranged in Five-Year Periods, 1894–1909   33
2–3   Bell Revenue, Expenses, and Earnings per Telephone,
        1885–1976   35
2–4   Summary of Assumptions Used in AT&T Financial
        Projection Scenarios   50
2–5   AT&T Financial Performance for Various Scenarios,
        1975–1985: Return on Equity—Historical   55
2–6   AT&T Financial Performance for Various Scenarios,
        1975–1985: Interest Coverage Ratio—Historical   56
2–7   AT&T Financial Performance for Various Scenarios,
        1975–1985: Debt-to-Capitalization Ratio—Historical   57
2–8   AT&T Telephone Rate Indexes for 1985 under Various
        Scenarios   60

2–9   Embedded Direct Cost Study Bar Chart
Representation of the Historical B Scenario Results
for the Year 1985   61

2–10   Embedded Direct Cost Study Bar Chart
Representation of the Historical C Scenario Results
for the Year 1985   62

2–11   AT&T Financial Performance for Various Scenarios,
1975–1985: Return on Equity—No Inflation and
Nominal   63

2–12   AT&T Financial Performance for Various Scenarios,
1975–1985: Interest Coverage Ratio—No Inflation
and Nominal   64

2–13   AT&T Financial Performance for Various Scenarios,
1975–1985: Debt-to-Capitalization Ratio—No
Inflation and Nominal   65

3–1   How Telephone Services Is Paid For   78

3–2   Toll Conversation Minutes, New York Telephone
Company, 1975   84

4–1   Costs and Benefits of Perfect Regulation   113

4–2   Strong Economies of Scale   114

4–3   Slight Economies of Scale   115

4–4   Economies of Scale Exhausted at Less Than Half the
Size of the Market   116

4–5   Monopoly Pricing with Low Elasticity of Demand   118

4–6   Monopoly Pricing with High Elasticity of Demand   119

4–7   Monopoly Pricing Constrained by Alternative
Technology   121

4–8   Illustrative Examples of Curves Fit by Statistical
Methods on Different Assumptions about the Process
Generating the Observations   133

4–9   Investment Cost for New Terrestrial Transmission
Systems   136

5–1   President's Task Force Estimates on Land Mobile
Radio Transmitters   188

D–1   Residuals from Regression   328

Preface

      This book is the result of a study undertaken by Charles River Associates Incorporated (CRA) beginning in the summer of 1977 for the International Business Machines Corporation (IBM). The goals were (1) to analyze the economic forces (competitive, regulatory, technological, etc.) at work in the telephone industry; (2) to analyze the impact, if any, of increased competition on both the telephone industry and telephone customers; and (3) to identify appropriate public policies for the industry.

      In pursuing these goals, the authors not only reviewed the literature and performed certain original analyses, as detailed in this book, but also met on a continuing basis with a small group of IBM technical personnel. They were particularly helpful in briefing the authors and providing them with information about various technical options and changes confronting the telecommunications industry; in their review of the draft manuscripts, they also brought an extra dimension of reality and insight to particular policy problems. In addition to the people from IBM, other regular participants in the review of the draft manuscripts were John McGowan and Robin Landis of CRA; they not only made many helpful comments and critiques of numerous early drafts, but also had primary roles at CRA in coordinating the entire study. William K. Jones, Milton Handler Professor of Trade Regulation at the Columbia University School of Law, also made helpful comments and critiques of earlier drafts. J. Walter Hinchman, former Chief of the Common Carrier Bureau of the Federal Communications Commission, reviewed in detail an early rough draft of the manuscript; unquestionably, this book is much improved because of his helpful suggestions and thoughtful analyses. Finally, John Lintner, Thomas McCraw, and Steven Breyer of Harvard University read early drafts and provided the authors with many useful comments and evaluations that, again, have been translated into substantive improvements in the book itself.

The conclusions and views expressed in this book, nevertheless, are strictly those of the authors. Indeed, even this statement may be too inclusive because many compromises and discussions among the authors were necessary to create the policy conclusions in the final chapter. In the final analysis, when choices had to be made among different policy options, they were essentially made by me, in my role as senior author. Accordingly, any blame, retribution, complaints, brickbats, or other paraphernalia of dissent should also be directed to the senior author. Certainly, none of the institutional participants in this study should be held accountable for any crime other than facilitating the researchers' task. I am sure, in fact, that some of the policy conclusions stated herein are not accepted by everyone at CRA, let alone IBM. In defense, I would only observe that, as with so many public policy debates, only a very innocuous set of conclusions or recommendations would have achieved a broad consensus.

While absolving my colleagues, as well as others, of blame for any shortcomings or policy disagreements, I nevertheless am very indebted to my coauthors for their many contributions. Robert Wilson was responsible for the pricing and many of the costing studies. Alan Baughcum assumed the leading role in developing historical, income incidence and market structure materials. Ellen Burton participated in several of the different research activities and was responsible for the basic drafts on econometric costing. Louis Caouette assumed primary responsibility for the financial and technological analyses.

In addition to all of these contributions, important support activity was rendered by several other people. In particular, Lindsay Noble provided research assistance for virtually every aspect of the work, sometimes at a level that closely approached actual authorship. In addition, John Horning, acting as a research associate, did significant original work in synthesizing information on telephone service problems of rural areas. Judy Chen and Bonnie Olson assisted in many areas, including making the calculations for various alternative rate structures in Chapter 3.

Numerous individuals assisted the authors and research staff in producing both intermediate and final drafts of this manuscript. Acknowledging gratitude is an insufficient reward for the many months of work their assistance represents. Eleanor Lintner and Sally Markham at Harvard typed numerous drafts and redrafts, hunted down citations and missing data, and generally oversaw "quality control" for the work. At CRA, Linda Carroll, Harriet Ullman, Priscilla Gebre-Medhin, and John Connolly provided patient and thorough editorial assistance. Karin Pawlowski supervised the production of at least four drafts of this book, typed by her and Susan Parker, Kathy Davenport, Denise ConSales, and Robert Scheier. Lynne Larson's typing and coordinating assistance was invaluable, in particular during the first draft of the manuscript. Joanne Doughty of CRA's Data Processing Department was of great assistance in helping to generate the data in the pro forma tables in Appendix B. Assistance in preparing the graphs and figures in this book (as well as additional work for various presentations of preliminary results) was ably provided by Sharon Nathan, Jean Fried, and Nicole Harris.

Although many pages are used in this book to "document" the case, the fundamental conclusion can be stated succinctly: Let a free market decide the truth of present speculations about the size, nature, and direction of the telecommunications industry; if these decisions are made in the marketplace, instead of in a regulatory proceeding, *both* the industry and its customers will almost certainly be better served.

Implementing such a policy, of course, may prove difficult. Specifically, for reasons explained in this book, transition problems of some magnitude may be encountered in moving from the structured and regulated world characteristic of most of the telecommunication industry today to the more open and competitive environment that looms on the horizon. This competitive environment, incidentally, is likely to occur in one form or another regardless of public policy. Indeed, the real policy question in this complicated, rapidly changing, technologically fascinating industry is not so much whether competition will reign, but rather in what form and how quickly.

John R. Meyer
Harvard University

Chapter 1

Introduction: The Industry in Perspective

ANTECEDENTS

The U.S. telephone industry is the product of many years of development and evolution. This description pertains not only to the industry's technological characteristics, its industrial and corporate relationships, but also to its pricing policies and regulatory status. From all this history, a set of relationships has emerged with important implications for public policy and for the future growth of such various services as communications, data processing, electronics, entertainment, and education.

The telephone network itself was started in the latter part of the last century with the creation of a number of independent community telephone systems.[1] After several individual community networks were solidly established, the next and quite natural step was to tie these together to offer a service that permitted customers in one community to speak to those in another. The Bell System saw the potential in such service very early, starting its "long lines" department in

1895; indeed, the creation of a department offering long distance service was also the occasion for creating American Telephone & Telegraph (AT&T) itself as a specific corporate entity. This early process of tying one community to another was apparently relatively expensive, given the technology then available. The rates charged for early long distance calls reflected these costs as well as the fact that the industry was largely unregulated and devoid of competition except from mail and telegraph, which were imperfect substitutes. Accordingly, the market for early long distance phone service was sufficiently robust to tolerate charges that were more than adequate to cover the relatively high costs of such service.

As the years went by, the rate of technological improvement, and therefore cost reduction, in long distance service proved quite remarkable. The economies achieved in long distance service have proceeded at a far faster pace than they have in local telephone service. Long distance rates have adjusted to this technological improvement, although only with a considerable lag. In essence, interstate rates, after some early reductions, have been more or less constant in current dollar terms over recent decades. Since other prices in the economy were generally increasing, this has meant that long distance telecommunication has become less costly in real terms and an increasingly "good buy" for consumers, as shown in Figure 1-1. Spurred on by these relatively low rates, long distance service has expanded and become an ever more important source of income to the telephone companies. Between 1950 and 1978, for example, toll service revenues increased almost 18 times, while local service revenues were up by a multiple of about 10, as shown in Table 1-1. This translates into an actual compound growth rate for toll services of 10.8 percent, compared with only 8.4 percent for local services.[2]

Some of the proceeds from long distance service have apparently been used to keep charges for local services lower than they might otherwise have been. Certainly, as shown in Figure 1-1, local telephone rates did not go up as rapidly as most other prices in the economy, although they did rise more rapidly than long distance rates and recently have shown a tendency to keep closer pace with general inflation. Indeed, the telephone industry has argued that had the positive contributions (i.e., revenues less "direct" costs) generated in vertical service (extension phones, PBX, Centrex) and toll service not been available for covering joint and common costs, residential and business local rates would have required a 70-percent increase as of 1973. Furthermore, the industry asserts that this percentage has been rising each year. The required percentage increase in local rates to cover

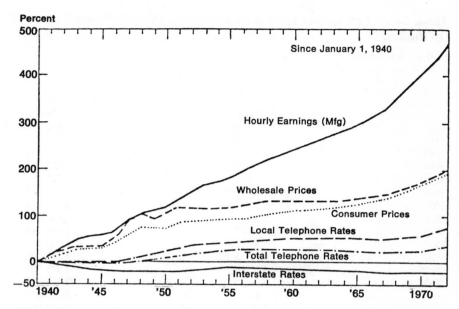

Figure 1–1. Telephone service rates versus price and wage changes, 1940–1972. *Source:* Paul J. Berman and Anthony G. Oettinger, "The Medium and the Telephone: the Politics of Information Resources" (Cambridge, Mass.: Harvard University Program on Information Technologies and Public Policy, June 1976). Taken from U.S. Congress, Senate Committee on the Judiciary, Subcommittee on Antitrust and Monopoly, Hearings on S. 1167, Part 2: The Communications Industry, July 30 and 31; August 1 and 2, Data submitted by American Telephone and Telegraph Company, p. 1721.

all joint and common costs would have been 72 percent by 1974 and almost 79 percent by 1975.

THE REGULATORY PROBLEM

A pattern may thus have emerged in which the substantial profits generated by long distance service have been used to keep tariffs low on local services. The exact extent and character of what some call "cross-subsidization" and others "contribution" from long distance to local service has been the subject of many intense debates

Table 1–1. Bell System Operating Revenues, 1950–1978 (thousands of dollars)

Year	Local Service Inter-state	Local Service Intra-state	Local Service Total	Toll Service Inter-state	Toll Service Intra-state	Toll Service Total
1950	—	—	$1,941,369	—	—	$1,184,655
1951	—	—	2,146,439	—	—	1,342,915
1952	—	—	2,397,552	—	—	1,470,267
1953	—	—	2,642,905	—	—	1,571,088
1954	—	—	2,836,958	—	—	1,720,742
1955	—	—	3,086,455	—	—	1,959,667
1956	—	—	3,368,608	—	—	2,176,241
1957	—	—	3,647,596	—	—	2,357,650
1958	—	—	3,944,443	—	—	2,490,649
1959	—	—	4,250,778	—	—	2,786,144
1960	—	—	4,547,409	—	—	2,996 436
1961	—	—	4,797,528	—	—	3,217,300
1962	—	—	5 088,488	—	—	3,471,787
1963	—	—	5,389,702	—	—	3,737,146
1964	—	—	5,633,732[a]	—	—	4,205,484[a]
1965	—	—	5 961,279	—	—	4,613,708
1966	—	—	6,354,655	—	—	5,274,390
1967	—	—	6,737,734	—	—	5,737,866
1968	—	—	7,184,077	—	—	6,341,158
1969	—	—	7,774,392	—	—	7,297,842
1970	—	—	8,455,967	—	—	7,874,069
1971	—	—	9,186,952	—	—	8,650,009
1972	10,393	10,351,645	10,362,038	6,323,472	3,447,208	9,770,680
1973	10,218	11,407,408	11,417,626	7,269,758	4,007,781	11,277,538
1974	10,749	12,801,059	12,811,808	7,923,505	4,536,307	12,459,813
1975	9,473	14,017,289	14,026,762	8,849,367	5,074,567	13,923,934
1976	10,781	15,597,002	15,607,783	10,049,091	6,014,917	16,064,008
1977	10,545	17,059,060	17,069,605	11,206,725	6,885,476	18,092,201
1978	—	—	18,684,609	—	—	20,770,263

Source: American Telephone and Telegraph Company, *Bell System Statistical Manual: 1950-1977,* May 1978, p. 202. Revenue Data for 1978 taken from American Telephone and Telegraph, 1978 *Annual Report.*

[a]Reflects reclassification prescribed by Federal Communications Commission of certain private line revenues from local to toll effective January 1, 1964.

before regulatory commissions and other forums, with the question of whether it is called a cross-subsidy or a contribution depending to a considerable degree on how direct costs are determined and allocated.[3] At one extreme (see **Figure 1–2**) AT&T claims that about $7 billion is contributed from long distance and vertical services to underwriting what they consider common and joint costs of all telephone services. Few outside the industry argue that no such contribution or cross-subsidization whatsoever occurs; many outsiders, however, insist that the pattern is a good deal more complex than that implied by the data in Figure 1–2. For example, it can be argued that the so-called common costs in Figure 1–2 are what most firms call overhead and that some of these costs vary in a meaningful and traceable fashion with different types of services rendered. Some analysts also contend that the joint costs (mainly for instruments and local loops) can be separated. For example, a purely local system could require less conditioning and in many locales would need only a four-digit switching capability; on this basis, it is argued, several billion dollars in joint costs should be assigned directly to long distance, considerably reducing any contribu-

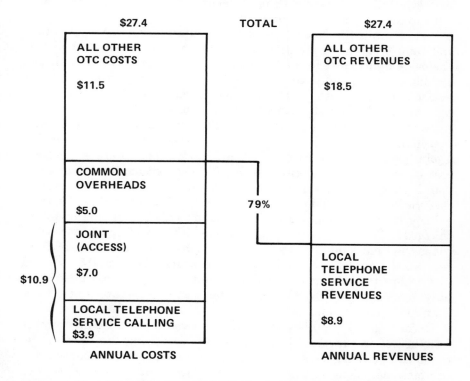

Figure 1–2. A Partitioning of 1975 Costs and Revenues at AT&T (billions). *Source:* American Telephone and Telegraph Company, "Embedded Direct Costs (EDC) Study, 1975," Third Supplemental Response to FCC Docket 20003, July 12, 1976, p. 6.

tion from this service.[4] Some critics of the telephone industry position, in fact, assert that long distance may impose substantial direct costs on local exchange networks that properly should be reimbursed from toll contributions. According to this reasoning, if all long distance services, including any competitive services, were obliged to compensate the local exchange network for all costs properly attributable to such services, the net amount might well equal or exceed the so-called toll contributions presently received. If this were true, no increase in local exchange rates would be required as a consequence of competition diminishing long distance revenues. Furthermore, some insist that important transfers or cross-subsidies occur within each of the categories (e.g., from high-density residential areas to lower-density suburban or exurban locations) and without knowing the specific details of these transfers no one can conclude about the scope, let alone the desirability, of these pricing patterns and any attendant contributions to covering common and joint costs.

Establishing the "truth" among this bewildering array of claims and counterclaims is greatly complicated by the fact that there are obviously many costs in the telephone network that are jointly shared by long distance and local services. The joint costs, particularly costs associated with creating and servicing the basic local loop between local central offices and residences or business locations, are troublesome to allocate in an economically efficient manner.[5] It is difficult therefore to assert with any certitude that a particular service sharing a joint production process really bears its "correct" share of the total cost burden. In such circumstances, the conventional economic wisdom is that prices charged should be related more to demand than to cost characteristics. The underlying principle is to use value-of-service pricing in which higher markups over directly assignable costs are charged to users who seemingly can least afford to forgo services (because the final amounts of services consumed will vary the least from services consumed at a level of usage in which rates equal "marginal cost").[6] For the telephone, this has meant in practice that businesses have tended to pay higher charges for a giving quantum of service than have most residential users.[7]

Unfortunately, application of this conventional wisdom may be complicated in practice by difficulties in identifying the relevant demand elasticities and cross-elasticities. An additional, and even more troublesome, long-running difficulty with all value-of-service ratemaking schemes is that customers being charged high markups do not, for easily understandable reasons, take kindly to this role. Customers bearing the higher markups obviously have strong incentives to seek and explore other alternatives for meeting their needs.[8] These explorations often prove fruitful in a real world of continuing changes in technology, market relationships, locational choices, trade patterns, and so on. In short, value-of-service rate making assumes static technology and unvarying tastes and market relationships; these assumptions are unlikely to be met in a constantly developing and envolving market economy. This limitation, incidentally, has always been prevalent. Value-of-service rate making, the basic pattern for railroad tariffs at the turn of the century, was a system under which so-called high-value goods were charged high markups over costs while low-value goods were either transported at a loss or charged very small markups; this arrangement was largely ended by the development of trucks and highways and the relocation of industry in the decades after 1920.

A new technological alternative that challenges established value-of-service rates may or may not be a truly low-cost technology. All that is necessary when price discrimination prevails is that the new entrants costs be lower than the value-of-service rates; it is *not* necessary that the entrant's costs be lower than the original vendor's costs. Of course, once the original vendor recognizes that the established business is being eroded by new entrants, a natural reaction is to attempt to cut rates so that they are brought more into line with costs. Often (as in some cases of rail-truck competition) the original vendor's costs may actually be lower than those of the new entrants; then, the original supplier may be able to drive off the new entrant through rate reductions if the vendor is unimpeded by regulatory or other market forces. Needless to say, if the new entrant has become established before the old vendor recognizes or understands the threat, and if there is governmental participation in the market through regulation or otherwise, the recent entrant generally will appeal to the government to prevent pricing retaliation by the original vendor. The entrant can base an appeal for regulatory or political protection on several different grounds. For example, the new might contend that the old vendor really does not have the lowest costs but is simply engaging in below-cost, "predatory" pricing to drive the new competition from the market; or the entrant may simply argue that the political process owes protection to new competitors since they made their investments in good faith on the assumption that regulation would maintain a certain "price umbrella" so that they could successfully recover their investment. As an illustration, acceptance of similar arguments has apparently led the Interstate Commerce Commission (ICC) to reject attemps by the railroads to use rate reductions to recover business from the trucks over the past two or three decades.[9]

This classical pattern of value-of-service rates being eroded by technological change and increased unwillingness of the customers paying the high markups to quietly accept their fate has clearly occurred in the communications field. Businesses, increasingly reliant on the communication of large amounts of information to conduct their activities, seem more and more intrigued by new technological possibilities for meeting these needs outside the established telephone system. As a result, the established rate and contribution patterns in the telephone industry are coming under attack and are increasingly difficult to sustain.

These difficulties with an increasingly out-of-date or technologically challenged rate structure may be further compounded for the telephone companies by the standard regulatory practice of determining "fair"profit for the industry on the basis of rate-of-return concepts. Under rate-of-return regulation, profits are considered adequate or fair if they equal a prescribed percentage of the investment or capital base of the industry. A great deal of time and effort has gone into defining what should be in the rate base and what constitutes a fair percentage or rate of return.[10] Unfortunately, although rate-of-return regulation may seem to give some coherence and theory to the arcane art of economic regulation, it can also set up incentives for the firms under such regulation to behave in inefficient ways.[11] In particular, since the level of profit is tied by rate-of-return regulation to the size of the rate base, and since rate bases are generally related in some fashion to the amount of capital investment committed to the industry, business firms under rate-of-return regulation can have an incentive to use too much capital relative to labor or other productive inputs not normally included in the rate base.[12] However, inflation, coupled with a continuing lag by regulators in responding to changes in costs and other circumstances, may have greatly dissipated these perverse incentives in recent years.[13]

Rate-of-return regulation may also create special efficiency problems when some markets are opened to competition while others remain franchised monopolies. A regulated multimarket firm may be able to successfully, and profitably, monopolize both its franchise monopoly market and the competitive market even when the technology employed in jointly supplying both markets is technically inefficient.[14] That is, joint technology may exist that is more costly than independent technologies for supplying each market. Since the joint technology uses common inputs, the "incremental cost" of serving the competitive market may be lower than the cost of using the independent technology. Thus, even if independently serving both markets results in lower total cost, the franchised monopolist may be able to substitute a more costly joint technology that allows the monopolist both to pass an incremental cost test and to undercut any other firm's price in the competitive market. Again, the incentive for undertaking such socially inefficient activity follows from rate-of-return regulation. If the allowed rate of return exceeds the cost of capital so that more than adequate profits can be drawn from a franchised monopoly market, a regulated firm may find it profitable to serve additional markets even if an unregulated firm would find it unprofitable. By so doing, the size of the rate base increases, allowing total profits to increase, and the regulated firms can cover any "losses" from serving the competitive market with additional revenue drawn from the monopoly market.

Additionally, rate-of-return regulation may have led to underdepreciation of telephone equipment[15] with consequent impacts on the regulated firm's incentive to innovate.[16] The estimates of this underdepreciation for the U.S. telephone industry in the mid-70s range from $5.5 billion to $11.5 billion; based on 1976 end-of-year figures, this would represent from 6 to 12 percent of net plant for the industry.[17]

Rate-of-return regulation may also discourage the use of the best technology. Specifically, under rate-of-return regulation, firms may be reluctant to make investments that save capital because such an act would reduce the rate base, thereby also reducing the permissible level of profits. In general, rate-of-return regulation may tend to favor investments that save labor but not capital.[18] The sluggishness of telephone companies in adopting satellite technologies may be such a case since satellites can sometimes provide more long distance communications capacity for a given level of investment than can terrestrial systems such as wire, microwave relays, and coaxial cables.[19] On the other hand, the telephone companies would point out that many considerations other than cost should enter any determination of whether to use terrestrial or satellite systems for long distance service; among these factors are security, exposure to interruption from climatic conditions, and general quality and reliability.

If regulation inhibits the rate of technological progress in the industry, as it well might under rate-of-return regulation, any disparity between rates and costs for a particular service can become a further source of aggravation. A new entrant under such circumstances may actually see a way of not only eroding the value-of-service rates but also of introducing a new technology that results in cost economies in the particular service offered.

Rate-of-return regulation may also create perverse incentives for pricing the services of a regulated industry. As already noted, rate-of-return regulation in effect can make capacity appear cheaper under certain circumstances than it should (or would) without regulation. Capacity needs, in turn, are largely determined in service industries like telecommunications by peak period demands. Accordingly, with rate-of-return regulation an incentive may exist to cut peak period rates below optimal levels (i.e., below what they should be to justify additional capacity); by so doing, more peak demand is generated, additional investment in capacity is made, and the permissible level of profits is increased.[20] As a corollary, little incentive may exist to "balance" loads over the daily cycle by lowering off-peak rates; for example, if lower off-peak rates attract business from the peak period, as they well might, then they could postpone the need for additional capacity in a growing market, thereby postponing justification of additional profits (via augmentation of the rate base) under rate-of-return regulation.[21]

THE POLICY ISSUES

Obviously, value-of-service rate making in combination with rate-of-return regulation that inhibits or slows the introduction of the best technology for certain services can result in service offerings that may not be well aligned with market demand and cost conditions. Certainly, it is a combination that brings certain aspects of regulation into question and can lead to a great deal of instability. Under such circumstances, many of the participants have extremely strong incentives to change the status quo. In telecommunications this results in an almost constant stream of proposals and hearings before the Federal Communications Commission (FCC) to introduce new services or rate concepts.

The essential issues are aptly illustrated by AT&T's "Multiple Supplier Network Study,"[22] which attempts to estimate how much costs of certain long distance services might be increased by having several rather than only one (monopoly) supplier. In this study, AT&T calculates that the cost of high-capacity interstate private line services from 1971 to 1991 could be as much as four times higher if provided by several private vendors rather than exclusively by the Bell network. AT&T's findings are summarized in Table 1-2, which has been taken from this study (and which includes the summary footnote reproduced as part of AT&T's original table).

Not unexpectedly, the other vendors, actual and potential, do not agree. They dispute AT&T's cost findings, in the first place arguing that AT&T might not achieve such low costs unless spurred by competition. Some potential competitors also assert that even if their costs might be somewhat higher than Bell's, their services would be more than sufficiently superior to compensate for any additional costs incurred (again reminiscent of truck/rail confrontations). Still other potential competitors suggest that the AT&T cost calculations, to the extent that they are valid, rely on achievement of scale economies that AT&T sees as available to AT&T alone, whereas the truth is that the market will be so large that other entrants will also be able to achieve these economies. In general, a common rebuttal to AT&T is that its costs are extrapolations or guesses about an unknown future in which both total demands (and therefore scale economies) and future technological changes are unknown.[23]

Table 1–2. AT&T Estimates of the Costs of Providing Interstate Private Line Service (High Capacity Transmission Facilities), 1971–1991 (present value in millions of 1971 dollars)

Item	Cost of Service on OCC Network[a] (A)	Cost of Service as Part of Single B-1 Network[b] (B)	Difference (A) — (B) (A) — (B)
Construction cost	$882	$199	$683
Income tax	215	65	150
Operating cost	311	77	234
Total cost[c]	$1408	$341	$1067

Source: American Telephone and Telegraph Company, "Multiple Supplier Network Study: The Cost of Multiple Intercity Networks Compared to a Single Integrated Network," submitted as Bell Exhibit 57, FCC Docket No. 20003, July 12, 1976.

[a]OCC is "other common carrier," i.e., other than AT&T.

[b]Single B-1 network is one such that all long distance service is provided by AT&T lines on one integrated system.

[c]The estimated cost of multiple supply is the difference of $1067 million shown above.

The cost of meeting private line demand on the OCC network is four times as much as it would cost to meet that demand on the Bell network, i.e., $1408 million versus $341 million.

The basic elements in the long, tedious, and extensive exchanges between AT&T and its challengers can be seen in broad perspective by simply juxtaposing Figure 1-2 with Table 1-2. AT&T asserts that it can do anything that its challengers can do more cheaply than the challengers (see Table 1-2), but that the "need" to maintain the present relatively low and flat rate structure for local services through substantial contributions from long distance (see Figure 1-2) prevents the telephone companies from reflecting their cost advantage in lower tariffs on long distance calls. AT&T's challengers, on the other hand, insist that AT&T would never achieve these cost economies unless challenged by competitors and that they surely would never provide a similar quality of service or level of innovation unless so challenged.

The usual way of reconciling large-scale differences in perceptions of future demand and technological developments in a market economy is, of course, to subject these differences to the test of the market itself. Thus, if AT&T is correct and its challengers are wrong, AT&T could clearly establish this point by simply reducing its prices and driving its new competition out of business. The market test would resolve important but necessarily nebulous differences about the relative qualities of the different services to be offered and, relatedly, differences about rates of innovation and product improvement.

The telephone industry's major (announced) objection to such a market test is that this might undermine the present pattern of rates and contributions. The telephone industry has long made the argument that anything that eroded the contribution of long distance service would necessarily increase the amount that would have to be charged for local services. As John deButts, former Chairman of AT&T, has said:

> However, if Local Telephone Service were required to cover all Joint (Access) costs by itself, similar to the situation that existed prior to the institution of the station-to-station separations procedure, present local service revenues would have to be increased considerably. If other service categories were restructured and repriced so as to only cover their directly attributable costs, Local Telephone Service rates would have to be increased about 78 percent, on the average, if overall revenue needs are to be maintained.[24]

This position that large profits from long distance service are necessary to maintain the rate structure of local service has many weaknesses. To begin with, and perhaps most basically, the particular pattern of rates that has historically evolved from the development of the private telephone industry in the United States is not necessarily socially desirable. Actually, no one really knows exactly who does pay in some ultimate or final sense for the phone service under the present complicated pattern of separations and contributions.[25]

The current debate on incidence, for example, is largely in terms of contributions among service categories, for example, interstate versus basic exchange. For purposes of welare analysis, it is necessary to look to the user of the service—and the prices paid by specific users are not clearly determinable from looking at generalized service categories. To illustrate, business is essentially a legal construct and cannot "pay" costs (or taxes) in any ultimate sense. The management of the firm collects revenues and disburses them among the factors of production. Whatever rate is charged by the phone company to businesses must be reflected in the price to the consumer or the compensation paid to labor, capital, land, or management. The actual incidence of the payment depends on a variety of considerations such as demand and supply price elasticities, market structures, industry cost conditions, type of charge, size of the entity imposing the charge, and whether or not there are unrealized rents to be exploited by the imposition of the charge.[26] In the not too long run, "business" may merely shift the burden of any telephone charge to customers or to factors of production (labor, capital, land). Given any shifting of telephone charges by businesses, the net price to the user consists of the direct charges for the user's own use *plus* the indirect business usage paid for in purchases of output or in diminished factor prices. A similar argument would apply to governmental use of telephone services. The government cannot "pay"; it can only pass along its charges to taxpayers or factors of production. The final *net* price to consumers would thus include imputed shifts of government charges as well as any business shifts of such charges.

AT&T takes the position that a service is not subsidized if its revenues cover directly attributable costs and at least some part of joint and common costs.[27] Using this approach, the Bell System concludes that for 1975 *all* of its major service categories covered their directly attributable costs, as shown in Table 1-3 (reproduced from an AT&T exhibit presented to the FCC).[28] Thus each service category, by AT&T definitions, "contributed" something to covering common and joint costs.

This result is at least partly due to the very size of nonattributable costs as defined by AT&T. Of total costs, 43.7 percent were not attributed to any specific service category. As already noted, a rigorous long-run incremental cost analysis might well reduce the proportion of costs that are nonattributable. Even more likely, much of the overhead cost now considered unassignable by Bell might be expected to vary with changes in various service outputs or activities. It seems difficult to believe that all management costs are totally unrelated to volume. Indeed, a Stanford Research Insititute report commissioned by AT&T makes a good case for allocating some of these costs.[29] To illustrate what might be possible under strong regulatory and court pressures, these influences have managed to promote analyses that in recent years have decreased the percentage of costs classified as "institutional" (nonattributable) by the U.S. Postal Service from a majority to a minority of total Postal Service costs.[30] Another method of generating pressures to know costs more precisely is, of course, the opening of a market to competition.[31]

Table 1–3. AT&T Summary of Results by Service Categories, 1976 (millions of dollars)

(1) Service Category	(2) Revenues	(3) Costs	(4) Contributions (2) - (3)
Interstate/intercity[a]	$6,840	$3,170	$3,670
Intrastate other than basic exchange	11,680	8,370	3,310
Basic exchange	8,950	3,920	5,030
Joint costs	—	6,990	(6,990)
Common costs	—	5,020	(5,020)
Total	$27,470	$27,470	—

Source: American Telephone and Telegraph Company, "Embedded Direct Cost (EDC) Study, 1975," Third Supplemental Response to FCC Docket No. 20003, Bell Exhibit 1–B, July 12, 1976.

[a]Interstate/intercity services: The directly attributable costs included are the traffic-sensitive costs related to usage of the plant provided by the operating telephone companies for interstate MTS and WATS and the directly attributable costs associated with interstate private line service.

Moreover, nothing is automatic or inevitable about a reduction in the high profitability of some services resulting in a sharp increase in charges for other services.[32] There are many alternatives. For example, a cut in long distance rates does not necessarily mean a cut in long distance revenues; if demand is highly elastic, a rate cut might induce an increase in traffic more than sufficient to offset the rate reduction,[33] or in an economy characterized by general growth and productivity improvement, some decline in the yield to the phone company on long distance services might be offset by market growth elsewhere. The inevitability of rate increases for local service follows only in a very static world.

Therefore, it is significant that entry of new competitors into long distance communication services probably will not significantly arrest the telephone companies' growth, at least in the foreseeable future. One of the most rapidly growing components of telephone services has been residential long distance;[34] this fragmented market would seem largely unattractive to the newly proposed business-oriented long distance services to be rendered by specialized or "other" common carriers. Of the new entrants or proposed entrants, none has announced any intentions to serve such markets, although they could obviously change their minds if the profit opportunities materialized.[35] Likewise, most newly proposed services would seem ill-suited to serve the needs of, say, small outlying establishments of even large multiestablishment firms or the needs of most professional offices or relatively short distance high-density communications now served by local private line services of the telephone companies.[36] For new entrants not using local telephone company switching, the market would be very much limited, at least initially, to high-volume point-to-point communications.

As shown in the next chapter, substantial shifts in revenues to services offered by new entrants and total *insensitivity* of joint and common costs to volume changes would be needed to require large rate increases for telephone users as a consequence of new competition. From a welfare standpoint, the impact should be even less since some of any telephone rate increase should be offset by reductions in charges for other goods, or in higher wages or increased dividends, and so on, if and as business pays less for its communications. Again, it is necessary to look at the impact of the *changes* in rates as they impose burdens on the users who must *ultimately* pay the costs.

Competition, moreover, results in long-run market prices close to the cost levels of the *new* firms entering the market. This does not imply that prices necessarily will drop to or below established telephone companies' direct costs. These costs may be lower than those of the competition,[37] thus providing a gap within which competing firms can operate without totally eliminating the contributions from the newly competitive service.

The question thus arises of whether the "contributions" that would remain (even with competition) plus any future growth in their level would be sufficient to cover any conceivable increase in joint and common costs. Clearly, the answer would depend on many factors, many if not most of which are unknown. For example, at what rate would local service itself grow? What costs would be associated with the augmentation of this local service? Would these costs be high because many of the new locations brought into the network are relatively remote and difficult to serve? Or would the costs be relatively low because the new locations were in already dense areas or were located where simple external switching and trunking or digitalization would minimize the costs of network augmentation? Would the new phones be at households or businesses where the use of toll services would be extensive so that they would generate enough demand for high-markup services to easily pay for themselves, or vice versa? Would those who utilized the local services also demand and use other, profitable telephone services such as Yellow Page listings, extra extensions, and so forth? Finally, how might future technological changes influence the costs of extending the network?

In general, the introduction of new competition may not reduce the growth in demand for existing telephone company services. By fostering a more rapid rate of innovation in both rates and products, new entrants might so increase the total market demand for communications that the telephone companies find themselves after the fact growing more rapidly with competition than without. The question would still remain of whether the new growth, if so generated, would be necessarily profitable. Establishing profitability, in turn, would depend upon exactly where the new growth occurred; whether, say, it was associated with augmentation of the present peak or, conversely, with filling in the off-peak periods; also of importance would be the particular services required to satisfy the augmented demands, the presence or absence of rate-of-return regulation, and so on. Strictly a priori, however, new competition could conceivably improve telephone company finances.

Furthermore, regardless of whether the growth rate of established telephone companies accelerates or declines or remains the same, future profitability in the different sectors of telecommunications depends also on future technological and cost trends. Although in the past the apparent pattern of technological change has been to produce economies more rapidly in long distance than in local service, this pattern may not be inevitable.[38] Certainly, technological changes can be expected in local services in the future. One of the more interesting possibilities would be enhancing the capacity of local loops and wires, so that the copper wire now used for connecting most households into the telephone system could be used for many other communications purposes beyond voice communications. The value of any such innovation depends, to a considerable extent, on identifying markets for the new capacity. Among the possibilities would be electronic settlement of household accounts and bills and connection of household microcomputers or terminals into larger computers. Meeting some of these demands might, in fact, provide an important source of growth for future telephone company revenues.[39]

Whatever form they might take, several impending or existing technological innovations might reduce the cost of rendering local telephone service.[40] To the extent that these innovations can be introduced faster than labor or other costs might be driven up by inflation (or other causes), such technological changes could also help reduce any need for cross-subsidization or contribution from long distance and other high-markup services. Whether or not technological improvement in local service can outrace general inflation remains a question for the future. Certainly the possibility cannot be ruled out in advance. The future in this, as in so many aspects of telecommunications, is unknown—and perhaps is almost unknowable because of the rapidity of technological and market changes.

Over and over, in fact, many imponderables about future developments assume a central role in deciding the wisdom, or lack therein, of pursuing certain regulatory policies in telecommunications. Technological change and its normal accompaniment, market growth, can and *will* have an important effect on all the central regulatory issues and policy tradeoffs—economies of scale versus competitive discipline to lower costs and innovate; the alleged need for long distance profits to keep local residential service rates low; and the extent to which rate-of-return regulation and other regulatory concepts may create incentives to be inefficient. It seems appropriate, therefore, to examine closely telecommunications growth patterns—past, present and future—as an indispensable prerequisite for understanding the advantages and disadvantages of different policy alternatives. Chapter 2 discusses in detail these growth issues and their relation to the major issues facing the industry.

As will often be noted in the forthcoming chapters, the alleged inevitability of having to increase local service rates as a result of predicted losses in long distance revenues is a major argument used against permitting freer entry. Growth, as discussed in Chapter 2, offers one challenge to this argument. An alternative solution that would avoid the seeming need to raise local rates is a restructuring of present telephone company pricing policies. Some hypotheses about such restructurings are explored in Chapter 3. The analysis of pricing issues in Chapter 3 begins with a discussion of the stated goals of regulation and how well the current pricing structure achieves these goals. Following that, alternative rate structures are analyzed as means of meeting public policy goals, including keeping local rates at reasonable levels while still keeping the telephone system financially viable; these analyses suggest that the goals of both the telephone companies and public policy might be better met by greater use of such alternatives.

Chapter 4 is concerned with one of the major issues of telecommunications regulation—that of natural monopoly. A major economic justification for telecommunications to be deemed a public utility requiring regulation has been the claimed existence of substantial economies of scale. The size of these scale economies is, however, an unanswered question, dependent on the predicted size and growth of the relevant market. In addressing these issues, Chapter 4 analyzes the available econometric and engineering evidence on economies of scale in telecommunications, including an evaluation of the impact of growth and technological change on such economies.

Any discussion of regulation and public policy alternatives must include an evaluation of the purported benefits of regulation and whether or not they have, in fact, been achieved. Thus, Chapter 5 sets out to examine two of the more salient benefits claimed for regulation in the telecommunications industry—the achievement of universal service and a greater rate of technological innovation. In general, regulation attempts to create a market structure preferable to that which would ostensibly exist under a free market regime. The applicability of this proposition in the case of the local loop—reportedly the "last bastion" of natural monopoly in telecommunications—is analyzed in the last section of Chapter 5.

Finally, Chapter 6 summarizes the arguments put forth in the preceding chapters and discusses in greater detail the public policy alternatives to regulation. This book concludes that more competition and less regulation offer better prospects for U.S. telecommunications than continued regulation. In light of this view, there are a number of crucial issues that need to be addressed to effectuate a successful transition to reduced or minimal regulatory involvement in the industry. Three of the more important of these issues are discussed in Chapter 6: entry, rate regulation, and predatory pricing.

NOTES

1. See *Investigation of the Telephone Industry in the United States,* Report of the Federal Communications Commission on the Investigation of the Telephone Industry in the United States, made pursuant to Public Resolution No. 8, 74th Congress. 76th Congress, 1st Session, House Document No. 340, 1939. In particular, see Chapter 5.
2. Growth rates computed in this book are annually, *not* continuously, compounded. The formula for the former computation is $r = annual\ growth\ rate = n\ \sqrt{future\ value/present\ value} - 1$, where n is the number of years over which change in value occurs.
3. Technically, a cross-subsidy exists only when revenues are less than all of the directly variable costs, fully and correctly estimated. AT&T prefers the word "contribution" to "cross-subsidy" on the grounds that local service revenues in toto cover local direct costs as defined by AT&T accounting procedures.
4. On this subject, see Walter Bolter, "The FCC's Selection of a 'Proper' Costing Standard After 15 Years—What We Can Learn from Docket 18128," in Harry M.

Trebing, ed., Michigan State University Public Utility Papers, *Assessing New Pricing Concepts in Public Utilities* (East Lansing: Michigan State University Division of Research, 1978), pp. 333–372. It might be argued, however, that even customers who make only local calls might be charged for access to the long distance network. Even if some customers never make a long distance call, they have obtained the ability to place and receive long distance calls, which is of value to them. This value is called option value and is discussed in Burton S. Weisbrod's "Collective Consumption Services of Individual-Consumption Goods," *Quarterly Journal of Economics* 78 (August 1964): 471–477.

5. This is not meant to imply that more meaningful allocations than those now reported by the telephone companies are impossible to achieve.

6. See Jules Dupuit, "On the Measurement of the Utility of Public Works," *Annales des Ponts et Chausées,* 2nd Series, VII (1844), and reprinted in International Economic Papers, No. 2 (New York: Macmillan, 1952), pp. 83–110; F. D. Ramsey, "A Contribution to the Theory of Taxation," *Economic Journal* 37 (1922): 47–61; and Marcel Boiteux, "Peak-Load Pricing," in *Marginal Cost Pricing in Practice,* edited by James R. Nelson (Englewood Cliffs, N.J.: Prentice-Hall, 1964), pp. 59–90.

7. It would appear, historically at least (see Chapter 3, first Section), that business demands have been less elastic than residential demands. Whether this will persist in the future, if and when competition and technologies proliferate, is obviously conjectural.

8. An examination of some of these issues has emerged in the literature (assuming static technologies and markets) under the rubric of "sustainability." See Chapter 6 for further discussion.

9. As a further corollary, this process can eventually lead the original low-cost vendor to "abandon" the previous cost advantage by systematically not investing in the technology required to compete in these markets. Since the original vendor is prevented from exercising a competitive cost advantage in these markets, there is little incentive to maintain competitive abilities there. This process, augmented by other technological and market changes, has apparently been at work in intercity freight markets; see R. Levin, "Allocation in Surface Freight Transportation: Does Rate Regulation Matter?" *The Bell Journal of Economics and Management Science* 9 (Spring 1978): 18–45; and Kenneth D. Boyer, "Minimum Rate Regulation, Modal Split Sensitivities, and the Railroad Problem," *Journal of Political Economy* 85 (June 1977): 493–512. From the social standpoint, of course, this systematic withdrawal of the low-cost competitor is a loss.

10. See, for example, Merton J. Peck and John R. Meyer, "The Determination of a Fair Rate of Return on Investment for Regulated Industries," in *Transportation Economics* (New York: National Bureau of Economic Research, 1965), pp. 199–244. Other good discussions of this subject are found in Thomas R. Stauffer, "The Measurement of Corporate Rates of Return: A Generalized Formulation," *The Bell Journal of Economics and Management Science* 2 (Autumn 1971): 434–469; and Stewart Myers, "The Application of Financial Theory to Public Utility Rate Cases," *The Bell Journal of Economics and Management Science* 4 (Spring 1972): 58–98.

11. Harvey Averch and Leland L. Johnson, "Behavior of the Firm Under Regulatory Constraint," *The American Economic Review* 52 (December 1962): 1052–1069; Alvin K. Klevorick, "The Graduated Fair Return: A Regulatory Proposal," *The American Economic Review* 56 (June 1966): 477–484; and William J. Baumol and Alvin K. Klevorick, "Input Choices and Rate of Return Regulation: An Overview of the Discussion," *The Bell Journal of Economics and Management Science* 1 (Autumn 1970): 162–190.

12. This possible incentive to use too much capital is called the "Averch-Johnson effect" after its authors. Three of the original and more notable of the articles on this subject are Averch and Johnson, "Behavior of the Firm Under Regulatory Constraint," pp. 1052–1069; Stanislaw H. Wellisz, "Regulation of Natural Gas Pipeline Companies: An Economic Analysis," *The Journal of Political Economy* 55 (January 1963): 30–43; and Elizabeth E. Bailey, "Economic Theory of Regulatory Constraint," Ph.D. Dissertation, Princeton University, September 1972.

13. Bailey and Coleman suggest that the existence of a regulatory lag often encourages cost minimization. See Elizabeth E. Bailey and Roger D. Coleman, "The Effect of Lagged Regulation in an A–J Model," *The Bell Journal of Economics and Management Science* 2 (Spring 1971): 278–292. Also, Leland Johnson ("Behavior of the Firm Under Regulatory Constraint: A Reassessment," *American Economic Review* 63 [May 1973]: 90–97) describes current regulation as caught in a "Type I" lag—defined as an inflationary spiral that leads to rate increases—where firms are unable to raise rates as fast as inflation is increasing costs. Such a situation should make it unlikely for a firm to overcapitalize.

14. This result is from Kenneth C. Baseman, "Open Entry and Cross-Subsidy in Regulated Markets," paper presented at the National Bureau of Economic Research, *Conferences on the Economics of Public Regulation,* December 15–17, 1977.

15. The underdepreciation allegedly results primarily from a desire to maintain a high rate base so as to justify larger returns.

16. On this subject see Howard Anderson, "AT&T and IBM: The Battle Continues," *Telephony* (April 10, 1978): 78–89. Also see "Response of the North American Telephone Association to Options and Papers Prepared by the Staff of the Subcommittee on Communications, Committee on Interstate and Foreign Commerce, U. S. House of Representatives," *Domestic Communications Common Carrier Policy* (July 11, 1977): 12–13.

17. See Bolter, "Selection of a 'Proper' Costing Standard."

18. Much has been written on this subject. See Fred M. Westfield, "Innovation and Monopoly Regulation," in *Technological Change in Regulated Industries,* edited by William M. Capron (Washington, D.C.: The Brookings Institution, 1971), pp. 13–43; Alfred E. Kahn, "The Graduated Fair Return: Comment," *The American Economic Review* 58 (March 1968): 170–173; William R. Hughes, "Comment," in *Performance Under Regulation,* edited by Harry M. Trebing (East Lansing, Mich.: Michigan State University, 1968), pp. 73–87; F. M. Scherer, *Industrial Market Structure and Economic Performance* (Skokie, Ill.: Rand McNally, 1970), pp. 529–537.

19. William G. Shepherd, "The Competitive Margin in Communications," in *Technological Change in Regulated Industries,* edited by William M. Capron (Washington, D.C.: The Brookings Institution, 1971), pp. 86–122. Also see the detailed discussion of these cost issues in Chapter 4.

20. Wellisz, "Regulation of Natural Gas Pipeline Companies: An Economic Analysis."

21. This depends on all other usual conditions for existence of the Averch-Johnson effect holding, a set of assumptions not always met in an inflationary economy. In particular, the permitted regulatory rate of return may not be greater than the cost of capital in an inflationary environment.

22. American Telephone and Telegraph Company, "Multiple Supplier Network Study: The Cost of Multiple Intercity Networks Compared to a Single Integrated Network," submitted as Bell Exhibit 57, FCC Docket No. 20003, July 12, 1976.

23. Stanford Research Institute, "Analysis of Issues and Findings in FCC Docket 20003," prepared for American Telephone and Telegraph Company, April 1977,

Bell Exhibit 65–A; and T+E, "A Project to Analyze Responses to Docket 20003," Contract #FCC–0163, September 24, 1976.

24. AT&T, "Embedded Direct Cost Study." The source of the 78 percent calculation is shown in Figure 1–2. The differences between the 79 percent shown in Figure 1–2, and the 78-percent in the quotation given above is apparently due to rounding.

25. These issues are explored in greater detail in Chapters 2 and 3 and Appendix A.

26. Again, further discussion of these problems can be found in Chapter 3 and Appendix A.

27. See AT&T, "Embedded Direct Cost Study."

28. Ibid, p. 009.

29. Stanford Research Institute, "Analysis of Issues and Findings in FCC Docket 20003."

30. See the Opinion and Recommended Decision of the Postal Rate Commission, Docket No. R77–1 (May 12, 1978), vol. 1, p. 169. The commission had been able to reduce the percentage of toal costs that could not be distributed from 51 percent in 1971 to 25 percent as of 1978.

31. See above footnote and T+E, "A Project to Analyze Responses to Docket 20003," Contract #FCC—0163, September 24, 1976.

32. As suggested above, by utilizing various methods of allocating the joint and common costs shown in Table 1–3, it is possible to "document" the existence of a wide variety of cross-subsidies. One can demonstrate, for example, by the appropriate manipulation of joint and common costs, that local telephone service is subsidized by nonlocal services; by not too dissimilar manipulations, one could demonstrate that local telephone service subsidizes nonlocal services. The latter argument (that basic exchange subsidizes other intrastate services) has in fact been made in two state regulatory cases. (On this subject see T+E, "A Project to Analyze Responses to Docket 20003: Deliverable G," Final Report prepared for the Federal Communications Commission, October 26, 1976, pp. 46–53. The T+E report refers to the New York Public Service Commission, "Proceedings on the Motion of the Committee as to the Service Provided by New York Telephone Company," Case No. 25290, 1969–1977; also, Massachusetts Department of Public Utilities, Docket 18210, 1974.) Thus, using more or less the same numbers, some regulatory commissions can find that current telephone rate structures disguise various cross-subsidies and contributions while AT&T finds that each of its service categories makes contributions and none is subsidized. The conclusions, in short, depend crucially on methodology and choice of numbers.

33. These possibilities are explored further in Chapter 3.

34. Leland Johnson, in his article, "Problems of Regulating Specialized Telecommunications Common Carriers" (Santa Monica, Calif.: Rand Corporation, 1976), Table 1, shows that for the years 1969–1974 residential long distance grew at a rate of 10.4 percent per year; business long distance grew at 8.9 percent per year. Residential long distance accounted for 51 percent of toll calls in 1974.

35. At the time this book went to press, two alternative offerings were available for long distance service to residential customers.

36. New services tied to the local switching network could compete for some long distance miscellaneous business messages; the main limitations would be local message unit or intrastate long distance charges to the local access point of the new OCC (other common carriers) network (since any such charges would be an offset to potential savings on the alternative long line service).

37. Compare the AT&T argument (in "Multiple Supplier Network Study") that its costs may be only one-fourth those of its potential competition. Of course, massive

diversions of traffic might lead to a loss of scale economies that could exert an upward pressure on AT&T's costs. As explained in the next chapter, any such massive diversion is extremely unlikely. And as explained in Chapter 4, there is some doubt or controversy about the extent of scale economies in the industry.

38. The underlying cost relationships suggesting such a pattern of technological change are explored in greater detail in Chapters 2 and 4.

39. These growth potentials are investigated further in Chapters 2 and 5.

40. A more detailed discussion of some of these possibilities can be found in Chapter 5.

Chapter 2

The Impact of Growth

The telephone industry is dynamic, growing rapidly, and undergoing continuing changes in its service offerings and its technologies. Since growth and change are so much a part of this industry, understanding its policy problems is impossible without understanding its growth processes and patterns.

In this chapter historical patterns of growth and competition in the telephone industry are evaluated. Attention is also focused on what future growth rates for the industry might be and what these rates might portend for the financial viability and general economic health of the industry. On the basis of these analyses, it will be shown that: (1) the telephone industry has displayed considerable ability in the past to handle competition; (2) the industry's potential for growth in the future is probably higher than in the recent past; and (3) AT&T's financial performance is not likely to be impaired significantly by competition, either recent or prospective.

HISTORICAL COMPETITION AND GROWTH IN THE TELEPHONE INDUSTRY

The telephone industry's history, for purposes of analysis, can usefully be broken into seven periods characterized by different growth characteristics and business strategies. These periods, like all historical taxonomies, are necessarily rough approximations. Nonetheless, they are helpful in understanding why the industry is what it is today. For easy reference, the seven periods are shown in simple tabular form in Table 2–1.

Table 2–1. Key Periods and Dates in Telephone Industry History

Period	Dates of Importance and of Demarcation	Significance
Uneasy monopoly (1876–1879)	March 10, 1876; January 30, 1877	Two Bell patents issued
	November 10, 1879	Bell-Western Union patent suit settled
Patent monopoly (1879–1894)	January 30, 1894	Expiration of second Bell patent
Open competition (1894–1907)	December 2, 1899	Widener withdraws financial support; independent long distance network consortium collapses
	May 1, 1907	Vail regains presidency of AT&T
Vail and financial acquisition of independents (1908–1913)	1910	Mann-Elkins Act applies ICC regulation to telephone companies
	1913	Minnesota rate cases establish "use of property" as basis for cost allocation leading to board-to-board separations
	December 19, 1913	Kingsbury Commitment: AT&T agrees to interconnect with independents
Post-Kingsbury agreement: interconnection of independents (1914–1919)	June 1919	Vail resigns as president of AT&T
Steady growth (1920–1967)	1930	Smith vs. Illinois Bell authorizes station-to-station separation of toll revenues
	1930–1933	Negative growth in number of Bell telephones (—6 percent/year)

Table continued on following page.

Table 2–1. (*continued*)

Period	Dates of Importance and of Demarcation	Significance
	1934	Creation of FCC
	1947	First formal separations manual
	1945–1949	Increase in growth rate for Bell telephones (10 percent/year)
	1959–1960	FCC liberalizes frequency use by private microwave carriers ("Above 890" Decision)
	1961	AT&T introduces Telpak
Incipient competition (1968–present)	June 1968	FCC issues Carterfone decision
	1969	MCI hearings and entry
	1969	AT&T introduces Series 11,000 Tariffs
	1971	FCC permits specialized carriers in intercity markets
	1971	AT&T proposes Hi-Lo tariffs
	1973	FCC authorizes value-added carriers
	1975–1976	AT&T proposes digital data service and dataphone switched digital service
	1976	FCC adopts fully allocated cost standard
	1977	AT&T proposes multiple schedule private line rates in place of other rejected tariff submissions

Monopoly Periods

The first period, that of incipient or uneasy monopoly, lasted from 1876 to 1879, and began with the issuance of two basic Bell patents. These patents, not surprisingly, were challenged in the courts by Western Union, the dominant communications firm prior to the telephone's development. A settlement was reached in 1879, and although some litigation lingered on, a relatively stable patent monopoly was established by Bell that lasted for almost 15 years thereafter. During this second period of patent monopoly, lasting from 1880 to 1894, Bell was in effect a monopolist in the marketplace, and it had the sole rights to develop the industry.

Open Competition

At the end of the Bell patent monopoly in 1894, a period of open competition began, but not until Bell first had tried to extend its patent control beyond 1894 with the so-called Berliner patent. However, the courts' interpretation of Bell's rights under this patent was so limited that, in effect, the market was free for entry.

During the ensuing period of open competition, dating from 1894 to 1907, independent telephone companies, rural telephone cooperatives, and various mutual companies entered local markets, sometimes in competition with Bell and sometimes in areas not previously receiving any telephone service. An incipient effort was also launched to compete with Bell in long distance service because of Bell's refusal in those years to interconnect its long lines with the local services of independent telephone companies. Some of the new independent telephone companies therefore sought financial support to set up a long line toll facility as an alternative to Bell. In 1899, however, one of the prime supporters of the independents' financial initiatives, Peter Widener, withdrew his financial support.[1] A few years later in 1907, J. P. Morgan financial interests took over the Bell System and installed Theodore Vail as president of AT&T. Actually, Vail had been president of the Bell System once before in the 1880s, but during that earlier period did not have the firm control over AT&T affairs that he was to exercise during his second sojourn in the office.

Acquisition of Independents

Beginning in May 1907, after Vail resumed the AT&T presidency, the company changed its policies dramatically. Specifically, it aggressively sought to buy out competitive independents. This policy reversed a previous emphasis on expansion that had often led to head-to-head competition between AT&T and the independents. It was also during Vail's second presidency (in 1910) that the ICC was given regulatory power over the telephone companies. The ICC apparently did not do much besides instituting a uniform system of accounts, as adapted from railroad antecedents; it has remained the basic accounting system of the industry for decades and is deemed by most observers as inadequate for making many of the analyses required for modern managerial and policy decisions.[2]

Kingsbury Commitment

In March 1913, AT&T under Vail also undertook to comply with some then recently enacted antitrust laws.[3] Specifically, in December 1913, Kingsbury, a vice president of AT&T at the time, negotiated an agreement with the U.S. attorney general whereby AT&T, among other things, agreed to interconnect with the independent telephone companies. With the "Kingsbury commitment" the strategy of the Bell System thus changed from trying to buy out the independent telephone companies to linking up with them through interconnection. Long distance revenues were divided or separated between the interconnected companies according to the "board-to-board principle," which essentially held that all the costs of providing local service should be recovered from local charges.[4] In the process of implementing these separation procedures and interconnecting, the cooperation of others in the industry was largely secured and the industry as it exists today was established. Vail presided over this reorientation, retiring from the presidency of AT&T in June 1919.

Steady Growth

In the years that followed, beginning about 1920, the Bell System experienced steady growth, with the number of phones in the system growing at about 4 to 5 percent each year. There were only two exceptions to this pattern. The first occurred from 1930 to 1933 when the Bell System, as a result of the Great Depression, experienced a decline of about 6 percent a year in the number of its telephones. On the other hand, from 1945 to 1949 a spurt in growth was experienced as demand pent up by World War II was unleashed. During these postwar years, the number of Bell telephones in use increased by about 10 percent a year, as opposed to the standard increase of 4 or 5 percent a year observed during most earlier and later years.

It was also during this period (in 1934) that Congress created the FCC. Unlike the ICC, the FCC began to apply some regulation to the telephone industry. Probably the most important development of this period, however, was the Supreme Court's action in the case of *Smith* v. *Illinois* (1930). In essence, separations on the basis of board to board were rejected, leading to the use of station to station as a method of recovering local service costs from telephone revenues. On the argument that local loops were used jointly by long distance and local services, the *Smith* v. *Illinois* decision made it mandatory to recover some of the local loop costs from long distance charges. However, a distinction was made between traffic-sensitive and nontraffic-sensitive costs;

the former were charged directly to a particular service and the latter (basically costs associated with joint facilities) were assigned to specific services according to arbitrary usage formulas. Since almost 50 percent of telephone investment came to be classified as nontraffic-sensitive costs, considerable discretion was introduced into costing and rate making by adoption of the station-to-station concept. As one financial officer in the industry has put it, "It is easy to understand why some regulators have said that the history of separations consists of simply agreeing upon a desired answer, calling a meeting to fit the formula, and then naming a new plan after the city in which the meeting was held."[5]

Incipient Competition

The most recent period or phase in telephone industry development dates from the Carterfone decision of 1968. With this decision, the industry entered a period of limited competition as the FCC opened up the terminal equipment market to new entrants. Actually, some competition in intercity markets was discernible as early as 1959 when the FCC allowed private microwave operators freer use of the frequency band. In the post-1968 period, however, incipient competition manifested itself in a series of proposals for new service or entry by those outside the industry and counterproposals to fend off these efforts. As shown in Table 2–1, AT&T has made many new intercity rate and service proposals in the years since 1968. The Bell System has also recently undertaken changes in its marketing and research and development strategies, apparently in response to this new competition.[6]

Significantly, none of this post-1968 competition has involved local service, which continues to be regarded as a natural monopoly. This contrasts with the period at the turn of the century when direct competition in providing local service was not uncommon. The continuation of telephone companies' monopoly position in local service manifests itself in, among other ways, a rising number of interconnection cases (i.e., situations where established or potential intercity competitors seek the right to interconnect with local telephone company services). The (1972–1973) Specialized Common Carrier hearings, Execunet, and ENFIA proceedings before the FCC and recent Canadian hearings[7] illustrate this increasing or renewed focus on interconnection. Any increase in competition in intercity services in the future could bring this local service monopoly issue even more to the fore.[8]

Relative Growth Rates

The different periods in telephone history have been characterized by different growth rates, as shown in Table 2-2. During the very early period of uneasy monopoly a very high growth rate of almost 118 percent per year was achieved; this might be expected since the beginning was from a base of zero phones in 1876. During the patent monopoly period the market growth rate tapered off to 15.9 percent per year.[9] Actually, by the time the patent monopoly was drawing to a close in 1893, the annual growth rate was only about 2 percent.[10] During the period of open competition, the growth rate in the market rose sharply to almost 27 percent per year. Under the policy of buying rather than competing with independents, instituted by Vail in 1907, market growth was at an annual rate of 8 percent. After the Kingsbury agreement and adoption of a policy of interconnecting with independents, market growth proceeded at an annual rate of 4.7 percent and then settled into a steady pattern at 4.5 percent for the next 47 years.[11]

Since Bell enjoyed a monopoly during the first two periods, the growth rate for Bell phones is also the market growth rate during those years. Market penetration during the early period of uneasy monopoly and initiation of the system was, as might be expected, slow, reaching only 0.06 telephone per 100 persons. The penetration rate increased slowly in the patent monopoly period, at the end of which there was 0.4 telephone per 100 persons.

Under the AT&T policy, pursued from 1907 to 1914, of acquiring independent telephone companies, the number of independent telephones not connected with the Bell System fell, never achieving the same size again. The growth rate of nonconnecting independents declined to a negative 7.8 percent per year in the years immediately after 1907. Bell's annual growth rate also dropped from 22.7 to 13.3 percent during this period. However, the penetration rate grew to just under 10 phones per 100 persons, still far short of market saturation. Decline of the nonconnecting independents accelerated further after the Kingsbury agreement; the growth rate for Bell also dropped to 6.4 percent per year, as penetration went to 12 telephones per 100 persons.

During the years of steady growth after 1920, penetration increased from 12 telephones per 100 persons to over 50 telephones per 100 persons. The growth rate for Bell (plus connecting independents) and the market, which were essentially synonymous after the Kingsbury agreement, dropped to 4.6 and 4.5 percent, respectively. For the recent

Table 2–2. Historical Patterns of Telephone Industry Growth

Period	Penetration: Telephones per 100 of Population at End of Period	Annual Percentage Growth in Telephones		
		Bell System[a]	Nonconnecting Independents	Total Market
Uneasy monopoly (1876–1879)	0.06	117.8%	—	117.8%
Patent monopoly (1879–1894)	0.41	15.5%	—	15.9%
Open competition (1894–1907)	6.96	22.7%	47.2%	26.6%
Vail and purchase of independents (1908–1913)	9.72	13.3%	−7.8%	8.0%
Post-Kingsbury and interconnections (1914–1919)	11.97	6.4%	−8.9%	4.7%
Steady growth (1920–1967)	51.93	4.6%	Not Applicable	4.5%
Incipient competition (1968–1976)	72.00	4.5%	Not Applicable	4.5%

Source: Data for the years 1876-1968 taken from U.S. Department of Commerce, Bureau of the Census, *Historical Statistics of the United States: Colonial Times to 1970.* Part 2 (Washington D.C.: 1975), Series R 1-12, pp. 783-784. Data for the years 1969-1976 taken from the United States Independent Telephone Association, *Independent Telephone Statistics,* Vol. 1, 1977 Edition; and Federal Communications Commission, *Statistics of Communications Common Carriers,* 1976.

[a]Bell and connecting independents after 1907.

period of incipient competition, growth in telephones has also been at a rate of 4.5 percent per year, for Bell and the market.

Figure 2–1 presents the growth experience of the telephone industry in graphical form, using the same data as in Table 2–2. The number of Bell System telephones is measured in natural log form to avoid problems of scale. The very large early growth rates of the initial period fall off during the patent monopoly period and then rise with open competition. As Bell strengthened its hold on the market during Vail's presidency of AT&T the growth rate declined, and with the Kingsbury agreement it fell still further. Over the next 40 or 50 years, the growth rate remained at about 4.5 percent per year (with the exception of the Great Depression and the post-World War II period). After the emergence of possible new competition in 1968, the growth rate changes little from that of the preceding years.

Clearly, not all variation in growth rates can be attributed to changes in competitive strategies, regulations, or legal milieu; there were also wars, depressions, and technological changes, among other influences. The steady decline in growth rates after 1907 reflects, of course, a not uncommon pattern in markets as they mature. Nevertheless, this market by most penetration measures was hardly mature in 1907—or in 1920, for that matter. The pattern of growth rates falling, rising, and then falling again, as competition sequentially did not exist, flourished, and was circumscribed in the years from 1879 to 1919, does not coincide merely with market growth but suggests competitive factors at work as well. Of course, part of the growth spurred by competition at the turn of the century represented duplication in service. AT&T's annual reports during the first part of the century assign a 15-percent figure to duplication in cities in which there was competition. (That is, 15 percent of businesses or households in exchanges with more than one telephone company had two telephones in order to connect with the available telephone system.)[12] At the same time, however, non-Bell telecommunications were growing rapidly elsewhere. In particular, much of the expansion in rural markets apparently was not attributable to Bell but to then newly created independents and farmer cooperatives.

Competition at the turn of the century (coupled with technological change and volume increases possibly leading to scale economies) also inspired considerable rate reductions, thus offsetting to some extent any social cost arising from duplication. Figure 2–2, taken from Bell's 1909 Annual Report, indicates the change in revenues per station in markets with and without competition. Starting in 1894, as the patent monopoly ended, the revenues per station were in the range of $70; by the end of the period (1909), the revenues per station had fallen to $31

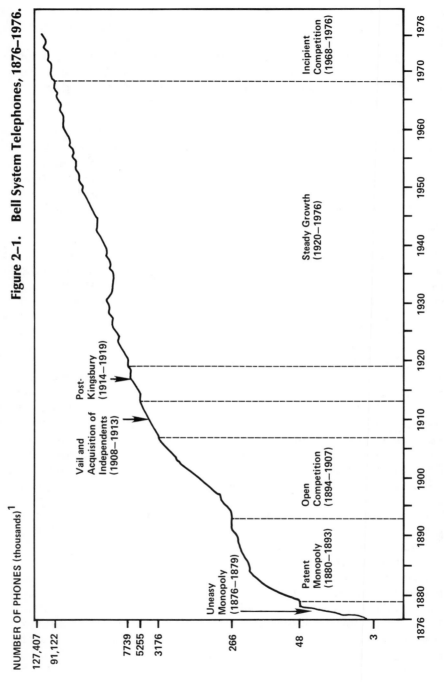

Figure 2–1. Bell System Telephones, 1876–1976.

NUMBER OF PHONES (thousands)[1]

127,407
91,122
7739
5255
3176
266
48
3

1876 1880 1890 1900 1910 1920 1930 1940 1950 1960 1970 1976

Uneasy Monopoly (1876–1879)

Patent Monopoly (1880–1893)

Open Competition (1894–1907)

Vail and Acquisition of Independents (1908–1913)

Post-Kingsbury (1914–1919)

Steady Growth (1920–1976)

Incipient Competition (1968–1976)

[1]This curve was calculated using the natural log of thousands of telephones.

32

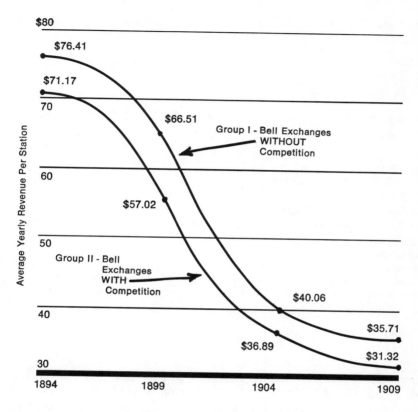

Figure 2–2. Reduction in the Average Yearly "Bell" Exchange Revenue Derived from Certain Groups of Cities Arranged in Five-Year Periods, 1894–1909 (January 1 of each year). *Source:* American Telephone and Telegraph, *Annual Report* (1909).

to $35 per station. The decline in revenue per station over the period 1894 to 1909 was about the same (in dollars and rate of decline) both in the exchanges without competition and in those with competition, the pattern of rate decline being roughly parallel throughout the entire period. This pattern is consistent with Bell moving to lower rates to preempt competition or to combat price cuts by actual competitors, as well as with the emergence of scale economies and improvements in technology.[13]

Table 2–3, taken from AT&T's annual reports of 1909 and 1913, shows that entirely different revenue experiences occurred in toll and local exchange services. From 1895 to 1905, exchange service revenue per station dropped from around $70 to about $30. During the same

Table 2–3. Revenues and Expenses of Bell Operating Companies, 1895–1913

Item	Dollar Amount				
Earnings per Exchange Station	1895	1900	1905	1909	1913
Exchange service	$69.75	$44.68	$33.31	$31.37	$30.45
Toll service	11.35	12.60	9.95	9.42	9.03
Costs					
Average plant cost per exchange station (including exchange and toll construction)	$260	$199	$145	$145	$141
Average cost per mile of pole line (toll), including wire	219	348	438	610	NA[a]
Average cost per mile of wire (toll), including poles and conduits	81	71	62	63	70

Source: American Telephone and Telegraph Annual Reports, 1909 and 1913.

[a]Not available.

years, toll revenues by station rose from $11 in 1895 to $12.60 in 1900 and then dropped to $9.95 in 1905. This pattern of toll service revenues suggests that competitive pressure may have been greater in the local exchange market than in the toll service market, or that major technological improvements in long distance telephone service did not occur until at least 1900.[14]

In fact, cost trends for this period show no clear downward trend that would totally explain the tremendous falloff in nontoll revenues per exchange station. The average plant cost per exchange station of $260 in 1895 dropped to $145 in 1905, a percentage decline (44 percent) smaller than that in revenues per exchange station (52 percent). While the average cost per mile of pole line went up substantially, average cost per mile of wire went down and then up over the period. The entire fall in revenue per exchange station would thus not seem to be explained by only technological changes or cost reduction during the period of open competition (though these cost and technological influences surely explain much of the fall in revenue).

Figure 2–3 presents an overview of Bell's development in real terms (revenues, expenses, and earnings per telephone deflated by the consumer price index) from 1885, just after the telephone system's inception, until 1976. In the patent monopoly period real revenues per

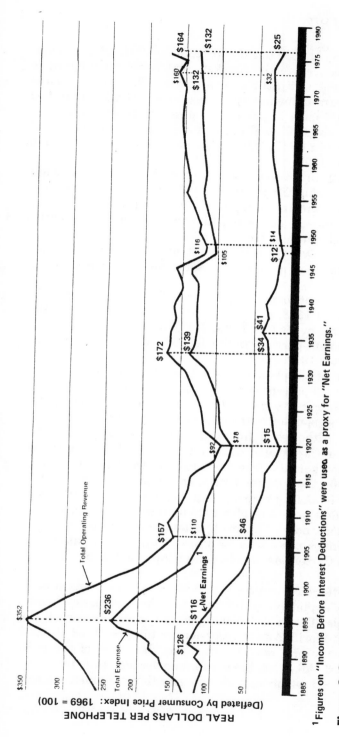

Figure 2–3. Bell Revenue, Expenses, and Earnings per Telephone, 1885–1976. *Sources:* Data for the period 1885–1935 taken from Federal Communications Commission, *Proposed Report; Telephone Investigation* (Pursuant to Public Resolution Number 8, 74th Congress), 1938, p. 149. Data for the period 1936–1939 taken from American Telephone and Telegraph Company, *Annual Report, 1936–1939*. Data for the years 1940–1975 taken from American Telephone and Telegraph Company, *Bell System Statistical Manual: 1940–1974* (May 1975) and *1950–1975* (May 1976). Data for 1976 taken from the Federal Communications Commission, *Statistics of Communications Common Carriers, year ended December 31, 1976*. Calculations by Charles River Associates.

[1] Figures on "Income Before Interest Deductions" were used as a proxy for "Net Earnings."

35

phone went from $250 in 1885 to $352 in 1895. By 1907, after competition had been introduced to the market, real revenues dropped to $157, or by 55 percent in a little over a decade. The rate of decline slackened in the years immediately thereafter but continued to $92 per station by 1920,[15] for a 74-percent drop in real revenues per station between 1895 and 1920. Real net earnings per telephone for Bell also fell from a high of $130 in 1892 to $46 per telephone in 1907 and $15 per telephone by 1920.[16]

After 1920, growth in revenues, expenses, and earnings per phone are all remarkably stable. From 1920 to 1976, annual real revenues per phone are largely within the range of $100 to $160. For net earnings, the range is mostly between $15 and $30 across the entire period. A couple of upward trends are observable in the 1920–1933 and post-World War II periods, but otherwise revenues and earnings per telephone are more or less stable when measured in real terms.

For the era of incipient competition that may have dawned after the Carterfone decision in 1968, it is interesting to inspect growth rates prior to and after this event (see Table 2–4). Total operating revenue increased by 7.4 percent per year in the decade prior to 1968, and 11.1 percent per year after 1968.

The growth rate of telephones was 4.8 percent per year between 1959 and 1968 and 4.3 percent per year from 1968 to 1977. The revenues per phone in nominal terms grew at 2.6 percent per year in the pre-Carterfone era as opposed to the post-Carterfone era growth of 6.6 percent annually. Although a higher growth rate in revenue per phone appears after Carterfone, there was also a different inflation experience in the two periods. From 1959 to 1968, the annual inflation rate as measured by the consumer price index was 2 percent, while from 1968 to 1977 it was 6.4 percent. Real growth rates in total operating revenue for Bell were thus about 5.3 percent per year before the Carterfone decision and 4.5 percent annually after the decision, whereas real revenues per telephone grew at an annual rate of 0.6 percent before the Carterfone decision and 0.2 percent annually after it.

Most of this difference in pre- and post-Carterfone real rates of growth is probably explained by differences in the business cycle during the two periods. The 1959 to 1963 period had one recession, the mildest of all post-World War II recessions, while the 1968 to 1977 period had two recessions, one of which was the worst of the post-World War II period. All in all, the Carterfone decision thus far seems not to have influenced telephone company growth rates much one way or another.[17]

Table 2–4.   Bell System^a Growth Rates Before and After the 1968 Carterfone Decision

	Annual Growth Rates (Percent)	
Item	1959–1968	1968–1977
Bell System nominal growth:		
Total operating revenue	7.4	11.1
Telephones	4.8	4.3
Revenue per telephone	2.6	6.6
Consumer Price Index	2.0	6.4
Bell System real growth:		
Total operating revenue	5.3	4.5
Telephones	4.8	4.3
Revenue per telephone	0.6	0.2

Source: *Bell Statistical Manual,* 1950–1975; AT&T, *1977 Statistical Report; Economic Report of the President, 1978.*

^{a}Bell System as measured herein does not include Cincinnati Bell or Southern New England Telephone.

In short, the telephone industry in this country has undergone quite different growth experiences during different periods of its history. These different growth experiences generally correspond to changes in Bell's business strategies. When Bell dominated the market through patent monopoly, growth was substantial but not dramatic. When Bell competed openly and intensely with the independents after the patents expired, a very large expansion occurred in both market penetration and growth rates. The expansion and penetration, of course, *might* have happened in these years even without competition, but it is at least suggestive that such growth rates were not achieved in the less competitive eras both before and after this period of open competition. Furthermore, during the period of open competition, residential and business consumers paid steadily lower rates. Although these benefits of lower rates and increased growth must be weighed against the costs of network duplication, it is not obvious that consumers were harmed by open competition at the turn of the century. Moreover, the lower rate structure that emerged in the competitive era was maintained in real dollar terms after open competition ceased, apparently as a matter of AT&T policy since regulation throughout much of the ensuing period was only nominal or sporadic.

POTENTIAL GROWTH FOR THE
TELEPHONE INDUSTRY

As the historical record demonstrates, the telephone industry experienced extended periods of dramatic growth, particularly from 1895 to 1907. More recently, however, the overall picture is of an industry characterized by considerable stability in all of its growth characteristics, ranging from financial to developmental to pricing strategies. Underneath (or alongside) this stability, however, volume and revenue growth for the telephone industry have significantly outpaced growth in the economy as a whole, as shown in Table 2–5. From 1970 through 1976 real GNP growth was 2.9 percent per year. The population grew at less than 1 percent per year while the number of households increased annually at 2.5 percent. At the same time, Bell's toll revenues increased by 12.6 percent per year; long distance rates increased by only 3 percent per year (as compared with 6.6 annual inflation), and thus toll volume implicitly grew at 9.3 percent per year. Similarly, but less dramatically, local exchange revenues for Bell increased at an annual rate of 10.8 percent during the 1970–1976 period; local rate index increases of 5.4 percent per year (again less than the annual inflation rate) resulted in the implicit volume in local exchange areas for Bell growing at 5 percent per year.

Data in Table 2–5 also indicate that the volume for all forms of electronic communications increased at a rate greater than that for nonelectronic media. For example, the number of newspapers in the United States actually declined during the years 1970 through 1976, and the number of pieces of mail carried by the U.S. Postal Service grew at less than 1 percent per year. The reasons for differential rates of growth among the communications media seem clear: the pickup, handling, and delivery functions required by paper communications are considerably more labor-intensive than are electronic communications, and labor costs have been inflating significantly (see Table 2–11). In general, electronic communications have benefited from a rapid rate of technological innovation that has lowered costs and provided new and better communications modes and services.

There is little reason to believe that the problems of labor-intensive communications media (in an era of widespread wage inflation) will soon be eased. Nor is it apparent that the rates of technological invention and innovation in electronic communications are likely to slow much. Developments on the horizon include optic fibers, expanding use of ever more sophisticated electronic switching systems, and more

Table 2–5. Recent U.S. Growth Experience, 1970–1976

Area of Growth	Annual Growth Rate (Percent)
Nominal gross national product	9.6
GNP deflator	6.6
Implicit real GNP	2.9
Population	0.8
Households	2.5
Bell System local exchange	
Revenues	10.8
Local rate index	5.4
Implicit volume	5.0
Bell System toll	
Revenues	12.6
Long distance rate index	3.0
Implicit volume	9.3
Other communications media	
Telephones	4.3
Cable television subscribers	15.7
Television sets[a]	5.8
Number of newspapers	−0.1
Pieces of mail	0.9

Sources: Television data from *Television Factbook* 47 (Washington, D.C.: Television Digest, Inc., 1978 edition). Bell System revenue data from American Telephone and Telegraph *Bell System Statistical Manual, 1950–1977,* May 1978. Rate index figures from *Telephone and Telegraph, 1976 Statistical Report.* All other data from U.S. Department of Commerce, *Statistical Abstract of the United States,* 1978.

[a]Television sets in use by end of year.

use of satellite transmission.[18] The slower growth rates in paper communications media than in electronic communications may thus persist.

The prospect of various industries, such as banking, finance, or securities, seeking to lower their total costs by increasing their use of electronic communications also points to the possibility of enlarged growth rates for the telephone industry.[19] Certainly, the current relatively high rate of growth for the industry in long distance communications seems easily sustainable. The question is whether even this high rate might rise markedly in the near future.

To assess this possibility, hypothetical growth scenarios for the years between now and 1990 can be considered, first for households (Table 2–6) and then for business (Table 2–7). The volume of telephone usage by households in 1977 was 1.1 billion household minutes per day, implying an average daily use of the telephone per household of approximately 15 minutes. Current consumption of paper communications media per household is estimated to average about one daily newspaper and one book and one periodical per month. The hypothetical development scenario outlined in Table 2–6[20] assumes that much of these papers, journals, and books will convert to electronic transmission into the home by 1990. It also assumes that 50 percent of the 1977 volume of mail received by the household will be transmitted electronically by 1990.

As a result of these assumptions, the volume of potential household electronic communications per day would grow to 4.8 billion minutes by 1990, implying an approximate average household use of the telephone of 48 minutes per day. Such a potential volume does not require, of course, that individual members of the household actually be on the phone for 48 minutes per day, but only that 48 minutes of electronic transmission would come into the home (some of it, for example, possibly being stored in terminals designed for such purposes).

The growth rate implicit in such a scenario is approximately 9 percent per year between 1977 and 1990 for the typical household's volume of telephone usage.[21] This scenario is *conservative* in that it does not project any new media that might arise as a result of the application of new technology to household communications.[22] Nor does it assume increased usage of the mails, books, or periodicals as a result of the new means of transmission. In particular, households do not receive individual two-way video transmissions (or any television transmission over telephone lines) in this scenario; the capacity required for such video transmission would explode growth rates in this industry.[23]

The importance of the scenario outlined in Table 2-6, with all of its limitations and qualifications, lies in the derivation from very conservative assumptions of a possible overall growth rate of 12 percent per year for telecommunications in the household sector. This is considerably higher than the 7 percent per year growth in total use of telephone communications experienced in the United States between

Table 2–6   Potential Growth in the Volume of Household Telephone Usage, 1977–1990,[a] Based on Usage by Households in 1977

Assumptions
Growth:
1. Households at 2.5 percent per year
2. Residential call volume per household at 4.5 percent per year
3. Paper communications per household at 0 percent per year

Household communication use in 1977:
1. Eight pages of mail per day
2. One book and one periodical per month
3. One daily newspaper
4. Telephone used 15 minutes per day

Communication transmission speed:
1. Mail at 3 minutes per page with 50 percent of household mail transmitted electronically
2. Nonmail paper communication at 3 seconds per page

Results
Annual growth rates for 1977–1990:
1. For household sector, 12.0%
2. Per household, 9.3%

Source: Calculated by Charles River Associates.
[a]Not a forecast. Results based on hypothetical assumptions outlined above.

Table 2–7.   Potential Telecommunication Volume Growth for Intercity Transmission in 1990[a]

Assumptions
1. Long distance voice transmission quadruples from 1977 to 1990.
2. Two million videophones in use by 1990 require a total capacity increment of 200 percent of current intercity transmission volume.
3. Data and facsimile transmission increases by an increment of 100 percent of present intercity capacity.
4. One-half of first-class mail travels between cities electronically. This requires an increment of 100 percent of present intercity capacity.

Results
Volume potentially increases 700 percent between 1977–1990, implying an annual growth rate of 17.3 percent.

Source: Calculations by Charles River Associates.
[a]Not a forecast. Results are based on hypothetical assumptions outlined above.

1970 and 1976.[24] On a per-household basis, the current 4.5 percent annual growth rate would more than double to 9.3 percent.

Table 2–7 indicates the impact of a similar scenario when applied to the intercity transmission market for *both* household and business demand. In this scenario it is assumed that long distance voice transmission grows at the same rate as that experienced between 1970 and 1976. This scenario further assumes that 2 million videophones are in use (mainly by businesses) by 1990 and that data and facsimile transmission increase by an increment of 100 percent of present total intercity volume. As in Table 2–6, it is also assumed that 50 percent of first-class intercity mail is transmitted electronically.

If these assumptions prove valid, the implied annual growth rate for volume of intercity transmissions for the 1977 to 1990 period would be 17.3 percent.[25] Such growth would obviously be substantially higher than the annual growth rates experienced during the 1970–1976 period. Such an increase, it should be noted, would come not from any increase in the voice transmission growth rate, but from the adoption of videophones by businesses and by increased diversion of communications from paper to electronic transmission.

Again, the lesson to be drawn from the scenario in Table 2-7, as from Table 2–6, is that reasonable descriptions of future possibilities for applying electronic transmission media could yield growth rates that are higher than those recently experienced in the telephone industry. In short, the telephone industry should continue to be a growth industry. A crucial question, of course, is whether the industry's growth potential will be achieved to the same extent without competition as it would with competition. A competitive industry in such a growth environment is likely to be characterized by new products, new services, and new technological innovations. In such a market, all firms could have expanding opportunities for sales and profits so that the sales and profits of one firm may not necessarily impinge directly on the sales and profit opportunities of any other firm. On the other hand, competition, while not a zero sum game for all the participants, might still result in considerable diversion of business for some, particularly the now dominant firm, AT&T.

COMPETITION AND DIVERSION

Much of the controversy about telecommunications policy has, in fact, focused on recent FCC and court decisions[26] that open the telephone industry up to competition in several of its traditional market areas, essentially beginning in 1968 with the FCC's Carterfone decision.[27] During the first few years after 1968, the telephone industry required that telephone company connecting arrangements be inserted between each item of customer-provided equipment and the network at a significant additional charge. This tariff requirement was declared unlawful by the FCC in 1975, when the commission established a registration program designed to ensure that all equipment be designed and manufactured with adequate safeguards to protect the telephone network and its personnel from electrical harm. Although full implementation of this decision was delayed nearly two years through legal challenges, it finally became effective in October 1977 when the Supreme Court refused to review a lower court ruling that affirmed the commission's action.[28] As a result, the selling of equipment to be used on the customer's premises, including equipment for residential telephones, is now open to competition.

Long distance transmission services for business subscribers also became a subject of potential competition in the 1960s and 1970s, mainly through entry of specialized common carriers (SCCs). This long distance competition essentially had its start in the *Above 890* decision of 1959, in which the FCC ruled in favor of allowing private point-to-point microwave communications systems. The commission's opinion was based on the belief that the number of frequency bandwidths in the microwave range was more than sufficient to meet present and future needs of both private point-to-point systems and the common carriers.[29]

The competition for business long distance private line services intensified in the early 1970s with the FCC's 1971 decision on specialized common carriers.[30] As with the Carterfone decision, the specialized carrier decision was again seen as an attempt to protect communications users' rights. The 1971 decision seemingly evolved from the continued refusal of AT&T and the other established carriers to provide shared and part-time private line service. Thus, the FCC apparently felt that consumers with special interstate needs, not presently being accommodated, would best be served by creating a more competitive environment.

In essence, new competition has developed in distinctly different sectors: customer premises equipment for households and businesses and long distance transmission services. They are discussed in turn below.

Customer Premises Equipment

Household Telephones. Consumer telephones is the most recent sector of telecommunications to be opened to competition. As already noted, the final barrier to competition was removed when the Supreme Court in 1977 upheld a lower court's ruling that allowed customers to attach their own phonesets without interface devices. Since then, there has been an increase in the number of suppliers for customer-owned equipment, rapid technological developments in this kind of equipment, and a proliferation of novel phoneset designs.[31]

Equipment with special features and new technology are expected to be the primary growth areas in the household telephone area. Devices such as automatic dialers, digital displays, call-timing, and built-in answering machines have been introduced at a rapid rate. In addition, the use of standardized modular jacks, greatly facilitating installation and removal, is seen as a major factor promoting the expected growth in this sector. *Business Week* [32] cites forecasts of $1 billion in total revenues by 1980 (more than three times 1978 revenues).

The new entrants in retail consumer phones have only recently been challenged by direct competition from AT&T. AT&T still maintains a lease policy[33] and historically has never been deemed particularly strong in marketing. This situation, however, may be changing as AT&T has begun to create an effective marketing department, gathering marketing specialists from such companies as IBM and Xerox.[34] Indeed, many industry followers, including AT&T executives,[35] feel that it is highly unlikely that AT&T will simply allow more and more of Western Electric's business in supplying residential phones to be eroded by competition.

The response of Western Electric will be, of course, a prominent factor conditioning activity and growth in these sales. Currently, Western Electric produces around 10 million phones worth approximately $200 million annually; direct marketing by Western Electric to the customer or through Bell operating subsidiaries clearly would greatly augment competition in these activities.[36]

Business Telephones

Competition in business telephone equipment and services "took off" in the early 1970s, thus recording an earlier start than in consumer instruments. Initially, however, there were many problems due to overly rapid expansion, poor financing, and the inadequate maintenance services provided by some of the new entrants. As early as about 1973, many of the weaker or more inefficient companies had already left the competition.

For the business terminal equipment market, total revenues for 1977 were approximately $700 million,[37] and the annual growth rate in revenues of PBX and KTS equipment for the years 1975 through 1977 was approximately 54.2 percent.[38] Predictions of future interconnect equipment revenues are varied, ranging from a 1975 estimate of $800.7[39] million in 1984 to a 1978 estimate of $1.1 billion in 1984.[40] Given that business customer premise equipment revenues had already reached $700 million in 1977, the $800 million estimate would seem to be too conservative.

Success of the interconnect equipment industry outside of Bell may greatly depend on future AT&T decisions. Some observers suggest that AT&T's comparative strengths may not lie in the provision of equipment; indeed, some studies such as one completed by the New York Public Service Commission[41] purport to show that the Bell companies are actually losing money on certain telephone equipment leases. Nevertheless, Western Electric, given its scale of production and budget available for advertising and R&D, could prove to be a very effective competitor. The extent to which competitive market thrusts, and counterthrusts, will stimulate or retard the development of this sector clearly remains an open question.

Long Distance Transmission Services

In long distance communications, companies other than the telephone companies that provide private line transmission services are usually referred to as specialized common carriers or SCCs. There are four basic SCC types: landline private service carriers; satellite carriers; packet-switched carriers;[42] and Western Union or telegraph type competitors. The landline carriers currently realize the greatest percentage of total revenue commanded by SCCs; the satellite carriers, although still very small, would seem to possess the largest potential for future growth.

The Ad Hoc Committee for Competitive Telecommunications (ACCT)[43] has estimated total landline SCC industry revenues to be $134 million per year as of the end of 1977. This amount is up from $30 million in 1972. Forecasts of expected future industry revenues include numbers such as $400 to $800 million by 1980,[44] $425 million by 1980,[45] and $800 million by 1985.[46] The forecast of $425 million by 1980 would imply a five-year (1975–1980) annual growth rate of 30 percent, while the estimate of $800 million by 1985 would imply an average annual growth of 34.9 percent for the ten-year period 1975–1985.

The 1970s were not profitable years for most of the landline SCCs, despite the rapid increase in corporate phone bills[47] and an increasing demand for innovative electronic services during these years. Datran, for example, has ceased operations. Others, such as SPCC and Telenet, have become profitable only recently.[48] MCI became profitable in 1977 with the offering of its Execunet service.[49] Reasons suggested for this poor financial experience range from the perception that AT&T's Long Lines provides excellent service to the observation that some of the landline SCCs may have been undercapitalized.[50]

Forecasts for satellite revenue growth are generally more optimistic than for other telecommunications services. *The Wall Street Journal* in September 1978[51] reported 1977 satellite carrier revenues of $250 million and predicted growth to $1 billion in revenues by 1983. This would require an annually compounded growth rate of 20 percent. A growth rate of 26 percent per year through 1983 may even be on the conservative side. A January 1979 *Newsweek* article commented, "Analysts expect the satellite business to grow to nearly $2 billion by the mid-1980s up from $100 million or so last year."[52] Such optimistic forecasts apparently assume that satellite transmission will not be limited to large businesses but will also include governmental users, time-sharing firms, and firms participating in resale and sharing programs. Services in which satellites could have a comparative advantage (again, especially over very long distances) would appear to include data transmission services, TV transmission, video conferencing, and direct TV broadcasting.

One of the major threats that satellite carriers may pose to existing common carriers is as wholesalers of telecommunications services. For example, by leasing or selling their services to the nation's major independent telephone companies, satellites could effectively bypass AT&T's Long Lines. Some satellite systems might also develop services competitive with AT&T's MTS service through interconnection with local telephone exchanges.

Future Developments

Any predictions made of future revenue growth for customer premises equipment and SCCs (including satellite) are obviously uncertain. Corporate phone bills have been increasing at a rate of 17 percent a year for the last few years as compared to a rate of about 5 percent per year in the mid-1960s.[53] Business telephone expenditures might be expected to increase further with the advent of proposed electronic communications services such as electronic funds transfer, electronic mail, word processing, and video services. Whatever the case, rising communications costs are causing firms to look for less costly and less limited alternatives to the services now provided by the established phone companies. And the interconnect suppliers, in both anticipation of and response to this demand, seem willing to introduce the products and services that may satisfy these needs.

Based on estimates publicly available and as reviewed above, the potential revenue growth for telephone industry competitors might be projected as shown in Table 2–8. These estimates are probably high, implying an overall growth rate for these areas of 25–36 percent per year. Sustaining such growth rates in revenues over a 10-year period (from 1975 to 1985) would require a good deal of financial and entrepreneurial strength on the part of competitors. Assuming that the telephone industry could and would develop these markets in the absence of competition—a very big assumption—the total potential direct loss of revenues (evaluated at the midpoint of the range) might be placed at $3.2 billion by 1985.[54] Since a $0.2 billion annual market was already realized by the competitors in 1975, the increased diversion under these assumptions would be $3 billion.

In addition to direct loss of revenues due to sales by competitors, competition could potentially also reduce telephone industry revenues indirectly, mainly as the result of rate erosion. As noted above, early competition in the telephone industry (from 1895 to 1907) was accom-

Table 2–8.   Current and Potential Revenues of AT&T Competitors

	Annual 1975 Revenues (millions)	*1985 Annual Revenues[a] ($billions)*	*Implied 1975–1985 Annual Growth Rate*
Interconnect customer premises equipment	143[b]	1.0–2.5	22–33%
Specialized common carriers	40[c]	0.4–0.8	26–35%
Satellite carriers	16[b]	0.5–1.1	41–53%
Total	199	1.9–4.4	25–36%
American Telephone and Telegraph	28,957	—	—

Sources: In addition to sources cited in footnotes a, b, and c, data for American Telephone and Telegraph were taken from AT&T's *1976 Statistical Report.*

[a]Charles River Associates judgments as to size of market from predictions cited in the text of this chapter.

[b]"The Economic Effects of Competition in the Private Line and Terminal Equipment Markets," Preliminary Report by the Common Carrier Bureau, Federal Communications Commission, 1975, Table 1.

[c]Ad Hoc Committee for Competitive Telecommunications, "The Case for Competition in Telecommunications" (Washington, D.C., 1975), p. 7.

panied by a reduction in rates of approximately 50 percent. Recent competitive experience has also seen Bell lower or try to lower all private line rates by nearly one-half.[55] A reasonable reaction to competition in one service, moreover, can be to lower the rates of several other services as well, especially services facing possible revenue loss due to cross-elasticities of demand or to potential entrance of new direct competition. Thus, Bell might choose to lower its rates on conventional long distance message service (MTS) as a way of limiting the danger of competitive entry into such service from firms who have already decided to compete with Bell in private lines.

Table 2–9 itemizes Bell revenues that might be assumed to be vulnerable to erosion from such rate reductions. In 1975, $12.2 billion of Bell's revenues were accounted for by interstate message toll service

Table 2–9. Bell Revenues Assumed Vulnerable to Errosion by Rate Reduction, 1975 (Billions of dollars)

Vulnerable Service	Revenues
Interstate message toll service (MTS)	6.9
Interstate outward WATS[a]	0.8
Interstate private line service (PLS)	1.0
Leased telephone equipment[b]	3.5
Total	12.2
All other OTC revenues	18.5
Total Bell revenues	27.4
Total of vulnerable revenues as percent of all other OTC revenues	65.9
Total of vulnerable revenues as percent of total Bell revenues	44.5

Sources: Federal Communications Commission Docket No. 20003, Bell Exhibit 20-A, "Historical Data Regarding Bell System and Non-Telephone Company Provided Terminal Equipment," April 20, 1976; Federal Communications Commission Docket No. 20003, Bell Exhibit 1-B, "Embedded Direct Cost Study 1975," Third Supplemental Response, July 12, 1976; *Monthly Report No. 4,* Bell System Operating Revenues, Report to the Federal Communications Commission, December 1975.

[a]Outward WATS is assumed to be 75 percent of interstate WATS. See Bell Response to FCC Docket 18128, Central Submission, Vol. 7: Market Analysis — Inward and Outward WATS, July 8, 1977, Table 3–1, p. 3–3. Only outward WATS was assumed to be vulnerable. Inward WATS depends critically on the access to large numbers of homes and businesses. It is not likely that Bell will be significantly challenged on such local exchange penetration for decades to come. PLS and WATS revenues from Bell System MR4 reports cited below.

[b]In equipment markets, Bell's revenues are assumed to be 100 percent vulnerable. This is clearly exaggerated since marketing of new equipment will concentrate for some time to come on large urban markets. Furthermore, current competition is largely for business equipment.

(MTS), interstate outward WATS, and interstate private line service and equipment leases. These revenues represented 66 percent of all other OTC (operating telephone companies) revenues and 45 percent of total Bell revenues in that year. Clearly, any assumption that all these revenues are vulnerable to erosion by rate reduction is an over-estimate.[56] Yet, as subsequent analysis will demonstrate, even with such large estimates of erosion to competition, Bell should have suffi-cient revenues to more than survive financially.[57]

FORECASTING THE IMPACT OF COMPETITION AND GROWTH ON AT&T's LOCAL SERVICE RATES AND FINANCIAL PERFORMANCE

It is often argued by those in regulated industries that deregu-lation or expanded competition would undermine their ability to at-tract financial capital. The trucking and airline industries have made this argument at various times and some telephone financial execu-tives have at least hinted at it. It has been suggested, for example, that increased competition in providing communications services might ultimately lower profitability to the point where the telephone system would have to increase local phone rates dramatically in order to attract needed outside capital, whether equity or debt.

As noted in the last section, increased competition could affect AT&T revenues negatively in two ways: (1) by taking away some share of volume; and (2) by pushing market prices down. These diver-sions, as the argument is made, could so reduce telephone company profits that the companies would not be able to issue the amount of debt and equity required to finance necessary increases in plant and equipment investment. Quality and availability of service would then decline, and the national objective of a universally available standard service for telecommunications would be less attainable unless local rates were increased substantially.

Financial Hypotheses and Scenarios

To investigate these possibilities, different scenarios about the finan-cial future of AT&T need to be investigated. These scenarios, their hypotheses, and a full discussion of the underlying structure of the financial models employed for their analysis are explained in detail in Appendix B. Generally, the scenarios use the revenue and cost catego-ries as in the AT&T embedded direct cost (EDC) studies, with some modifications, and the same balance sheet categories as in the AT&T annual reports; the projection period runs from 1975 to 1985. Figure 2–4 provides a simple diagrammatic summary of the assumptions used and scenarios investigated.

Three basic "historical" scenarios are analyzed. In the *Historical A Scenario,* it is assumed that the general economic conditions prevalent in the 1973–1975 period[58] continue through the projection years. Revenue and cost items thus grow in this scenario at the rates experienced in the 1973–1975 period (see Table 2–10) when the inflation

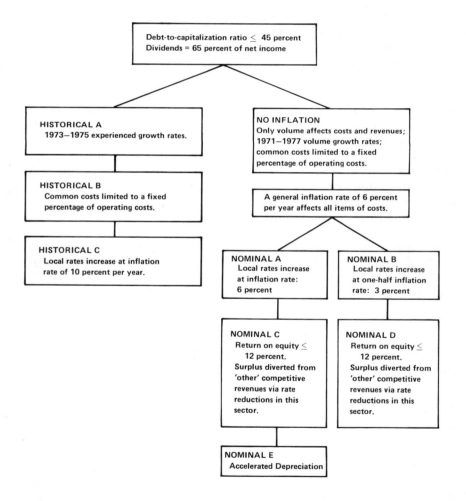

Figure 2–4. Summary of Assumptions Used in AT&T Financial Projection Scenarios.

rate averaged around 10 percent per year (see Table 2–11). Dividends are paid at the historically observed rate of 65 percent of net income (after taxes) in this as well as in all other scenarios. Similarly, debt is constrained in this and all other scenarios to never exceed 45 percent of total capitalization (i.e., new equity is issued to maintain a 45-percent debt ratio); this seems consistent with AT&T financial targets and maintenance of top ratings for AT&T debt. The *Historical B Sce-*

Table 2–10. Annual Bell System Growth Rates, 1973–1975[a] (percent)

Service Category	Annual Revenues	Annual Costs
Interstate (operating telephone companies (OTC) portion)	10.9	7.2
Intrastate excluding local telephone service	12.5	11.3
All other OTC revenues and costs (subtotal)	11.9	10.1
Local telephone service (calling)	9.5	10.0
Joint (access) costs	—	9.9
Common overhead costs	—	16.5
Total OTC revenues and costs	11.1	11.1

Source: FCC, "Embedded Direct Cost Study, 1975"; Bell Exhibit 1-B; FCC Docket No. 20003.

[a]Rates are compounded annually.

Table 2–11. General Economic and Telecommunications Industry Growth Rates, 1973–1975

Item	Annual Growth Rate (percent)
Consumer Price Index	10.1[a]
Index of adjusted hourly earnings, total private nonagricultural (current dollars)	8.5[a]
Bell local rate index	6.0[b]
Bell long distance rate index	3.0[b]
Number of telephones	3.6[b]
Local conversations	4.0[b]

[a]From the *1978 Statistical Report of the President.*
[b]American Telephone and Telegraph Co., *Statistical Report,* 1976 and 1977.

nario is identical to *A* except that the growth of "overhead" costs (called "common" costs by AT&T), is not projected to grow at the rate experienced from 1973–1975, but is instead limited to the 1975 ratio that these costs were of total operating costs. In short, it is assumed that the growth rate of these costs was exceptionally high in the 1973–1975 period and that in the long run these costs will represent a fixed percentage of total costs. The *Historical C Scenario* is the same as *B* except that local phone rates are assumed to increase at the same rate as that assumed for general inflation (10 percent in the base years of 1973–1975 and therefore, by assumption, in the projection years 1975–1985 as well).

In the *No-Inflation Scenario,* a noninflationary environment is assumed so that costs and revenues grow at "real" rates, that is, only as volume grows. No price adjustments and no diversions of revenue to competition are hypothesized. In the *No-Inflation Scenario* and all following (i.e., nominal) scenarios, volume is assumed to grow at the rates observed in the 1971–1977 period, except for common costs, which are kept at a fixed proportion of total operating costs. This longer time period was selected so that the *No-Inflation* and *Nominal* scenarios would not be unduly influenced by any short-term aberrations (see Table 2-12).

In all the *Nominal* scenarios, a general inflationary environment of 6 percent a year is assumed. This inflation is assumed to affect all

Table 2–12. Experienced Growth Rates for Various Elements of AT&T, 1971–1977 (percent)

Total revenues	12.0
Operating expenses	11.9
Number of conversations	4.5
Number of phones	4.2
Local revenues	11.0
Rate index	5.2
Volume	5.5
Toll revenues	13.1
Rate index	2.6
Volume	10.2

Source: American Telephone and Telegraph Co., *Statistical Report,* 1976, and 1977. Calculations by Charles River Associates based on data from source cited above.

items of cost. In *Nominal A,* local phone rates grow with the rate of inflation so that in real terms local rates are held constant. Revenues for "other" services, however, grow only with increases in volume; that is, the original rate structure is, by implication, held constant for this sector so that real rates decline with inflation. The *Nominal B Scenario* is basically the same as *Nominal A* except that local phone rates increase at half the inflation rate. In the *Nominal C Scenario* the same assumptions are used as in the *Nominal A,* except for two items: (1) the return on equity is constrained not to exceed 12 percent a year (which is the target level recently ordained by the FCC);[59] and (2) any potential surplus in net income (i.e., in excess of the 12 percent) is used to reduce proportionately "other revenues" from what these would have been had they grown at the rate assumed for the *Nominal A Scenario.* Thus, in the *Nominal C Scenario,* return on equity is held to a level of 12 percent a year by implicitly cutting rates in the other revenue category (which is otherwise expected to be under pressure from competition).

In the *Nominal D Scenario,* the same general assumptions are used as in the *Nominal C Scenario,* except that local phone rates increase only at half the inflation rate (as in *Nominal B*). Given that local phone rates increase more slowly, the amount allowable for potential diversion by rate cuts in "other" revenues will be smaller than in the *Nominal C Scenario.* In the *Nominal E Scenario,* the same general assumptions are used as in the *Nominal C Scenario.* The only modification is that the depreciation rate for plant and equipment is increased by one percentage point every year for the first five years of the projection and then stays at that higher level for the following years, effectively doubling AT&T's present depreciation rates. Return on equity is again constrained not to exceed 12 percent a year; here again, any potential surplus in net income is used to reduce proportionately "other revenues."

Results of the Scenarios

Table 2–13 reports for each scenario three measures of financial performance for the final year of the projection (i.e., 1985): return on equity, debt ratio, and an interest coverage ratio. Charts that compare the year-to-year progression of these ratios for each scenario are also shown; also plotted on these charts are the actual results for the years 1975 to 1978, a period when AT&T's net income after taxes rose by more than 69 percent.

In the *Historical A Scenario,* showing what could happen to AT&T if the 1973–1975 (EDC study) growth rates were extended to 1985, re-

Table 2–13. Impact of Competition on AT&T's Financial Performance in 1985

Scenario	Brief Description	Return on Equity (percent)	Interest Coverage Ratio	Debt Ratio (percent)
1975 actual results	Base year comparison	9.7	3.3	47.9
Historical A	1973–1975 experienced growth rates	12.0	3.4	45.0
B	Realistic common cost growth	19.2	4.9	45.0
C	Local rates at full inflation rate	29.6	7.2	44.1
No-Inflation	Only volume affects cost	16.4	5.2	44.8
Nominal A	Inflation, local rates at full inflation rate	24.2	6.5	44.9
B	Local rates at half inflation rate	18.2	5.1	45.0
C	Limit on return on equity; local rates at full inflation rate	12.0	3.7	45.0
D	Limit on return on equity; local rates at half inflation rate	12.0	3.7	45.0
E	Accelerated depreciation	12.0	3.8	44.0

Source: Calculations by Charles River Associates.

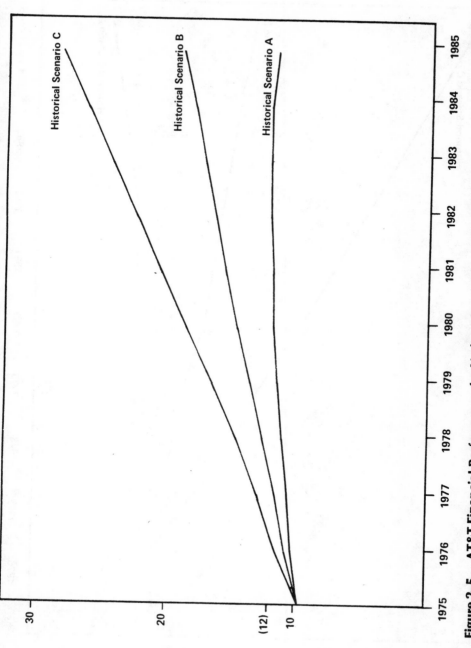

Figure 2–5. AT&T Financial Performance for Various Scenarios, 1975–1985: Return on Equity—Historical.
Source: Reproduced from data appearing in Appendix B.

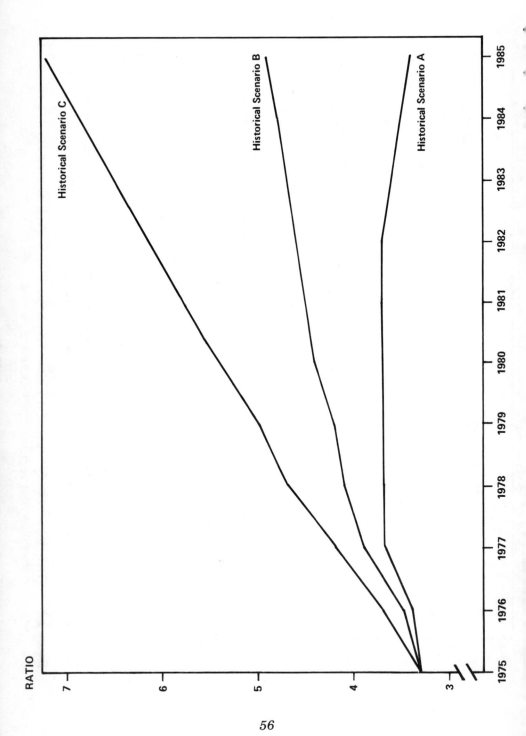

RATIO

Historical Scenario C

Historical Scenario B

Historical Scenario A

1975 1976 1977 1978 1979 1980 1981 1982 1983 1984 1985

7

6

5

4

3

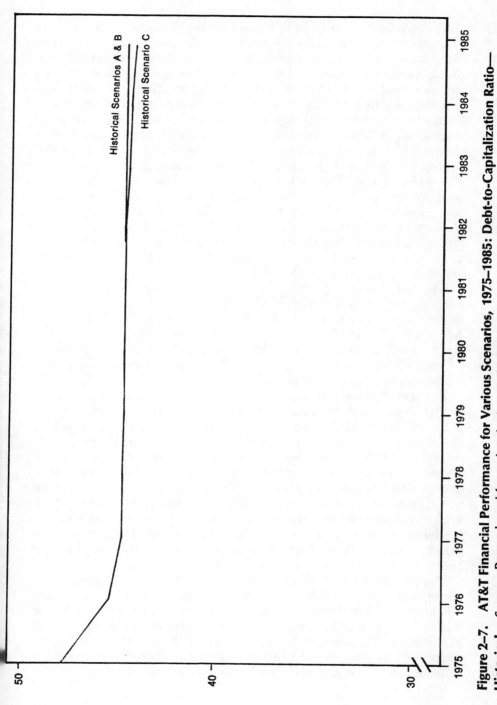

Figure 2–7. AT&T Financial Performance for Various Scenarios, 1975–1985: Debt-to-Capitalization Ratio—Historical. *Source:* Reproduced from data in Appendix B.

sults improve initially, only to deteriorate gradually in later years (Figures 2–5, 2–6, and 2–7). As can be seen by comparing the *Historical A Scenario* with *Historical B,* this deterioration is essentially due to the excessive growth of common or "overhead" costs; if common costs are held at a fixed percentage of total operating costs, as in *Historical B,* the financial measures improve every year, with return on equity topping 19 percent in 1985. *Historical C* shows that financial results would be even better if local phone rates were allowed to increase at the full inflation rate. The return on equity would reach 29.6 percent by 1985, and the debt ratio would slowly move below the 45 percent maximum. For the local phone service customer, this *Historical C Scenario* would mean that local rates would remain constant in real terms at their 1975 level.[60] By way of comparison, it was implicitly assumed in the *Historical A* and *B* scenarios that local rates would fall in real terms over the years. These and other price experiences under the different scenarios are summarized for 1985 in Table 2–14 and shown graphically in Figure 2–8.

As another way of illustrating these financial results, Figures 2–9 and 2–10 reproduce the *Historical B* and *C* results for 1985 in terms of bar charts similar to charts AT&T has utilized in its EDC studies. AT&T's EDC studies were done for only three years (1973 to 1975), and the cost and revenue breakdowns cannot be derived independently from published financial statistics. The EDC categories also do not conform to common financial practice.

Because of these differences and difficulties, only limited relevance should be attached to extrapolations based on the EDC methodology. If the EDC methods are accepted for the sake of argument, however, the assumptions underlying the *Historical B Scenario* imply that AT&T would have extra revenues of $10.9 billion in 1985 over and above revenues needed to generate a return on equity equal to that obtained in 1975 (see Figure 2–9). Using a higher return on equity would reduce this amount somewhat, but it would still leave AT&T with significant revenues to meet competition by cutting rates or losing business to competitors. For example, a cost of equity of, say, 15 percent would still mean extra revenues of $4.8 billion for the year 1985.

The assumptions underlying the *Historical C Scenario* lead to even larger surpluses by 1985. As shown in Figure 2–10, there are excess revenues of $23.0 billion over the 1975 rate of return on equity. If a return on equity of 15 percent is assumed, the extra revenues equal $16.9 billion. These amounts appear unrealistically large in comparison to 1975 net income ($3.1 billion), and under the existing regulatory structure, they probably would not be permitted to occur. The

Table 2–14. Telephone Company and General Economic Rate Indexes for 1985 under Different Financial and Development Scenarios (1975 = 100)

Scenario		Brief Description	General Price Level	Local Rates Index		"Other" Services Rates Index	
				Nominal	Real	Nominal	Real
Historical	A	1973–1975 experienced growth rates	259	167	65	131	51
	B	Realistic common cost growth	259	167	65	131	51
	C	Local rates at full 1973–1975 inflation rate (10%)	259	259	100	131	51
No-Inflation		Only volume affects costs	100	100	100	100	100
Nominal	A	Inflation, local rates at 6% inflation rate	179	179	100	124	69
	B	Local rates at half inflation rate (3%)	179	134	75	124	69
	C	Limit on return on equity; local rates at 6% inflation rate	179	179	100	96	53
	D	Limit on return on equity local rates at half inflation rate (3%)	179	134	75	110	61
	E	Accelerated depreciation	179	179	100	108	60

Source: Calculations by Charles River Associates.

figures do show, however, that, under plausible assumptions (growth rates of costs and revenues similar to those in the 1973–1975 period, a constraint on the growth in common costs, and local phone rates keeping pace with inflation), AT&T's financial position would be strong.

The *No-Inflation Scenario* indicates what would happen to AT&T in an inflation-free environment where unit costs are kept stable. Measures of financial performance improve every year, with return on equity reaching 16 percent by 1985 and the debt ratio also falling below 45 percent in the last few years (see Figures 2–11, 2–12, and 2–13).

The *Nominal A Scenario* represents the case where there would be little increase in competition for AT&T in an environment of 6-percent general inflation. Local rates in this scenario are projected to

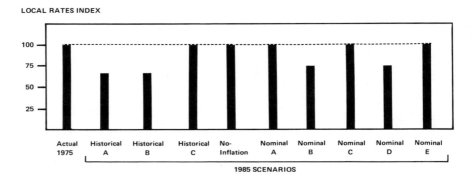

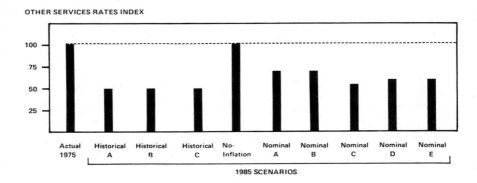

Figure 2–8. AT&T Telephone Rate Indexes for 1985 under Various Scenarios (in real terms, where 100 is equal to 1975 rates). *Source:* Reproduced from data in Table 2–14.

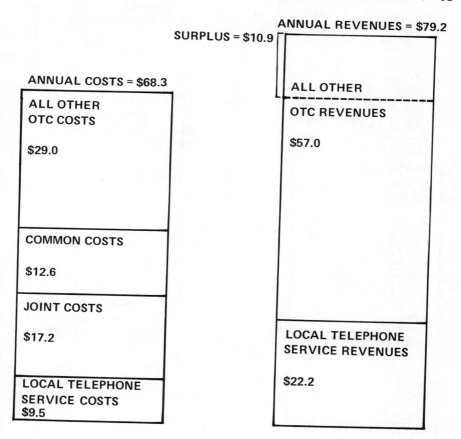

The figures shown are derived from the Historical B Scenario results appearing in Appendix B.

Revenue items are similar to those in Appendix B.

An allocation for return on investment before interest and after taxes was added to each of the four cost items, in conformity with the EDC study methodology. This allocation is in proportion to the relative investment in plant and equipment. The return on investment is assumed to be the same as in 1975.

The surplus represents the difference between annual costs (including the allocated return on investment) and annual revenues.

Figure 2–9.   Embedded Direct Cost Study Bar Chart Representation of the Historical B Scenario Results for the Year 1985 (billions of dollars).

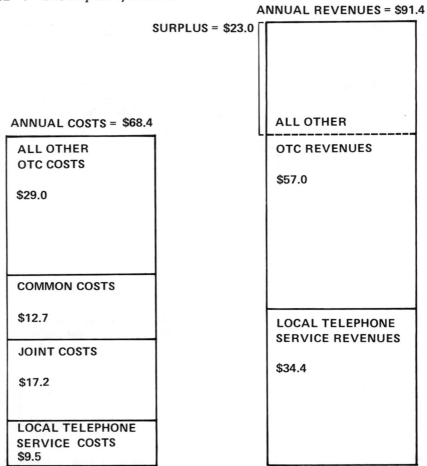

ANNUAL REVENUES = $91.4

SURPLUS = $23.0

ANNUAL COSTS = $68.4

ALL OTHER
OTC COSTS

$29.0

COMMON COSTS

$12.7

JOINT COSTS

$17.2

LOCAL TELEPHONE
SERVICE COSTS
$9.5

ALL OTHER

OTC REVENUES

$57.0

LOCAL TELEPHONE
SERVICE REVENUES

$34.4

The figures shown are derived from the Historical B Scenario results appearing in Appendix B.

Revenue items are similar to those in Appendix B.

An allocation for return on investment before interest and after taxes was added to each of the four cost items, in conformity with the EDC study methodology. This allocation is in proportion to the relative investment in plant and equipment. The return on investment is assumed to be the same as in 1975.

The surplus represents the difference between annual costs (including the allocated return on investment) and annual revenues.

Figure 2–10. Embedded Direct Cost Study Bar Chart Representation of the Historical C Scenario Results for the Year 1985 (billions of dollars).

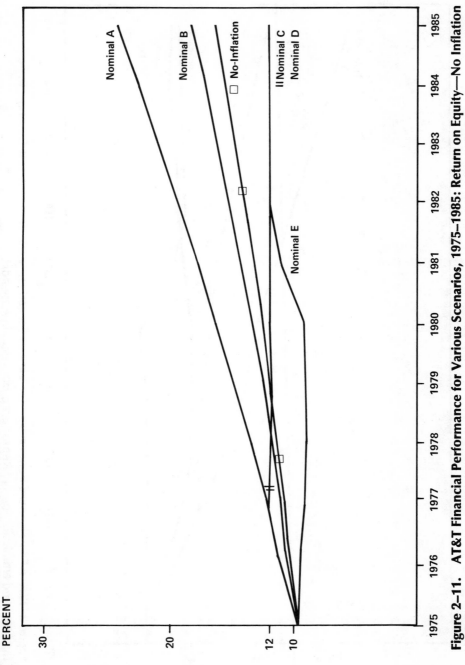

Figure 2–11. AT&T Financial Performance for Various Scenarios, 1975–1985: Return on Equity—No Inflation and Nominal. *Source:* Reproduced from data in Appendix B.

PERCENT

Nominal A

Nominal B

No-Inflation

Nominal C
Nominal D

Nominal E

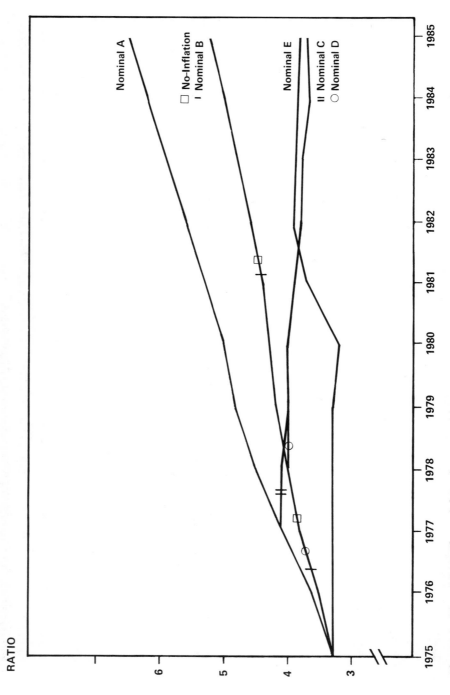

RATIO

Nominal A

No-Inflation
Nominal B

Nominal E

Nominal C
Nominal D

1975　1976　1977　1978　1979　1980　1981　1982　1983　1984　1985

Figure 2–12. AT&T Financial Performance for Various Scenarios, 1975–1985: Interest Coverage Ratio—No Inflation And Nominal. *Source:* Reproduced from data in Appendix B.

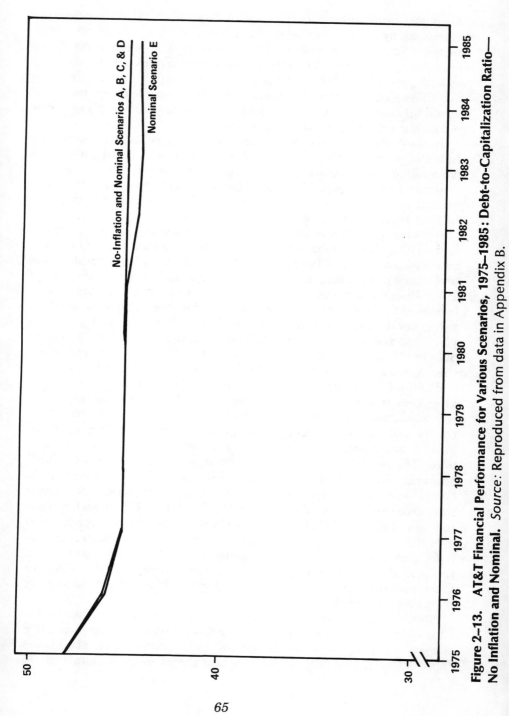

Figure 2–13. AT&T Financial Performance for Various Scenarios, 1975–1985: Debt-to-Capitalization Ratio—No Inflation and Nominal. *Source:* Reproduced from data in Appendix B.

65

keep pace with inflation while nominal rates on "other" services are increased by 24 percent over the 10-year period, which effectively means a "real" cut in rates for these services of 31 percent. As can be seen from the charts, this would gradually drive return on equity up to 24.2 percent in 1985, and the interest coverage ratio to more than six times. The 45-percent maximum debt ratio is maintained, but seemingly not so as to require equity issues that seriously jeopardize achieving a favorable return on equity.

In the *Nominal B Scenario,* where local phone rates increase at only half the general inflation rate and all other rates are adjusted as in *Nominal A,* the measures of financial performance again gradually improve over the projection years. Because local phone rate increases are cut to half of what they were in *Nominal A,* the return on equity in 1985 is cut from 24.2 percent to 18.2 percent.

Nominal C Scenario incorporates inflation, competitive diversion, and slimmer profit margins for AT&T. This "worst case" scenario nevertheless results in a better financial picture for AT&T in all of the projection years than was actually experienced in the base year of 1975. Inspecting Figures 2–11, 2–12, and 2–13, a good financial performance for *Nominal C Scenario* occurs while local phone rates follow the general price level in the economy (i.e., increase at 6 percent a year) and nominal prices for all "other" services are cut by 4 percent over the 10-year period, as compared to their 1975 level. In real terms, these price reductions on all "other" services amount to 47 percent.

Nominal D Scenario differs from *Nominal C Scenario* mainly in its allocation of so-called growth dividends between local and "all other" services. Except for the first four projection years when the return on equity and interest coverage is slightly higher in *C* than in *D,* all other years present virtually similar measures of financial performance. The return on equity constraint imposed on *Nominal C* and *D* scenarios, in conjunction with their high financial needs and traditional depreciation policy, forces the interest coverage ratio down over the years from a peak of 4.1 reached in *C* for the 1978 projection year to a low of 3.7 in 1985; this 3.7 ratio is nevertheless an improvement over the 1975 ratio of 3.3. The surplus available for diversion in *Nominal C* and *D* scenarios exceeds by a wide margin the estimates made in the preceding section for revenues of competitors in 1985. Thus, in these scenarios, AT&T's financial performance would not be hurt by increased competition.

Nominal E Scenario shows the impact of combining a higher depreciation rate with a constraint on return on equity and the possibility of a competitively induced diversion of traffic and revenue. The higher depreciation rate effectively cuts in half the average economic life of

AT&T's plant and equipment, from around 20 years to 10 years; such a shorter life might be needed in a competitive environment where technological innovation and new product and service introduction would proceed at a faster pace.

The measures of financial performance in *Nominal E,* despite some transitional difficulties in the early years, do show improvement over the decade. Return on equity, after a slight decrease in the initial years, soon hits its constrained maximum of 12 percent; interest coverage increases to nearly four times, while the debt-to-capitalization ratio decreases to 44 percent. Interest coverage, moreover, is understated by the EBIT[61]/interest ratio for this scenario, since depreciation funds are *not* included in EBIT and in this scenario, by definition, depreciation is accelerated.

Furthermore, net debt and equity issues are much smaller under this *E Scenario* than under all the others; this result is made possible by the higher cash flow generated by the higher depreciation charges and compensating higher revenues (as explained fully in Appendix B). Equity issues would be required only in the early years of the projection (because of the assumption that debt must be cut to 45 percent of total debt and equity), but return on equity and dividend maintenance would be somewhat strained (see Appendix B).

The obvious way out of such difficulties in the real world setting would be to keep the debt-to-debt-plus-equity ratio temporarily at its 1975 level of about 48 percent for a few years (say until 1980) and only subsequently reduce it to the target 45 percent ratio. This would reduce the need for new equity issues in the early years and thereby alleviate pressures on dividends and return on equity.

In *Nominal E Scenario,* customers again experience local phone service rates equal in real terms to those of 1975, while rates for other services are reduced 40 percent overall in real terms; here again, the surplus available to meet new competition seems comfortable, given the estimates of competitive growth outlined in the preceding section.

Conclusions

Clearly, if the preceding scenarios are a reasonable representation of the range of development possibilities confronting AT&T in the near future, potential competition should have little adverse effect on AT&T's financial performance in the years from now to 1985. All three selected measures of financial performance are generally better in the projection years than in 1975. Moreover, in none of the scenarios do local rates increase any faster than prices in general, and in some of them local rates fall relative to the general price level. Final-

ly, the general trend is toward improvement in the financial performance ratios, rather than deterioration, as the decade ends; accordingly,
extrapolation beyond 1985 should not profoundly alter the
conclusions.

SUMMARY

The telephone industry is no stranger to competition. During
the period following the expiration of Bell patents at the turn of the
century, competition was widespread at the local level among Bell and
non-Bell providers of telephone services. While this competition no
doubt resulted in some duplication of services and in "wasteful overinvestment," during the period rates also dropped and the size of the
telephone industry increased.

In the second decade of this century, competition effectively ended
because of the interconnection of the independent telephone companies with Bell. Since then, the growth in the industry has been generally slow to moderate[62] but steady. For example, measured by the
growth in the number of telephones per year, the industry has expanded annually at a 4.5-percent rate since 1920.

Beginning in the 1960s, the telephone industry was once again confronted with the possibility of competition, but this time in long distance rather than local service. In particular, limited competition has
emerged in private line services and in long distance services among
selected major metropolitan areas over the last decade. In the mid-
1970s, this competition has spread beyond long distance service, as
new entrants into the market offered customer premise equipment to
both businesses and residential customers. As earlier, competition
may very well be associated with increased growth rates for the industry,[63] augmented by impending technological changes creating new
services and demands for telecommunications.

Estimates by industry observers of the potential revenues that competitors might achieve in the near term as a result of competition with
AT&T are, moreover, quite small compared to current and prospective
AT&T revenues. Examined against AT&T's financial outlook under a
variety of development scenarios, these diversions do not seem to pose
a serious threat to AT&T. Indeed, the "growth dividend" generated by
the development of the telecommunications market in a technologically advancing milieu should mean that enough potential revenues will
be available to AT&T so as to effectively meet price competition while
still not jeopardizing AT&T's financial capability.

NOTES

1. It is alleged that J. P. Morgan assisted Widener with some streetcar financing difficulties in New York in exchange for Widener's withdrawing from the effort to compete with Bell long lines. See Richard Gabel, "The Early Competitive Era in Telephone Communication, 1893–1920," *Law and Contemporary Problems* 34 (Spring 1969): 350; see also FCC, "Report on the Investigation of the Telephone Industry in the U.S." (Washington, D.C., 1939), p. 131. For a contemporaneous account, see *Commercial and Financial Chronicle* 69 (1899): 1151.
2. For these and other reasons, the FCC is considering modifications in its accounting requirements. See FCC, "Revision of the Uniform System of Accounts and Financial Reporting Requirements for Telephone Companies," FCC Docket No. 78–196, Notice for Proposed Rule-Making, 1978.
3. This was just after J. P. Morgan's death and may or may not have been coincidental. Historical judgments on this point seem inconclusive or in conflict.
4. The Minnesota rate cases for 1913 apparently provided the legal basis for this approach. See *Minnesota Rate Cases,* 230 U.S. 352.
5. David C. Mitchell, Controller of the Rochester Telephone Corporation, in "The Impact of Competition in Communications on Settlements and Separations," *Assessing New Pricing Concepts in Public Utilities* (East Lansing: The Institute of Public Utilities, Michigan State University, 1978), p. 377.
6. A good summary of these activities can be found in Manley R. Irwin, "Implementing Competition in Intercity Communication Services," paper prepared for the *Ad Hoc Committee for Competitive Telecommunication,* mimeo, September 1977. Also see Bro Uttal, "Selling is No Longer Mickey Mouse at AT&T," *Fortune* (July 17, 1978): 98–104.
7. Docket 78—2, Canadian Radio-Television and Telecommunications Commission, in Re: Canadian Pacific Limited and Canadian National Railway Company vs. Bell Canada. Application dated June 14, 1976.
8. Further discussion of this point can be found in Chapters 5 and 6.
9. Table 2–2 shows slightly differing growth rates for the Bell System versus the market due to the inclusion of 1894 in the period of patent monopoly. The monopoly ended during the year; by the end of 1894, there were 15,000 nonconnecting independently owned telephones. Total telephones at the end of 1894 numbered 285,000.
10. Part of this decline in the growth rate in the early 1890s was, however, almost surely due to generally depressed business conditions in these years.
11. One must be careful about making comparisons between growth associated with early competition in local exchange service and growth associated with current competition in such areas as private line services. The earlier period was one of low penetration *in numbers of instruments* while *telephone services* at present are much more numerous and accessible. Nevertheless, the record clearly shows that telephone competition in the past has not been associated with retarded growth.
12. *AT&T Annual Report* (1909), p. 28. Also see J. H. Ainsworth and G. R. Johnston, *A Discussion of Telephone Competition* (Columbus: The Ohio Independent Telephone Association, February 1908), p. 29.
13. Disentangling and assigning particular causative roles to these various influences would go beyond the present analysis, and is probably almost impossible to do. An indication, though, that Bell's price reductions followed rather than led competition comes from analyzing the 1894–1899 period separately. In this initial competitive period, Bell's prices dropped 15 percent while competitors' prices declined by 20 per-

cent. The fact that competitors' prices were initially lower and fell at a faster rate than Bell's is suggestive of at least some competition, not just technological change.

14. Bell's acquisition of the loading coil in 1900 impacted toll development while the decline in total operating revenue per station may have reflected (among other causes) the expiration of basic Bell patents. See "Report of the Federal Communications Commission on the Investigation of the Telephone Industry in the U.S.," House Document No. 340, 76th Congress, 1st Session, 1939, pp. 132–133.

15. The Bell System was taken over by the government during World War I, and revenues per station may have dropped "artificially" due to wartime price controls.

16. There is little indication that the Bell System's profits were unduly limited by competition. Data on net income per dollar of book value of capital stocks show rates of return between 10 and 14 percent for the 1900–1920 period (see FCC, *Proposed Report; Telephone Investigation* [Pursuant to Public Investigation No. 8, 74th Congress], 1938, p. 608) with the exception of years during World War I (see preceding note). Annual rates of inflation for this period averaged 3 to 4 percent.

17. This finding of little relationship between AT&T performance before and after Carterfone is, of course, not particularly surprising if one views the decision as of little consequence until augmented by the FCC's registration program of 1977. If 1977 is regarded as the effective date for "feasible" or "true" competition, it is still too early (as of 1978–1979) to evaluate the consequences of any new or potential competition.

18. More discussion and documentation on these points can be found in Chapters 4 and 5.

19. For several industries telecommunications costs are a significant proportion of total expenses. One report itemized telecommunications costs as a percentage of total operating expenses for the airline industry at 4 percent, for banking and finance at 1.5 percent, for the insurance industry at 2 percent, for manufacturing at 0.5 percent, and for the securities industry at 10 percent. Cf. James Martin, *The Wired Society: A Challenge for Tomorrow* (Englewood Cliffs, N.J.: Prentice-Hall, 1978), p. 198.

20. It should be stressed that the scenario sketched in Table 2–6 is a hypothesis, not a forecast.

21. The use of the year 1990 as the target for conversion of present paper media to electronic communication is, of course, arbitrary and is not intended as a prediction. Allowing 13 years for the development of the possibilities in the Table 2–6 scenario does not imply, moreover, an overnight transition; nevertheless, it is not at all certain that by 1990 such a scenario will actually occur. The exercise outlined in Table 2–6 is intended more as a sensitivity analysis than as a firm forecast of the future.

22. However, it may be optimistic on the prospects for developing terminals to store information electronically at the household.

23. Video transmission by wire presently requires the equivalent of 1,200 one-way voice channels. Were all communication in this scenario to involve two-way video, the implied annual growth rate in the household market would be approximately 100 percent per year. However, increased demand for such services would probably stimulate research to lower video bandwidth requirements.

24. During the 1970–1976 period, annual growth in total telecommunications usage ranged between 5 percent for local exchange volume and 9.3 percent for toll telecommunications volume, implying an approximately 6.5 percent annual growth in usage during those years. Specifically, 7 percent is an estimate based on the 5 percent growth in local exchange use and 9.3 percent growth in toll volume as shown in Table 2–5.

25. Again, 1990 is an arbitrary date and the scenario in Table 2–7, as in 2–6, is not meant to be a forecast.

26. A history of the recent legal developments can be found in the Memorandum of Evidence (of Herbert E. Forrest) submitted by CNCP Telecommunications (April, 1978) in Re: Canadian Pacific Limited and Canadian National Railway Company vs. Bell Canada. Application dated June 14, 1976.

27. *Carterfone*, 13 FCC 2d 420 (1968), recon. denied, 14 FCC 2d 751 (1968).

28. See *First Report and Order Docket No. 19528*, 56 FCC 2d 593 (1975) and *Second Report and Order in Docket No. 19528*, 58 FCC 2d 736 (1976).

29. Federal Communications Commission, *In the Matter of Allocation of Frequencies in the Bands Above 890 MC*, Memorandum Opinion and Order, 29 FCC 825 (1960), FCC Docket No. 11886.

30. *Specialized Common Carrier Decision*, 29 FCC 2d 870 (1971).

31. For example, in anticipation of growth in the demand for this equipment, General Telephone and Electronics has increased its phone production by 20 percent. ("Classy Phones: A Grab at Ma Bell's Market," *Business Week*, June 19, 1978: 92B–92I.) Northern Telecom, Incorporated, projected that by 1978 at least 10 percent of the total consumer telephone market would be sold through retail outlets with sales mounting to $212 million. Northern Telecom also boosted its production by 168 percent, increasing output from 207,500 units in 1977 to 556,000 units in 1978. John Crudele, "Phone Manufacturers Seek Retail Outlet at CES," *Electronic News* (June 19, 1978). Northern Telecom sales revenues reported in *Electronic News* (May 8, 1978): 1.

32. "The New New Telephone Industry," *Business Week* (February 13, 1978): 68–78.

33. Bell now sells the exterior of "design" phonesets while retaining ownership of its interior mechanical components.

34. For further discussion on AT&T's attempt to become more consumer oriented, see Uttal, "Selling Is No Longer Mickey Mouse at AT&T," pp. 98–104.

35. See Crudele, "Phone Manufacturers Seek Retail Outlet at CES."

36. Ibid.

37. Ad Hoc Committee for Competitive Telecommunications (ACCT), "The Case for Competition in Telecommunications," May 1978; and Harry Newton's article, "1978: The Year Telco Competition Becomes Real for the Non-Telco's?", *Telephone Engineer and Management*, January 15, 1978, pp. 91–95.

38. Note that even moderate increases in unit volume appear as large percentage increases, given a small initial base. Revenue data are taken from the Newton article cited above.

39. Stanford Research Institute, "Market Forecasts for Terminal Equipment Areas," March 21, 1975. Prepared for American Telephone and Telegraph Co., Bell Exhibit 19, FCC Docket No. 20003.

40. Reported in *Telecommunications Report*, May 1, 1978. Also see National Association of Regulatory Utility Commissioners, Report After Investigation, *An Investigation Into the Economics and Quality of Service Impact on Telephone Service Subscribers Resulting from the Interconnection of Subscriber-Provided Equipment to the Public Switched Telephone Network, and from Competition by the Specialized Common Carriers in the Provision of Telecommunications Services, May 15, 1974.*

41. New York Public Service Commission, *The Review of Cost Impact of Interconnection Within the Service Area of New York Telephone Company*, January 1975.

42. Packet-switching carriers allocate bandwidth to a customer only when a block of information is ready to be sent, and only for sufficient time for that one block to

travel over one network link. In this way, many users can share the same transmission line previously dedicated to only one user.

43. Ad Hoc Committee for Competitive Telecommunications (ACCT), "The Case for Competition in Telecommunications."
44. Arthur D. Little, Inc., *Business Communications 1975–1978* (Cambridge, Mass.: May 1979, p. 29).
45. Newton, "1978: The Year Telco Competition Becomes Real for the Non-Telco's?"
46. ACCT, "The Case for Competition in Telecommunications."
47. See the *Business Week* article of February 13, 1978. Moreover, it was reported that total business telephone service expenditure exceeded $25 billion for 1977.
48. Telenet has now merged with General Telephone and Electronics, Inc.
49. About one-half of MCI's revenues, according to the company's annual reports in 1977 and 1978, were generated by its Execunet shared service. By comparison, revenues from AT&T's interstate MTS currently equal seven times the revenues generated by its interstate PLS (private line service) offerings (see Table 2–5).
50. The future financial health of the landline segment of the SCC industry would seem to depend heavily upon court decisions about MCI's Execunet service, which provides a service very much like AT&T's wide area or WATS offering. These court decisions could allow specialized carriers to offer direct-dialed long distance service in competition not only with AT&T's WATS, but possibly with AT&T's message toll service (MTS) as well. See Stephen A. Caswell, "Specialized Carriers in the Balance," *Computer Decisions,* January 1978, pp. 52–54.
51. *The Wall Street Journal,* September 8, 1978.
52. "Pie in the Sky," *Newsweek* 93 (January 22, 1978): 54–56.
53. "The New New Telephone Industry," *Business Week.*
54. Such a direct loss of revenues should, of course, also reduce AT&T's direct costs.
55. Newton, "1978: The Year Telco Competition Becomes Real for The Non-Telcos?" p. 91.
56. Alfred E. Kahn, *The Economics of Regulation,* Vol. II (New York: John Wiley and Sons, 1971), p. 136, notes that the FCC staff in Docket 18920 estimated that only 2 to 4 percent of Bell's existing business would be subject to competition. These estimates, though, did not contemplate subsequent court decisions that invalidated FCC rulings that the specialized common carriers should compete only in private line business.
57. Services deemed vulnerable to rate erosion by competition were based primarily on judgments as to whether these services were substitutes or complements. Since current competition in the private transmission market is on long-haul routes, only interstate revenues were deemed vulnerable. To the extent that certain intrastate routes (particularly in large western states) are vulnerable to competition, such a judgment understates potential revenue erosion. Countering this assumed immunity of intrastate routes is the assumption that *all* interstate MTS and PLS routes are vulnerable to rate erosion.
58. The 1973–1975 period was selected, both because it covers the years for which AT&T made EDC studies available and because it was not a strong financial era for the Bell System. During this period the overall economy was inflating at double-digit rates; telephone costs, particularly for local services, were rising faster than the operating telephone companies could recoup through rate increases granted by the regulatory commissions.
59. FCC Docket 20376, 57th volume, FCC Reports, 2nd series (February 5, 1976), p. 972. Of course, certain parts of AT&T are regulated by state agencies, which may, and effectively often do, allow a different rate of return.

60. Economists make a distinction between "nominal" and "real" units. In the present case, "nominal" rates refer to rates expressed in terms of current monetary units; when they are corrected for the change in the general price level over time, these rates are then referred to as "real" rates.
61. Earnings before interest and taxes.
62. As compared to the telephone industry growth rate in number of telephones during the earlier competitive period, growth has been slower in recent decades. As compared to revenue growth in other industries, the telephone industry has done moderately well or has even been a growth leader in many respects.
63. Growth in recent years is probably better measured by diversified services and their associated revenues than by increases in numbers of telephones.

Chapter 3

Pricing

GOALS AND CRITERIA

An analysis of telecommunications pricing issues seems best begun by defining what society seeks through imposing regulations on public utility pricing policies. In this regard, the standard literature in economics and law suggests at least eight basic goals identifiable with regulatory rate making: (1) universal service; (2) static efficiency in resource allocation; (3) equity for different kinds of users and services; (4) financial self-sufficiency (total revenues equal to total cost); (5) prevention of uneconomic entry; (6) consistency with expected technological change; (7) administrative simplicity; and (8) historical continuity. A discussion of the rationale and implications of each of these eight goals follows.[1]

Universal Service

Probably the most widely proclaimed goal of pricing policy in the telephone industry is to achieve universal or widespread connection with the system.[2] The basic notion of achieving universality for communications is very deeply embedded in American folklore, politics, and legislation. It is, moreover, in the interest of both the telephone company and the public to have as many households and businesses attached to the phone system as possible. Universality of connections enhances values, both private and public; in particular, certain externalities (in the economic sense) can be gained from universal service.[3] A desire for a ubiquitous service has also had much to do with the historical pattern of rate development in the industry and particularly the reliance on relatively flat or average rates applied across broad geographic areas.

The public policy concern with universality has recently been reaffirmed by increased interest in what are generally called "lifeline" rates. Simply put, the lifeline concept is that a relatively low charge should be made for attaching to the system so as to provide everyone with an opportunity to be in contact with the outside world (i.e., maintain a lifeline connection to the outside world). This goal is deemed especially desirable for households headed by older people or by others with special physical limitations or handicaps.

Static Allocative Efficiency

Pricing, even in a regulated public utility, is expected to achieve a reasonably efficient utilization of the facilities involved. Thus, even under regulation, prices should reflect the extent to which resources are used to render the service being consumed. Without such an approximation, consumers would be led to either overconsume the service if prices were set too low relative to costs or to underconsume the service if prices were set too high relative to costs. In particular, prices are generally deemed to be a good means of inhibiting consumers from the wasteful use of a resource. Experience has taught that a zero price on almost anything of value — whether it be a pharmaceutical prescription in a nationalized medical plan[4] or a transit system in a central city [5]or use of a high performance urban expressway[6]—almost invariably leads many people who really do not place much or any value on such consumption to consume. This, in turn, may often interfere with or even prevent consumption by others who really do value the service or product. Very modest changes in such circumstances often greatly improve the satisfactions derived from the consumption of such goods and the ability of producers to supply them. Rates that reflect marginal costs more accurately should also help people to make better economic decisions "downstream"; for example, they should help telephone users locate efficiently relative to their demands for telephone service.

Another efficiency goal often sought through pricing is to establish a better balance between the location or timing of demands and the ability or capacity to meet such demands. For example, it has long been recognized in electric utilities that it is usually advantageous to charge lower rates for use of electricity during off-peak periods than during peak periods; in simplest terms, such differential pricing usually more closely reflects costs, provides a reduced need for financing, and improves companies' ability to determine total demands for investment or capacity in the industry.[7] Similar observations have long been made (but infrequently applied) to the use of urban transportation facilities.[8] The same concepts, however, often show up in the pricing policies of unregulated, private business activities, for example, higher charges for so-called high season use of resort hotels or higher greens fees on golf courses on the weekend.

Equity

The notion that rates should not be too inequitable in their distribution of the cost burdens between different groups in a society has also influenced rates in regulated industries throughout much of U.S. history. These equity concepts can take many forms (some of which may be inconsistent and not sustainable under close scrutiny). For example, rural users of the telephone service in the United States have traditionally experienced lower monthly connect charges than have urban dwellers,[9] as seen in Figure 3–1. Among other justifications, it might be deemed to be in the national self-interest, of both urban and rural dwellers alike, to have people at remote locations in contact with the more centrally or densely populated sectors of society (cf., universality goal cited above). However, a lower charge for rural phone service could also be inversely related to the costs incurred since rural phone loops are longer and commonly alleged to be more expensive than urban phone loops. Acceptance of a lower rate for rural dwellers therefore may need to be justified by some broader public policy goals, such as equity and universality considerations.

Businesses also generally pay more for their service than do residential users, as again can be seen from Figure 3–1. Value of service, or the probability that businesses use their phones more on the average than do residences, might justify such rates. However, notions of equity may also be involved, based on the view that businesses are "better able to pay" than ordinary residential users. Of course, this overlooks the possibility that in the long run residential users may pay for higher business telephone charges in higher prices, lower wages, lower dividends, and so forth.

Equity, of course, can be a very ambiguous concept. One man's equity is not necessarily another's. About the best that often can be done as a guide to public policy is to suggest that people in roughly the same situations should pay about the same charges.[10] Equity considerations may also accentuate the contradictions hidden in regulatory rate setting, as equity can often conflict with efficiency or financial self-sufficiency goals.

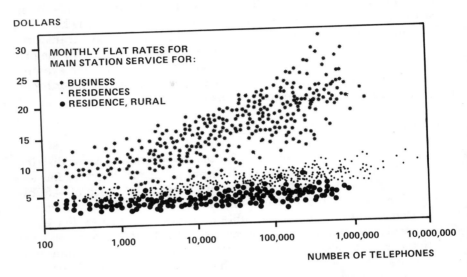

[1] Monthly rates are shown for basic flat rate main station service for individual (one-party) business and residence telephones and for rural service that extends into the hinterlands with up to eight parties on a line but no charge for mileage.

Rates vary with the number of telephones in the subscriber's service area, i.e., the number of phones that may be reached without additional charges. Through price-averaging, the variation within any one service may or may not reflect cost variation. The concept of "value-of-service" pricing is most clearly illustrated by the large differences between the prices and the variances for business and residential services that use essentially the same facilities. In some of the largest service areas only message-rate service is offered, especially for businesses.

The data show rates for Bell Companies. Current data are not available for most independent companies, since the U.S. Independent Telephone Association stopped compiling such figures ca. 1970. Data for California, Missouri, New York and Texas are not included.

Data compiled from National Association of Regulatory Commissioners, *Exchange Rates in Effect June 30, 1974* (Washington, D.C.: NARUC, 1975).

Figure 3–1. How Telephone Service Is Paid For. *Source:* Paul J. Berman and Anthony G. Oettinger, "The Medium and the Telephone: The Politics of Information Resources" (Cambridge, Mass.: Harvard University, Program on Information Technologies and Public Policy, 1976).

Financial Self-Sufficiency

Another major goal of the pricing policies of private telephone companies is to yield sufficient total revenues to cover total costs, including sufficient return on investment to attract the financing needed to maintain and develop the industry's network. This total revenue requirement has been codified in the regulatory structure through a determination that telephone companies be permitted rates that achieve a so-called fair rate of return on invested capital. Unfortunately, such an approach to regulation may also have led the industry into practices that lower its efficiency.[11] It is quite evident, however, that as long as the telephone company remains a private enterprise operating within the context of a market economy, failure of its prices or rates to yield total revenues equal to total requirements, including revenues needed for financing, could result in an unfortunate withdrawal of capital, and thereby service, from the industry. The U.S. railroad industry's experience is illustrative of the mechanics of how such a withdrawal can occur when regulation results in inadequate returns.[12]

The requirement that total revenue cover total cost has a long history in economics, going back to Ramsey in the English literature[13] and Dupuit on the Continent even before that.[14] The basic idea is that if total revenues do not cover total costs, the activity may not be worth the undertaking. In a market economy, moreover, the total revenue coverage of total costs is a prerequisite for purely private sector activities, and the relaxation of this requirement for regulated or public sector activities could create an overallocation of resources to regulated or public enterprises.[15]

Prevention of Uneconomic Entry

Another goal of pricing policy in regulated industries is to preempt wherever possible entry of competitors whose presence would erode overall industry efficiency. A social justification for such preemption can be developed in many instances, particularly where a natural monopoly is believed to exist in at least part of the activity.[16]

The possibility of uneconomic entry is especially likely when widespread use of value-of-service rate making opens up large discrepancies in the markups applied to different portions of the business. Clearly, if price is well above the cost of rendering a particular service and entry is likely, it normally pays the established vendor to reduce price to any level that remains above cost and yet keeps the business. Many of the problems of the U.S. railroads, for example, are attributable to their failure to recognize this fact promptly and to their permitting much highly profitable (so-called high-value) traffic to be diverted from rail to truck competition, particularly during the 1930s and 1940s.[17]

The same principle may be at work in telecommunications in AT&T's response to Datran's offer of competitive services in long lines; specifically AT&T responded to Datran proffered competition by quoting rates that AT&T insisted were fully compensatory but which Datran found so low as to be intolerable, to the point of forcing Datran into bankruptcy and eventual absorption by Southern Pacific Communications. From a slightly different perspective, looking back at Table 1-2, according to AT&T's own figures it might be possible for AT&T to charge rates as much as four times higher than their prospective costs for rendering private line services from now until 1991 and still preempt entry by others into these services.[18] Furthermore, if AT&T's cost estimates are correct, it would be in the public interest to let AT&T preempt entry by others since under these circumstances AT&T would be the low-cost vendor. Great regulatory debates are generated, however, over the issue of whether pricing action to preempt entry is really cost-justified or simply "predatory" pricing (i.e., at levels below cost) pursued by financially strong, large enterprises to prevent competitors from disturbing what is deemed to be an otherwise advantageous status quo.[19]

Consistency with Technological Change

A rate structure reasonably compatible with probable future technological changes in an industry is also usually deemed desirable. This can be particularly true in a regulated environment because of "regulatory lag." In virtually all regulatory situations legal challenges can result in lengthy hearings. As a consequence, there will be a lag between seeking a change and its implementation. In an evolving situation, whether moved by technology or other forces, those in a regulated industry can thus be well advised to anticipate developments.

All else being equal, a price structure that is more or less in line with technological and cost developments in an industry is easier to protect or sustain than one that becomes increasingly inconsistent with such trends. Of course, where prices do not rise as costs rise, this is acutely obvious — and often ends with bankruptcy. A lag in bringing prices down when costs decline, although a profitable strategy for the short run, may also prove very troublesome or counterproductive in the long run. For example, new competition may well be encouraged, including uneconomic entry. Relocation is yet another way for users to avoid high markups if prices become increasingly out of touch with costs. Use of entirely different kinds of communications, transportation, or production technologies is still another possibility.

In all these cases the end result of not keeping prices in phase with change can be quite unfortunate for the original regulated enterprise, and it is very likely to be unfortunate for society as well since incentives can be created to use inefficient alternatives (including wrong downstream investment or location decisions by those who use the service). Development of inefficient alternatives, in turn, can make it more difficult subsequently to unravel the mistakes of regulation. That is, regulation can create a set of vested or committed interests outside the regulated industry itself whose only "crime" may have been to respond intelligently to the price signals provided by the regulators.[20]

Administrative Simplicity

Very practical considerations for achieving simplicity of structure will condition the pricing policies of almost any firm, whether it is regulated or unregulated. For example, simplicity helps keep billing and bookkeeping costs low. Simplicity may also help achieve better consumer acceptance or understanding, which can sometimes be crucially important in a regulated industry.

Historical Continuity

It is generally considerd desirable to have prices evolve over time rather than move abruptly. Sharp changes in prices can be seriously distorting in various ways, as aptly illustrated by the recent sharp rise in petroleum prices engineered by OPEC in 1974-1975. Generally, slow evolutionary price adaptations can be accommodated by slow evolutionary adaptations in consumer or production practices; abrupt changes in prices, by contrast, can necessitate traumatic adjustments.

Historical continuity can also help with public acceptance. If rates change abruptly, investments made on the assumption that a certain set of rates would hold may be jeopardized, creating what are perceived to be inequities by some consumers or producers.[21] In general, very sharp unpredicted changes in a rate structure can be very disconcerting in a regulatory framework, with appeal to the political process for redress quite common.

THE PRESENT STRUCTURE OF TELEPHONE TARIFFS

An obvious question is the extent to which the present structure of telephone tariffs satisfies the eight criteria listed above. Overall the present telephone rate structure would seem to do only moderately well on this test. Moreover, not all of the various criteria have been equally well satisfied. This is hardly surprising since some of the goals are clearly conflicting. For example, historical continuity almost by definition will be difficult to reconcile with better anticipation of technological change; the equity and efficiency criteria are also not always compatible.

The telephone companies also may not accept this entire list of eight goals as legitimate or possible. The eight criteria, after all, devolve primarily from public policy considerations, and telephone companies are private concerns with interests that may differ from goals enunciated by public policy.[22] It is not clear, moreover, who or what— if anything—should bear the blame for any gaps between current rate practice and the goals listed. A long history is involved (going back to the nineteenth century) with many public and private participants.

Achievement of Universality

The first of the listed goals, universal service, is usually taken to mean that every household has at least one telephone. By this standard, the conventional estimate is that something like 94 percent of all U.S. housholds have main stations.[23] This compares well by international standards; even the high-income countries of Northern Europe rarely do better. Nevertheless, the number may be slightly misleading or overly optimistic in its implications if not carefully evaluated.

The reported percentage of 94 percent is arrived at by simply taking the ratio between the number of residential main stations in the United States divided by the number of households. Such a number does not actually tell how many households have telephones. Moreover, it clearly overstates the degree to which the telecommunications industry has achieved universal service.[24] In 1975, the number of housing units occupied year-round amounted to 72,523,000. However, census data indicate an additional 1,534,000 housing units for seasonal and migratory workers. Homes held for occasional use (second homes) accounted for an additional 1,050,000 units. The category of "other vacant" (1,246,000) would also appear to contain some seasonally occupied and second-home types of housing units. Depending on how many of these other housing units might possibly be deemed to have phone service, the total number of U.S. housing units could rise to a maximum of 76,353,000. With residential main station telephones estimated to be slightly more then 68 million in 1975, the estimate of telephones per household could be placed at anywhere from 89 or 90 percent to 94 percent.[25] Whatever the exact percentage, this figure is still a respectable achievement by most standards of comparison.

Efficiency Performance

The current telephone rate structure is immediately suspect when it comes to achieving the second goal, efficiency in resource allocation. It is a simple proposition of economic theory that rate structures largely invariant with intensity of use and time of day will not yield efficient utilization of capacity in systems characterized by such variations. The pattern shown in Figure 3-2, taken from data for the New York Telephone Company in 1975, seems fairly typical of the daily variation in telephone usage.[26] Three peaks (one occurring around 11 A.M., another around 3 P.M., and one in the evening) and a tremendous decrease in demand late in the evening, lasting through the nighttime and early morning hours, can be identified in most telephone systems.

In recognition of the inefficiencies implicit in such dramatic peaks and valleys, the idea of setting different peak and off-peak rates for local telephone calls is slowly spreading. Metering of the time used on a call is also somewhat more common today than it was a few years ago. Nevertheless, flat charges for local service are still basically the rule rather than the exception in the United States.[27] By contrast, usage metering and time-of-day differentials are more widespread in many foreign telephone systems.[28]

Another example of inefficiency in the current telephone industry rate structure may be the surcharge on touch-tone dialing. To the extent that touch-tone dialing permits more efficient utilization of the system, a touch-tone surcharge could be counterproductive in the long run. Similarly, to the extent that what were formerly junk mailings have become solicitations over the phone that tie up the system at peak hours, another symptom of inefficiency is present. Some recipients of solicitation calls might even argue that these calls should not be priced as though they are totally costless at the margin (as they are under flat rates) regardless of their timing, since they can obviously impose costs on recipients as well as originators. Another possible source of inefficiency in the present rate structure could be the FCC's alleged refusal to permit rates below fully allocated costs, even if the rates seem well above available or long-run incremental costs. Finally, the inability of the telephone companies to implement special charges for special services (such as exceptional use of information calls) further suggests that regulation sometimes inhibits development of a more efficient rate structure.

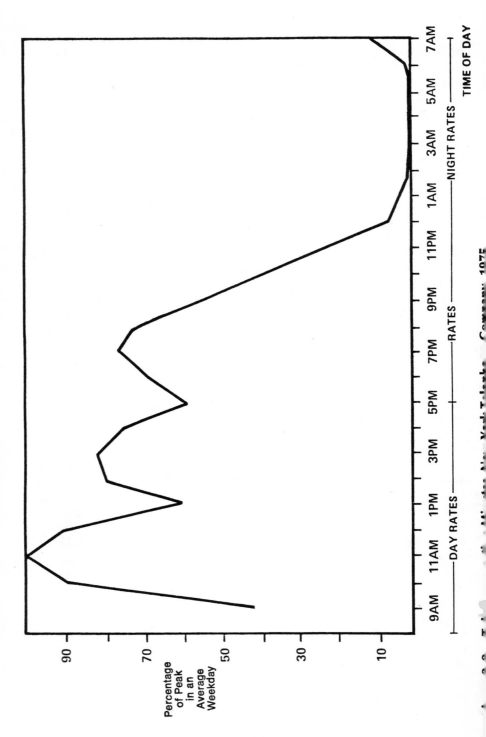

Equity in Theory and Practice

Some prevailing notions about the third goal, equity, are perhaps revealed by Figure 3–1, which shows the difference in rates to businesses and residential customers by size of exchange area. In small exchanges, businesses are charged twice as much as residential users. But in larger exchanges, the ratio of business to residential charges rises sharply, to something on the order of three to one. Even if businesses should be charged more than residential users (say, on value of service or intensity of use considerations), it is not clear what the proper ratio should be or why the ratio should vary with the size of the exchange area.

An attempt to identify equity concepts might also be made by scrutinizing the current rate structure. To this end, rates were sampled from the *Rate Handbook* [29] for exchange areas of comparable size from contiguous states. As shown in Table 3–1, significant differences were found in the rates charged by exchange areas of similar size located in different albeit contiguous states. In a comparison of small exchanges on opposite sides of the Florida/Georgia state line, for example, the Florida exchanges had business monthly charges of $19.20, whereas the Georgia exchanges charged $11.10; for residential users the charge in Florida was $8.10, while in Georgia it was $5.75. In large nearby exchange areas in Illinois and Indiana, the monthly charge in Illinois was $20 for business while it was $34 in Indiana. All of the companies involved in these comparisons, incidentally, are in the Bell System, and these differences in rates occur despite similar size exchange areas with presumbly similar value of service.

A variety of reasons, nevertheless, might account for these differences. Thus, it might be that rates are higher in Florida than in Georgia because Florida's telephone system is much newer on the average

Table 3–1. Sample of Bell Local Exchange Rates

Number of Telephones in Local Exchange	State	Monthly Rates	
		Business	Residential
0–1,000	Florida	$19.20	$ 8.10
0–1,200	Georgia	11.10	5.75
360,000–480,000	Illinois	20.35	7.90
380,000–460,000	Indiana	34.00	11.25

Source: National Association of Regulatory Utility Commissioners, "Exchange Service Telephone Rates in Effect June 30, 1976." Washington, D.C.: 1977.

and therefore was built in an era of inflating costs; these costs, in turn, are being recouped through higher connect charges. That Indiana's rates are higher than those of Illinois might be attributable to differences in costs or regulatory procedures in the two states, although these differences are not readily discernible.

Another interesting nonuniformity occurs in the surcharge for touch-tone service in different areas. The highest touch-tone surcharges are at least 2.5 times as high as the surcharges in some of the lower areas.[30] It would seem difficult to explain this discrepancy in terms of costs or value of service. In general, nonuniformities in local telephone rates can be discovered almost at will, and it is difficult to explain many of the disparities by cost differences or density differences or other characteristics.

Rate discrepancies or variations that are not readily explained by cost or value-of-service justifications also raise questions about the empirical validity of the sustainability argument that is sometimes advanced.[31] Under this line of reasoning, entry may be efficient for the entrant, but at the same time it may undermine a rate structure that is otherwise socially optimal (i.e., Ramsey-efficient in that it meets the conditions that total revenues equal total costs but has optimal allocative properties within this constraint).[32] By implication, some telephone company spokesmen sometimes seem to suggest that the existing rate structure actually possesses these optimality properties. The real rate structure, however, would appear to be a random construct of telephone history and various regulatory practices in which it is difficult to discern many formal or systematic properties.

Financial Health

The fourth goal, to have total revenues cover total costs, seems to have been reasonably well met by the telephone industry in recent years. Recent rates of return on Bell System equity have been on the order of 12 percent while that of the independent telephone companies have been in the 13-percent range.[33] Few financial analysts would argue, moreover, that telephone companies will encounter serious trouble marketing capital in the foreseeable future. AT&T itself and virtually all of the Bell subsidiaries still have triple A bond ratings, although a couple of Bell operating telephone companies have been downgraded recently to double A (which still seems somewhat short of financial jeopardy).

The only major indentifiable danger to financial self-sustainability for the telephone companies might reside in a claim made by some suppliers of telephone equipment. They argue that in the near future a write-off of millions of dollars of obsolete telephone equipment will be required because of historically low depreciation rates in the face of rapid technological change.[34] If so, lower tariffs induced by competition could exacerbate any problem encountered in raising the financial capital needed by the industry. However, as the financial analyses in Chapter 2 indicate, the difficulties can be exaggerated. Futhermore, to the extent that technological change can save on capital by reducing replacement costs on old equipment as it wears out, any depreciation shortfall will be alleviated.[35] Continuing steady growth in the capital stock also helps alleviate any such shortfall by providing an ever-expanding capital stock against which depreciation charges are assessed, and in advance of actual replacement needs.[36] Finally, if substantial debt financing is used, inflation in replacement costs for assets is at least partially offset (or shifted to lenders) by depreciation in the real value of the debt.[37] Nevertheless, if inflation persists and is not offset by these other considerations, even a 12- to 13-percent return might be deficient for the telephone companies, as for many companies in an inflationary economy.[38]

Prevention of Uneconomic Entry

With regard to the fifth goal, preemption of uneconomic entry, the telephone companies historically have relied mainly on either regulation or lawsuits rather than on utilizing the rate structure to discourage entry. For example, during the period of patent monopoly ending in 1893, the Bell System, not unreasonably, fought several lawsuits to protect its patent. However, the economic incentives to enter the industry provided by high rates of return became apparent with the expiration of the patent monopoly, for after 1894 a tremendous development of the independent telephone industry took place.[39] By 1907, in fact, the independents had nearly as many telephones as the Bell System did.[40] After that, with the changeover to a policy of acquisition of independents by AT&T, the Bell System outdistanced the competition and within 10 or 20 years had a share close to where it is now, that is, on the order of 80 to 85 percent of the market.[41] In recent years, with the FCC allowing some entry into long lines and terminal equipment, the fact that firms have entered the market again suggests that the current rate structure does not totally discourage entry, whether it is economic or uneconomic.[42]

To some extent, however, AT&T has shown signs recently of using rate policy more than in the past as a competitive tool. Certain WATS, private line, and telex rates that have been proposed recently (though often not allowed by the FCC) can be construed as attempts to prevent what AT&T envisions as uneconomic entry. In general, the argument has been made that when regulatory or legal remedies have been exhausted, the Bell System is quite capable of responding to entry with reduced rates.[43]

Achieving Consistency with Expected Technological Change

An evaluation of the technological consistency goal is perhaps best approached by considering recent changes in relative prices for various phone services. Looking at indices for the local rate structure from, say, 1960 to 1975 and deflating for the overall change in prices, local rates fell approximately 25 percent. During that same period, intrastate rates have fallen about 45 percent in real terms.[44]

These rate revisions would appear to be only partially aligned with corresponding developments in technology and cost. Local loop technology has developed less rapidly in recent years[45] than long distance technology. The available evidence would suggest that local loop costs remained constant or increased somewhat in real terms over recent years while intra- and interstate costs both declined, probably more or less by the same amounts.[46]

Achieving Simplicity

The present basic structure of residential rates is quite simple. Most householders know what their monthly charge is and what it has been. In most cases it is a flat monthly charge, plus or minus a bit in some instances for a few message units and toll calls.

It is substantially more difficult to argue that the existing rate structure for businesses is administratively simple. In fact, several consulting firms make a good living by reading business telephone bills to determine whether a business has the right services for its needs.[47] Some large firms even give in-house training courses just to teach people how to read telephone bills! Many large firms operating in different states also complain that different codes are used in different states for identical equipment. Furthermore, when it comes to new installations, it is often deemed a major undertaking to determine which of many alternatives is most economical.

The Historical Record for Continuity

Telephone rates could well be among the most stable prices in the U.S. economy as illustrated by Figures 1–1 and 2–3. Of course, some of this stability may have been "purchased," as just suggested, by retarding adaptation of the rate structure to changes in technologies and costs for particular services.

With regard to the almost inevitable conflict in rate making between continuity and adaptation to change, the telephone industry has clearly opted for continuity. Indeed, as already suggested, much of the current policy problem in the communications industry arises from rates on long distance service being kept relatively high even as the cost of rendering such service systematically declined over time. This gap has become particularly pronounced at longer distances since telecommunication costs are increasingly insensitive to distance while tariffs remain tied, though ever more loosely, to distance.

Goals versus Achievement: An Overview

In sum, the current telephone rate structure stemmed perhaps as much from historical developments, regulatory quirks, and accidents as from any grand design. It may help promote universal service at the local level by keeping local rates relatively low through averaging, but perhaps not quite to the extent sometimes claimed. Certainly, too, claims of allocative efficiency have to be modest for a pricing structure that employs a fixed monthly charge for most use of the system. Finally, the existence of substantial differences in business and residential phone rates, which are not easily explained by cost or value-of-service differences, further undermines any claims of static allocative efficiency. And if the ultimate goal is to achieve some sort of redistributive equity, whatever redistribution that is achieved would appear to be very rough justice indeed and more a product of private deliberations or accident than of public policy.

The possibility thus cannot be ruled out that while the present structure of telephone rates has some desirable features, it still might be improved. Specifically, pricing policy could seemingly be revised so as to better meet most of the acknowledged goals of rate making in a regulated industry while at the same time better serving the telephone industry's own needs (e.g., providing additional revenues to offset potential declines generated by new competition).

ALTERNATIVE RATE STRUCTURES

Desired Characteristics

Characteristics of alternative rate structures that improve on the existing situation are fairly easily identified. These rate structures may not necessarily be optimal in some comprehensive sense.[48] However, given the conflicts among goals, as discussed earlier, it may very well be impossible to assert that a single rate structure is optimal—even abstracting from the question of how to define and implement an optimal structure given the multitude of telephone companies and regulatory bodies that exist.[49]

One of the more obvious ways in which existing telephone rates might be improved is to adopt usage-sensitive pricing for local service; that is, have a measured service in which a charge is incurred per unit of service, such as one call or, say, a connection of five minutes' duration, and so on. Such a usage charge would accomplish several things. It would bring price closer to marginal cost and thereby improve resource allocation. It can also provide a new source of revenue to help cover joint and common costs and, in so doing, may even allow other charges (such as the basic monthly connect charge or long distance toll rate) to be lower. By way of calibrating the scope of such a change, approximately 50 percent of business phones nationwide are presently subject to some form of usage-sensitive pricing while only 10 percent of residential phones are metered for local calls.

A second desirable characteristic for a reformulated rate structure would be more use of peak load pricing. In situations where the same physical plant can be used to produce a service in different time periods and where the demand characteristics differ in these time periods, prices generally should not be the same at different times of day. Telephone companies, of course, already engage in peak load pricing on long distance services, but some excess capacity for handling long distance calls apparently still exists. This excess capacity might be priced so as to improve resource allocation and at the same time provide new revenue.

For local calls today there is very little peak load pricing, in spite of variations in calling patterns according to the time of day similar to those observed in long distance.[50] The only major local peak load pricing experiments under way today are in New York and in Illinois. These experimental pricing structures are similar to those presented below in that they have broadly defined peak, shoulder, and off-peak periods.[51] Whether these simple structures or more complicated approaches such as instantaneous signals[52] will ultimately be more desirable remains to be seen. Nevertheless local peak load pricing, if pursued sufficiently vigorously, should improve capacity utilization[53] as well as provide an alternative source of revenue. A transition to peak load pricing can be expected to require education of consumers and will naturally face resistance from very high-volume users that benefit from flat rate pricing.

A third characteristic that might be usefully introduced into the telephone rate structure would be distance-insensitive toll rates. This would represent a departure from the current system in which interstate rates increase gradually with distance and a considerable difference exists between intrastate and interstate tolls. The decision whether to adopt rates that are completely insensitive to distance requires a balancing of cost sensitivity with administrative simplicity. Although distance insensitivity aligns well with some new technology, such as satellites, it may be some time, if ever, before costs are completely insensitive to distance. However, distance-insensitive rates are simple for companies and regulators to administer and for customers to understand.[54]

Another move toward greater consistency with expected future technology would be to adopt generally lower toll rates than presently exist. Since current long distance rates seemingly exceed direct costs by a substantial margin, lowering toll charges should improve static resource allocation and reduce the probability of uneconomic entry.

Total revenues, of course, should continue to cover total costs if the telephone system is to continue as a private enterprise. As a proxy for total revenue covering total cost in the analyses that follow, any revenue decreases (e.g., from lower toll rates or lower monthly basic service charges) will be balanced by increased revenues from new sources (such as usage-sensitive pricing or peak load pricing).

Finally, in the examples analyzed below rates will be set to be uniform nationwide. These rates are administratively simple and also easy to analyze. Uniformity already exists for interstate rates, but intrastate rates and local rates currently vary considerably. There would, of course, be many practical problems in implementing uniform nationwide rates given the number of companies and state regulatory bodies that would have a voice in the process. Alternatively, therefore, the example rates can be viewed as nationwide averages from which individual states, for instance, could depart according to their own goals and needs.

Relation to Public Policy Goals

On the assumption that universal service is best achieved by low-cost connection to the telephone system (rather than, say, by achieving low-cost usage overall) several of the rate structure alternatives to be analyzed will include usage-sensitive pricing to achieve lower cost connection to the system. In addition, usage-sensitive pricing, peak load pricing, and generally lower rates will be suggested so as to help bring rates closer to cost and thereby improve resource allocation.

Equity might also be served by uniform nationwide rates. Equity, however, is an obviously subjective concept, and by certain measures the proposed alternative rate structures may not perform quite as well as the existing structure.[55]

The goal of achieving revenues sufficient to cover total costs would be met by maintaining present revenue levels, since the telephone companies now seem financially viable and self-sufficient. In addition, other features of the alternative rate structures, particularly usage-sensitive pricing and peak load pricing, provide new sources of revenue with which to help cover potential increases in total costs.

Lower toll rates and distance insensitivity should discourage uneconomic entry and would appear to be reasonably consistent with probable future and current technological developments, such as optic fibers,[56] microwave transmission, and electronic switching.

Distance-insensitive and uniform nationwide rates would also be administratively simple. Distance insensitivity, for example, should reduce the amount of information that has to be recorded when a long distance call is made. In addition, with lower toll rates people may use person-to-person service even less, thereby reducing operator and related administrative costs. Possibly, too, if toll rates are lower, complaints and claims for calls gone wrong will be reduced. On the negative side, however, usage-sensitive and peak load pricing at the local residential level will be more complicated than systems now employed in most areas.

Finally, the proposed alternatives definitely represent a change from the current system, and in that respect they are not historically continuous. However, these alternatives do maintain historic levels of revenue and thus cover total historical cost rather than just prospective costs; in that sense, they are more continuous than some of the other commonly proposed alternatives.[57]

Table 3-2 shows a breakdown of telephone industry revenue sources for 1975 by several categories of service. The companies represented in these totals account for about 99 percent of total industry revenues. The revenue sought from the alternatives to be analyzed equals the first two lines in Table 3-2, that for local subscriber and domestic message toll services. These revenues amounted to more than $26 billion in 1975, out of a total industry revenue in that year of nearly $35 billion. PBX and Key telephone system revenues are excluded from the pricing analysis to avoid the complexity of pricing these more complicated terminal equipment items. Since terminal equipment is subject to competition it may be reasonable to assume that these categories are approximately covering their costs and that their exclusion therefore does not distort the relationship between the remaining revenue and total cost. The most important additional exclusions are private line services and WATS, which together account for about $3 billion of revenue. Again, sufficient data are not available to include these services in the rate structure analysis adequately. The lack of knowledge about the demand interactions between private line,

WATS, and message toll service makes it difficult to say how much repricing of message toll service will affect these other two categories. On the other hand, WATS and private line service may already be more competitively priced than either message toll services or local services,[58] and for this reason repricing may not be of too much concern.

Local Rate Alternatives

Results that might be expected from adopting various alternative local rate structures, including the volume of local calls and total revenues, are summarized in Table 3–3.[59] For comparison, estimates of actual nationwide average charges and volume in 1975 are also shown on the top line of Table 3–3. Structures A through C are hypothetical alternative local rate structures whose effects have been estimated on various assumptions. Structure A, for example, involves a $3 monthly residential access charge and a $1 charge for residential phones; a $10

Table 3–2. Telephone Revenues, 1975 (millions of dollars)

Service	Independents	Bell System	Total
Local subscriber revenue	$2,182	$10,989	$13,171
Domestic message toll	2,342	11,018	13,360
PBX and KTS revenue[a]	227	2,803	3,030
Toll private line and WATS	187	2,707	2,894
Other	356	2,064	2,420
Total operating revenue	$5,294	$29,582	$34,876

Sources: Bell System data from Federal Communications Commission, *Statistics of Common Carriers,* Washington, D.C., for year ended December 31, 1975, pp. 23, 30. Independent Telephone Company data from United States Independent Telephone Association, *Independent Telephone Statistics,* Vol. I, 1976 (for year 1975). Total PBX and KTS revenue from Federal Communications Commission, *First Report,* FCC Docket No. 20003, Table III-I, Sept. 23, 1976. Bell PBX and KTS Revenue figure from "Historical Data Regarding Bell System and Non-Telephone Company Provided Terminal Equipment," Second Supplemental Response, FCC Docket No. 20003, Bell Exhibit 20–A, 1976.

[a]Total of $3.06 billion reduced by 1 percent to make coverage comparable to FCC plus USITA totals for other services. Independents figure obtained by subtracting Bell figure from total.

monthly access charge for business, plus a $1 charge for business and Centrex phones; and a $0.06 charge for calls made during a 9 A.M. to 5 P.M. peak, a $0.04 per call charge for the shoulder period of 5 P.M. to 11 P.M., and a per call charge of $0.02 for the rest of the day.[60]

As shown in Table 3-3, Structure A implies a reduction in the number of local calls from around 193 billion to about 165 billion per year. Demand for calls is reduced by the imposition of a local usage charge. To the extent that this reduction occurs in peak periods, telephone capacity is freed for future growth.[61] Structure A also results in a total revenue of $13.4 billion, which is slightly more than the initial level of total revenue from local service.[62]

Structures B and C result in higher levels of local revenue than the original $13.2 billion on the premise that these levels may be needed to offset declines in toll revenue if toll rates are reduced. Structure B differs from Structure A in having a lower monthly access charge for business. By assumption, this does not affect the total number of calls made by businesses, but it does decrease revenues by about $0.5 billion. Peak and shoulder call charges, however, are higher than in Structure A. Although the higher call charges induce a lower volume of calling, call revenues increase by $1.67 billion—thus increasing total local revenues to $14.5 billion. Structure C increases the business monthly access charge to $12 per month, and increases the charge per call to $0.10 during the peak period, $0.06 during the shoulder, and $0.04 during the off-peak period. The net result under Structure C is a decline in calls to 143.3 billion but, owing to the higher per call charges, an increase in revenues to $17.0 billion.

Structures A through C all have higher monthly charges for business than for residential phones. At least two justifications for this can be suggested. The first is historical continuity. The second is that businesses appear to receive roughly twice as many calls as residences do. In other words, on balance, residences are calling businesses. Some advocates of usage-sensitive pricing suggest it would be fairer to impose a charge on both the originator and the receiver of calls. Others, however, would disagree on the grounds that the originator controls the option of whether or not to call. On the grounds that businesses usually seek incoming calls more than residences, a higher monthly business connect charge is incorporated into Structures A through C to distribute more of the usage burden to businesses as willing recipients of incoming calls.

The usage charges in Structures A through C are per call rather than for duration of the call.[63] Since costs depend on both linking up and holding time,[64] some sensitivity of prices to duration would be desirable. However, given the limited data on costs and on the respon-

Table 3–3. Alternative Rate Structure Parameters and Expected Revenue for Local Telephone Service

Rate Structure	Monthly Charges			Call Charges			Total Calls Per Year (in billions)	Total Revenue[e] (in billions)
	Residence Access[b]	Business Access[c]	Phones[d]	Peak	Shoulder	Off-Peak		
Initial[a]	$7.43	$21.00	$1.00	$0.06	$0.04	$0.02	$193.1	$13.2
A	3.00	10.00	1.00	0.08	0.05	0.02	165.3	13.4
B	3.00	6.00	1.00	0.08	0.05	0.02	154.5	14.5
C	3.00	12.00	1.00	0.10	0.06	0.04	143.3	17.0

Source: Appendix C, Tables C–3 and C–7. Calculations by Charles River Associates.

[a]Initial charges represent estimated averages for 1975.

[b]Structures A–C represent average charge per residence main station access line.

[c]Structures A–C represent average charge per business main station and PBX trunk

[d]Phone category includes residential main stations, business main stations, total extensions, and Centrex phones but excludes Key telephones and PBX phones. See Appendix C, Table C–3.

[e]Total revenue includes installation charges and surcharges for equipment such as luxury extensions. If data were sufficient to show these items separately, other charges could be reduced accordingly.

siveness of duration to pricing, the simpler alternative of charges per call is used here.

In Table 3-4, the number of residence telephones existing under the different rate structures is shown, as well as the average number of calls per station.[65] In Structure A, with a $4 total monthly charge, the number of telephones demanded increases over the original or 1975 base level by about 6 percent.

In Structures B and C, meanwhile, the number of telephones increase by smaller amounts, 4 and 2 percent, respectively, even though the monthly charge is the same as in Structures A and B. This is because the increased call charges under Structures B and C dampen demand.[66]

Long Distance Rate Alternatives

Table 3-5 summarizes the differences among five alternative long distance rate structures analyzed. The initial charges shown are weighted averages of estimates of what the interstate and the intrastate charges were in 1975. The alternative toll rate structures are analyzed under two alternative sets of assumptions about elasticities[67] and capacity available for toll calls.

Independent studies by economists such as Dobell and Waverman and by Bell Telephone researchers have produced a wide range of elasticity estimates for long distance service.[68] For daytime service, these elasticity estimates range between −0.2 and −0.9. An elasticity of −0.2 would mean that a 10-percent descrease in price would result in only a 2-percent increase in quantity demanded, while an elasticity of −0.9 would mean that a 10-percent decrease in price would spur a

Table 3–4. Local Residential Demand

Rate Structure	Total Residence Main Stations (millions)	Calls per Residence Main Station (monthly)
Initial[a]	66.7	120.0
A	70.7	93.0
B	69.7	88.4
C	68.1	83.7

Source: Appendix C, Tables C–3. Calculation by Charles River Associates.
[a]Initial calls per residence main station represents an estimate.

Table 3–5. Alternative Rate Structure Parameters and Expected Revenue for Long Distance Telephone Service

Rate Structure	Assumptions		Charge per Minute					Total Revenue (in billions)
	Elasticity	Excess Capacity	Morning	After-noon	Evening	Night	Weekend	
Initial[a]			0.28	0.28	0.18	0.10	0.10	$13.4
1	Low	Low	0.25	0.19	0.14	0.02	0.06	12.1
2	Low	Moderate	0.13	0.11	0.11	0.01	0.05	9.6
3	Low	Low	0.27	0.27	0.14	0.02	0.06	13.4
4	High	Low	0.26	0.23	0.15	0.02	0.06	13.5
5	High	Moderate	0.19	0.17	0.12	0.02	0.05	13.2

Source: Appendix C, Tables C–4 through C–6e. Calculations by Charles River Associates.
[a]Initial monthly charge represents estimated averages for 1975.

9-percent increase in consumption. The range in empirical estimates for evening and night elasticities is from about −0.5 to −1.9; thus, a 10-percent decrease in price could result in anywhere from a 5- to 19-percent increase in consumption. Since these ranges are quite large, two values (referred to as high and low in Table 3–5) were selected within each range for analysis. The high elasticity is −0.7 in the day-time and −1.3 at night or in the evening, and the low elasticity is −0.3 in the daytime and −1.1 in the evening.

Two alternative assumptions are also made about the capacity available for toll calls. Capacity is of more concern in the long distance analysis than in the local because the number of toll calls increases under the suggested alternative rate regimes whereas the number of calls in the local analysis decreases or remains about the same. At the low end, capacity is assumed to be available at a level about 5 percent above the current average peak period usage (i.e., in the morning). The assumption of moderate capacity is that available capacity is about 30 percent above average peak usage. Figure 3–2 shows the distribution of toll conversation minutes for the New York Telephone Company in 1975 by time of day; these figures were hypothesized to be typical of nationwide usage and thus deemed appropriate for analyzing the rate structure alternatives. As can be seen, the morning peak is the highest, and thus capacity is a greater constraint during the morning than at other times of day.

In analyzing the alternatives shown in Table 3–5, the same basic procedure was used except with Structure 3. Specifically, the price in each period was lowered until usage was at or near the assumed capacity. This means, of course, that the price can be lowered less in the morning peak than in the afternoon, but it can be lowered more at night and still more on the weekends.

As a result of this procedure the bulk of increased long distance usage occurs during periods in which capacity is currently underutilized. Therefore the cost of additional facilities can be excluded from the analysis.[69] If existing peak usage approximates available capacity, the results in Structures 1, 3, and 4 are appropriate. If there is some moderate excess capacity at present, the results from Structures 2 and 5 are relevant.

As shown in Table 3–5, for Structure 1, with elasticity and capacity both assumed to be low, prices are $0.25 in the morning, $0.19 in the afternoon, $0.14 in the evening, $0.02 at night, and $0.06 per minute on the weekend; estimated total long distance revenue declines to about $12.1 billion.

In Structure 2, with moderate capacity (i.e., with capacity assumed to be 30 percent above the current peak usage), price can be lowered much more before usage approaches capacity. This results in lower prices than Structure 1; specifically, $0.13 in the morning, $0.11 in the afternoon and evening, $0.01 at night, and $0.05 on weekends; revenue declines still further, to $9.6 billion.

In Structure 3, instead of operating to the capacity constraint, revenue was constrained to remain at its initial level of $13.4 billion. This suggests how much rates could be cut, still keeping long distance revenues intact. Within this constraint, rates can be cut substantially only when demand elasticity is greater than −1.0, that is, in the evening and at night. By contrast, rates can decline only about $0.01 per minute during the daytime.

Structures 4 and 5 incorporate high elasticity assumptions. With low capacity as in Structure 4, usage approaches capacity in each period at the relatively high prices of $0.26 in the morning, $0.23 in the afternoon, $0.15 in the evening, and $0.02 and $0.06 at night and on the weekends. In this case, total revenue actually increases slightly to $13.5 billion. In Structure 5, with moderate capacity, the price can be cut more before the capacity constraint is encountered but revenue declines slightly to about $13.2 billion.

In assessing these calculations, it must be remembered that these hypothetical rate structures treat interstate and intrastate calls the same; that is, all toll charges are independent of distance. Since current interstate rates differ from intrastate rates, the impact of rate changes on both these types of calls has to be estimated separately. Initially, the average interstate rate would have been about $0.25 per call as shown in Table 3-6. By comparison, Structure 5 with high elasticity and moderate capacity, results in a weighted average toll rate of

Table 3–6. Comparison of Initial and Alternative Prices and Volumes of Message Toll Minutes Assuming High Elasticity and Moderate Capacity

	Interstate		Intrastate	
	Weighted Average Price per Minute	Annual Volume (billion minutes)	Weighted Average Price per Minute	Annual Volume (billion minutes)
Initial (estimated)	$0.25	32.2	$0.17	30.6
Alternative assuming high elasticity and moderate capacity	$0.11	75.5	$0.11	45.1

Source: Appendix C, Tables C–4 through C–6e. Calculation by Charles River Associates.

about $0.11 and an increase in interstate volume from 32 billion minutes a year to about 75 billion minutes a year. On the other hand, the initial intrastate average price was lower than the interstate price at about $0.17. At the universal average rate of $0.11, Structure 5 results in a smaller increase in intrastate volume, up only about 50 percent from 30 billion to 45.1 billion minutes a year.

The alternative long distance structures in some cases yield less revenues than today's toll rates while the hypothetical local rate structures have been designed to yield as much or more than current charges. Combinations of these structures can thus be identified that would yield total revenue more or less equal to initial 1975 revenues, that is, $26.6 billion a year. Table 3–7 shows these combinations. Local Structure A, for example, when combined with any of the toll structures 3 through 5, yields approximately the initial level of total revenue. Local Structure C, which yields $17 billion in revenue, can be matched with the worst toll revenue cases (where revenue declines to about $9.6 billion a year) and still achieve the original total revenue yield of $26.6 billion. Finally, local Structure B could be matched with toll Structure 1 and yield the original $26.6 billion. In short, there are a number of ways of repricing telephone service to better meet some of the goals outlined previously and still keep total telephone revenues intact. The great uncertainty in all this analysis, however, is the elasticity assumptions, for as noted previously, the empirical work done on elasticities has resulted in a large range of estimates; as shown in Structures 1 and 2, elasticities in the low part of the range can result in declines in toll revenue.

A few other studies have also looked at telephone pricing, using crude cost data in an attempt to estimate optimal prices (in an eco-

Table 3–7. Total Revenue under Selected Rate Structures (in billions of dollars)

Rate Structure	Local Revenue	Toll Revenue	Total Revenue
Initial	$13.2	$13.4	$26.6
A–3	13.4	13.4	26.8
A–4	13.4	13.5	26.9
A–5	13.4	13.2	26.6
B–1	14.5	12.1	26.6
C–2	17.0	9.6	26.6

Source: Appendix C, Table C–7. Calculations by Charles River Associates.

nomic sense).[70] Results from these studies are shown in Table 3–8 together with ratios comparing the results of these studies to selected combinations of the alternative rate structures analyzed here.[71] In spite of the differences in approach of the different studies there are some important similarities. In particular the results from the Littlechild and Mitchell studies, in terms of the ratios shown in the last two columns, fall more or less in the middle of the range of results from the present study.[72] These similarities are all the more striking in that these other studies have calculated optimal prices based on a much narrower range of public policy goals than in the present analysis.

SUMMARY

At least eight basic goals commonly have been identified with regulatory rate making: (1) universal service; (2) static efficiency in resource allocation; (3) equity for different kinds of users and services; (4) financial self-sufficiency (total revenues equal to total cost); (5)

Table 3–8. Comparison of Selected Price Structures with Littlechild and Mitchell Studies

Price Structure	Residence Monthly Rate[a]	Average Local Call Charge[b]	Average Interstate Price[b]	Ratio of Interstate to Average Local Price	Ratio of Monthly to Average Local Call Charge
Initial	$7.43	—	$0.75	—	—
A–5	4.00	$0.05	0.33	6.6	80
B–1	4.00	0.06	0.23	3.8	67
C–2	4.00	0.08	0.23	2.9	50
A–4	4.00	0.05	0.41	8.2	80
Littlechild (1970)	—	0.02	0.12	6.0	—
Mitchell (1976)	2.19	0.03	—	—	73

Source: Appendix C, Table C–8. Calculations by Charles River Associates.

[a]Residence monthly rate represents the sum of the monthly access charge per main station plus charge for terminal equipment. In local Structures A, B, and C this equals $3 access plus $1 equipment charge.

[b]Per three-minute call.

prevention of uneconomic entry; (6) consistency with expected technological change; (7) administrative simplicity; and (8) historical continuity.

The current telecommunication rate structure serves at least some of these regulatory goals well. For example, universal service appears to be nearly achieved in the United States, and the industry seems to be financially healthy. Residential rate structures, largely based on flat monthly charges, are quite simple. The industry also has met the goal of historical continuity by maintaining stable prices for many years. It could be argued that telephone rates are equitable in that businesses are charged more than residential customers by roughly a ratio of two to one, although the choice of this particular ratio seems arbitrary. On the other hand, variation in flat monthly rates to similar customers in nearby local exchanges of similar size indicates that similarly situated users can pay very dissimilar rates.

The current telephone rate structure does not accord well with the standard goals in other ways. It is not, for example, consistent with expected technological change and lacks simplicity in business rates. Moreover, the current practice of charging flat monthly rates for most local services with little usage-sensitive pricing is almost surely inefficient. The present structure also seems to be attracting entry into some segments of the business, entry that could (or could not) prove uneconomic in the long run.

It is therefore not difficult to identify simple alternative rate structures with characteristics that might improve existing telephone tariff schedules in many ways. Indeed, several sets of such alternatives are identified in this chapter. Such alternatives, among other advantages, could promote even more universality of service than now exists by lowering monthly connect charges. The alternatives could also improve resource allocation by moving rates closer to cost through greater use of usage-sensitive pricing, peak load surcharges, and generally lower long distance tolls. By bringing rates closer to costs, these alternative strutures should also discourage uneconomic entry and reduce unnecessary or frivolous use of the network. The alternatives are, in addition, more consistent with expected future technologies, being largely distance-insensitive, and on balance are probably easier to administer than the current system, even allowing for some added complexity of adopting usage-sensitive and peak load rates at the local level. As a final advantage, the alternative rate structures include the attractive feature of making it possible for existing common carriers to compete effectively with new entrants while maintaining their financial viability.

It would be quite premature to assert dogmatically that any of the rate structures outlined in this chapter should be adopted. The point, moreover, is *not* to assert that any particular rate schedule is *the* correct schedule. Rather, it is to suggest the possibility that new or revised schedules with the characteristics proposed in this chapter might well serve the basic goals and purposes of *both* the telephone companies and public policy better than do present tariffs.

NOTES

1. A similar list of pricing criteria for regulatory commissions is developed by William J. Baumol, "Reasonable Rules for Rate Regulation: Plausible Policies for an Imperfect World," Chapter 8, in *Prices: Issues in Theory, Practice, and Public Policy*, edited by A. Phillips and O. Williamson (Philadelphia: University of Pennsylvania Press, 1968), pp. 108123. Baumol's list was developed in a somewhat narrower context, namely that of evolving rules to avoid the worst aspects of Averch-Johnson effects distorting the behavior of regulated firms. Thus, Baumol's list of seven criteria would mostly map onto three of the criteria listed here; specifically, his criteria are mostly concerned with static efficiency, equity, and total revenue sufficiency. On the other hand, Baumol's list would not include or incorporate most of the more dynamic or historical concerns listed above. Baumol also does not include administrative simplicity as a criterion in his list, although he has raised it by implication in other writings, specifically as a coauthor with Otto Eckstein and Alfred E. Kahn of a paper on "Competition and Monopoly in Telecommunications Services," prepared for AT&T and reprinted by AT&T, dated November 23, 1970.
2. Although many observers suggest that universal service was proclaimed as a public policy goal in the Communications Act of 1934, the exact words are difficult to find. The act does, however, contain the phrase: "To make available so far as possible . . . a rapid, efficient . . . communications service with adequate facilities at reasonable charges . . ." *The Communications Act of 1934*, as amended, 50 Stat 189, Section 1.
3. Externalities occur when one person's consumption of a good or service affects another person's well-being. Competitive behavior will generally result in prices that fail to reflect these external benefits (or costs), and therefore some government correction may be required. The case for pricing telephone services to reflect externalities can, however, easily be overstated. In the case of telecommunications there are two potential types of consumption externalities: those associated with receiving a call and those associated with adding a new connection to the system. It can be argued that the call-related externalities can easily be internalized since, for instance, the parties to a call are generally either involved in a transaction or engage in reciprocal calling over a given period of time. Similar arguments can be applied to externalities associated with connection to the system. Analogies with transportation systems, where pricing to reflect externalities is not deemed necessary, can be made. See Charles Jonscher, "Benefits, Costs and Optimal Prices in Telecommunications Services," draft working paper (undated). For more traditional viewpoints see S. C. Littlechild, "Two-part Tariffs and Consumption Externalities," *Bell Journal of Economics* 6 (Autumn 1975): 661670; and J. Rohlfs, "A Theory of Interdependent Demand for a Communications Service," *Bell Journal of Economics* 5 (Spring 1974): 1637.
4. Martin S. Feldstein, *Economic Analysis for Health Service Efficiency* (Amsterdam: North-Holland Publishing Co., 1967).
5. Thomas Domencich and Gerald Kraft, *Free Transit*, a Charles River Associates Incorporated Research Study (Lexington, Mass.: Lexington Books, 1970).

6. William S. Vickrey, "Optimization of Traffic and Facilities," *Journal of Transport Economics and Policy* 1 (May 1967): 123136.

7. M. Boiteux, "Peak Load Pricing" (1949), pp. 5990, and "Marginal Cost Pricing" (1956), pp. 5158, translated and reprinted in J. R. Nelson, ed., *Marginal Cost Pricing in Practice* (Englewood Cliffs, N.J.: Prentice-Hall, 1964), pp. 5158.

8. Vickrey, "Optimization of Traffic and Facilities," pp. 123126.

9. If fewer people can be reached without incurring toll charges in rural areas, the value of the service in such areas might be deemed to be less, thus justifying lower charges. This, of course, overlooks the strong probability that the value of a communication capability also depends heavily on many circumstances, not the least of which may be remoteness itself (e.g., an ability to call a doctor for help or advice may be worth more at a remote farmhouse than in a city apartment located next door to a general hospital with an emergency service). Similarly, the value of a phone in relieving loneliness could depend on a number of considerations, ranging from misanthropic tendencies to accessibility of locale, etc. Finally, it is possible that most people's "community of interest" is much smaller than and largely invariant to size of calling area (e.g., a person's calling desires may be limited to 30 or so other people in the local area, whether that area is a town of 500 or a city of 1 million).

10. Usually referred to as "the principle of horizontal equity."

11. See, for example, the discussion in Chapter 1 of Harvey Averch and Leland L. Johnson, "Behavior of the Firm Under Regulatory Constraint," *American Economic Review* 52 (December 1962): 10521069; and Stanislaw H. Wellisz, "Regulation of Natural Gas Pipeline Companies: An Economic Analysis," *Journal of Political Economy* 55 (January 1963): 3043. Again, the incentives to pursue inefficient policies because of these effects will be greatly attenuated, or even reversed, if inflation or other effects create a situation in which the cost of capital exceeds the permitted "fair" rate of return.

12. John R. Meyer and Alexander L. Morton, "The U.S. Railroad Industry in the Post-World War II Period: A Profile," *Explorations in Economic Research* 2 (Fall 1975): 449501.

13. F. D. Ramsey, "A Contribution to the Theory of Taxation," *Economic Journal* 37 (1922): 4761.

14. Jules Dupuit, "On the Measurement of the Utility of Public Works," *Annales des Ponts et Chaussées,* 2nd Series, VIII (1844), and reprinted in *International Economic Papers,* No. 2 (New York: MacMillan, 1952), pp. 83110.

15. S. Margolin, "The Opportunity Costs of Public Investment," *Quarterly Journal of Economics* 77 (May 1963): 274289; E. J. Mishan, "Criteria for Public Investment: Some Simplifying Suggestions," *Journal of Political Economy* 75 (April 1967): 139146; J. Hirshleifer and J. Milliman, "Urban Water Supply: A Second Look," *American Economic Review* 57 (May 1967): 169178; E. J. Mishan, "Criteria for Public Investment: A Reply," *Journal of Political Economy* 78 (January-February 1970): 178180; and Robert Dorfman, *Prices and Markets* (Englewood Cliffs, N.J.: Prentice-Hall, 1978), pp. 170196.

16. The "natural monopoly" issue is discussed in the next chapter, and to some extent in Chapter 6. Preemption of uneconomic entry also arises as part of the "sustainability" argument sometimes used to justify continued or extended regulation. See W. J. Baumol, "On the 'Proper' Cost Tests for Natural Monopoly in a Multiproduct Industry," *The American Economic Review* 67 (December 1977): 810. This subject is also addressed in the Bell System's comments on the options paper "Domestic Communications Common Carrier Policy," Comments of American Telephone and Telegraph Company on Behalf of the Bell System, July 18, 1977. John

DeButts makes further reference to this issue in his letter and attachments to Senator Harrison H. Schmitt (May 9, 1977) in connection with the options paper mentioned above.

17. Given developments in highway and rail transportation in recent years, uneconomic traffic diversion may be declining See Richard C. Levin, "Allocation in Surface Freight Transportation: Does Rate Regulation Matter?" *The Bell Journal of Economics and Management Science* 9 (Spring 1978): 1845. In large part, though, this very narrowing of the gap in relative costs may represent management adaptation to the inefficient signals provided by regulation (e.g., rail managements have little incentive to invest in methods to regain high value traffic if regulation prevents railroads from getting that business).

18. AT&T's Multiple Supplier Network Study purports to find that the cost of providing the entire private line network by a separate entity could be as much as four times higher than the cost for the Bell System, if the new entrant has to start from scratch. As shown in Chapter 4, however, doubts exist about some of the assumptions underlying this study.

19. See the discussion in Chapter 6 on predatory pricing.

20. For example, people investing in suburban homes served by a commuter rail line may see their investments as unfairly devalued if the rail service is subsequently withdrawn.

21. Again, suburban commuters facing loss of their rail service is an example in point.

22. The list of eight goals outlined above can be compared with a list of five goals or desirable characteristics for a rate structure recently advanced by a Telecommunications Industry Task Force in "The Dilemma of Telecommunications Policy" (1977). The first goal stated by the task force was to have low rural rates. Clearly, this may be another way of stating universal service. The task force's third stated goal was nationwide rate averaging. This would seem to be an expression of some kind of equity goal, although the exact kind is not obvious. Rate averaging can be tied into universality aspirations as well. The task force's fourth goal was to achieve "fair competition" in pricing. Going somewhat beyond the label, the task force was apparently worried about uneconomic entry occurring, although other worries may also be involved. On balance, though, the most sensible interpretation of the industry's use of the phrase "fair competition" would seem to be a desire by the telephone companies to reflect promptly any true cost advantages that they might have so as to prevent higher cost suppliers from entering the market. The task force also explicitly lists economic efficiency as a target, which matches well with the static efficiency characteristic listed above. In fact, by stressing economic efficiency in general, not just static efficiency, the task force report can be interpreted as trying to incorporate some of the other goals listed above that go beyond static efficiency, such as consistency with expected technological change and having sufficient revenues to cover total costs. Of interest is the fact that neither administrative simplicity nor historical continuity are mentioned by the task force. Of course, historical continuity might be subsumed under equity, although historical continuity embodies other aspects as well. Simplicity may have been assumed to be so obvious as not to require separate mention.

23. Federal Communications Commission, *Statistics of Communications Common Carriers,* for Year Ended December 31, 1975.

24. The Bell System understands these deficiencies in the simple ratio as a measure of universality; in a Bell exhibit presented in Docket 20003 (Lewis J. Perl, "Economic and Demographic Determinants of Telephone Availability," prepared by NERA, April 15, 1975; FCC Docket Number 20003, Bell Exhibit 21) census data were used to indicate that pockets of deprivation may exist where telephone service coverage drops to as low as 20 to 30 percent in certain poor rural areas.

25. In 1975, AT&T reported 52,460,000 residence main stations not including Southern New England Telephone Co. and Cincinnati Bell. These two companies had 1,733,686 main stations in 1975. Breaking this down by the Bell System proportions of business main stations to residence main stations indicates that the two companies had 1,465,055 residence main stations (*Bell Statistical Manual 1950–1975,* May 1976, p. 506). Reporting independents owned 13,425,946 residence main stations. Nonreporting independents owned 1,275,000; assuming 80 percent of those phones were residence main stations implies an additional 1,020,000. Thus, a reasonable estimate of total residence main stations is 68,371,000 phones. Data on housing units area available from Bureau of the Census, *Annual Housing Statistics,* 1975, Current Housing Reports, Series H–150–75A. Data on main station (residential) are taken from AT&T's *Annual Statistical Report,* the FCC *Statistics of Communications Common Carriers,* and USITA's *Independent Telephone Statistics.*

26. P. J. Berman and A. C. Oettinger, "The Medium and the Telephone: The Politics of Information Resources" (Cambridge Mass.: Harvard University Program on Information Technologies and Public Policy, P–76–75, June 1976). Figure 3–2 pertains only to toll usage. Public information on diurnal usage patterns for local service is not readily available but, if it were, would probably indicate that the daytime peaks are less pronounced and the evening peak more prominent.

27. According to L. Garfinkel ("Usage Sensitive Pricing: Studies of a New Trend," *Telephony* [February 10, 1975] : 24), over 89 percent of residential local exchange service is priced on a flat-rate basis.

28. See Bridger M. Mitchell, "Telephone Call Pricing in Europe: Localizing the Pulse" (Santa Monica, Calif.: The Rand Corporation, January 1979).

29. National Association of Regulatory Utility Commissioners "Exchange Service Telephone Rates in Effect June 30, 1976" (Washington, D.C., 1977).

30. As of December 1978, for instance, calls to telephone company business offices indicated the following surcharges for standard touch-tone service: Boston $1.10, Buffalo $2.85, Dallas $1.30, San Francisco $1.45.

31. John C. Panzar and Robert D. Willig, "Free Entry and the Sustainability of Natural Monopoly, "Bell Laboratories Economic Discussion Paper Number 57, May 1976.

32. Ramsey, "A Contribution to the Theory of Taxation."

33. Figures taken from *The Wall Street Journal,* February 2, 1978.

34. See Howard Anderson, "AT&T and IBM: The Battle Continues," *Telephony* (April 10, 1978): 78–89; and "Response of the North American Telephone Association to Options Papers Prepared by the Staff of the Subcommittee on Communications, Committee on Interstate and Foreign Commerce, U.S. House of Representatives," *Domestic Communications Common Carrier Policy* (July 11, 1977): 12–13.

35. See Merton J. Peck and John R. Meyer, "The Determination of a Fair Rate of Return on Investment for Regulated Industries," in *Transportation Economics* (New York: National Bureau of Economic Research, 1965), pp. 199–244.

36. Evsey Domar, "The Case for Accelerated Depreciation," *Quarterly Journal of Economics* 67 (November 1953): 492–519; and Robert Eisner, "Accelerated Amortization, Growth and Net Profits," *Quarterly Journal of Economics* 66 (November 1952): 533–544.

37. See Franco Modigliani and Robert A. Cohn, "Inflation and the Stock Market," *Financial Analysts Journal,* March/April 1979, pp. 24–44; also Philip Cagan and Robert Lipsey, *The Financial Effects Of Inflation (Cambridge, Mass.: Ballinger Publishing Co. for National Bureau of Economic Research, 1978).*

38. For a further discussion, see Solomon Fabricant, "Accounting for Business Income Under Inflation: Current Issues and Views in the United States," *Income and Wealth* Series 24, No. 1 (March 1978): 1.

39. See Chapter 2.
40. The Bell System had 3.1 million stations or telephones compared to the Non-Bell total of 3.0 million. See U.S. Bureau of the Census, Special Report, "Telephones: 1907" (Washington, D.C.: U.S. Government Printing Office, 1910), p. 21, Table 8.
41. *Investigation of the Telephone Industry in the United States,* Report of the Federal Communications Commission on the Investigation of the Telephone Industry in the United States made pursuant to Public Resolution No. 8, 74th Congress. 76th Congress, 1st Session, House Document No. 340, 1939.
42. As yet, a good market test remains to be done to determine whether or not recent entry is economic or uneconomic.
43. William G. Shepherd, "The Competitive Margin in Communications," pp. 86–122, in *Technological Change in Regulated Industries,* edited by William M. Capron (Washington, D.C.: The Brookings Institution, 1971).
44. See "Economic Data of Regulation — Telephones," NARUC, July 12, 1976 (revised September 26, 1976), p. 97; and *Economic Report of the President,* Transmitted to the Congress, January 1976, p. 220, Table B–42.
45. Although it might have more nearly kept pace according to some critics by, for instance, installing "clip" and "phone store" concepts earlier and more widely to save on installation labor.
46. Specifically, local loop and instrument installation are both high in labor intensity and, therefore, sensitive to labor cost increases. Some of the increase in local loop costs might be due to more installations in less dense areas. On the other hand, housing statistics on multiple- versus single-family starts do not suggest much, if any, decline in the density of new housing constructed in recent years. On balance, then, local installation costs have probably been marked by cost increases with possibly some small offset in real operating costs. The goal of regulatory or telephone company pricing policy has apparently been to keep local rates steady in nominal terms, creating a decline in real terms. There is no apparent economic reason, however, why local telephone rates should have been protected from real cost increases during these years.
47. For example, Economics and Technology located in Cambridge, Massachusetts, or Communications Analysis Associates in Boston, Massachusetts.
48. The alternative rate structures discussed here will deal only with local exchange and message toll service. Private line service and WATS are excluded.
49. A promising approach to the analysis of pricing structures is the use of game theory to examine the conditions under which coalitions of producers and consumers may voluntarily band together. See for instance, J. Sorenson, J. Tschirhart, and A. Whinston, "A Theory of Pricing Under Decreased Costs," *American Economic Review* 68 (September 1978): 614–624. The results in this article, however, are not necessarily applicable to the telecommunications situation since the authors assume a single product produced under conditions of declining average cost.
50. Predominately residential exchanges may have less pronounced daytime peaks, or even have the highest peaks in the evening. Although the rate structure examples presented below incorporate peak charges during the day, exceptions should be made in individual cases to prevent shifting of the peak to evening hours.
51. A recent experiment with time-of-day pricing did not result in significant load shifting or improved capacity utilization. However, the experiment was probably not a good test of the approach since only off-peak rates were adjusted downward while peak rates remained unchanged. If peak rates were designed to more closely reflect costs of additional capacity, different results might have occurred. For a discussion of the experiment and its weaknesses, see J. K. Hopley, "The Response of Local Telephone Usage to Peak Pricing," and J. H. Alleman, "Comments," in *Assessing*

New Pricing Concepts in Public Utilities, edited by Harry Trebing (East Lansing: Institute of Public Utilities, Graduate School of Business Administration, Michigan State University, 1978), pp. 12–26 and 40–43, respectively.

52. For instance, the use of tones or lights has been suggested as a means of instantaneously indicating the level of capacity utilization and the corresponding charge that would be levied on a call made at that particular time.

53. See Bridger Mitchell, "Optimal Pricing of Local Telephone Service," *The American Economic Review* 68 (September 1978); 517–537.

54. Given the complexities invovled in making a public policy choice as well as in presenting hypothetical alternatives, the rate structures analyzed below include only a single rate and may be viewed as either distance insensitive or as averages of slightly more complicated rate structures (which could be distance sensitive).

55. A discussion and analysis of this point are in Appendix A.

56. Optic fiber costs are, of course, distance sensitive but probably at quite a low level as a portion of total telephone cost. In other words, optic fibers may drive line costs down to such a low level that even though they increase with distance, they are an inconsquential portion of the total cost of making a call (including switching, local loop, instrument, etc.), and thus distance becomes insignificant as a cost factor with fiber optics.

57. See, for example, the optimal rate structure suggested by S. C. Littlechild, "Peak-Load Pricing of Telephone Calls," *The Bell, Journal of Economics and Management Science* 1 (1970): 191–210.

58. Canadian National/Canadian Pacific Telecommunications strongly argued this case as part of a recent hearing on interconnections in Ottawa. (Canadian Radio-Television and Telecommunications Commission, "Re: Canadian Pacific Limited and Canadian National Railway Company vs. Bell Canada." CRTC 2–1978.)

59. For a closer examination of the impact of alternative rate structures on residential customers' telephone bills, calling volumes, and welfare, see Appendix A.

60. In Structure A, call charges are the same for businesses and residences. Since 50 percent of businesses and 10 percent of residences are currently on some kind of measurement system that involves usage-sensitive pricing, their revenue is already included in the $13.2 billion total revenue from flat rate and message unit service in 1975. Message unit service in 1975 amounted to about $2 billion of revenue.

61. A possible objection to all usage-sensitive pricing schemes can be that people will be paying more and getting less. However, usage-sensitive pricing tests the value of a call; those who continue to call place at least some minimal value on making the call. Moreover, to the extent that busy signals were reduced by eliminating overly long or unwanted calls, total satisfaction from the system could be enhanced.

62. Structure A is designed to offset a slight decline in toll revenue in toll rate structure 5 below. A lower figure for local revenue is needed to match other toll structures. This can be achieved through various adjustments. For instance, changing the business access charge in structure A to $8 per month while leaving other charges the same yields $13.2 billion.

63. Some figures are available on the average length of a local call, which appears to be about three minutes. Little is known, though, about the distribution by duration or how pricing would affect the duration of the calls. Metering costs would also be higher when charging for duration. By contrast, Schedules A through C are very simple systems, wherein only the number of calls must be recorded.

64. New York Telephone Company studies show that 35–40 percent of the costs of a local call are for holding time. On this point see John K. Hopley, "The Response of Local Telephone Usage to Peak Pricing," in *Assessing New Pricing Concepts in Public Utilities,* Proceedings of the Institute of Public Utilities Ninth Annual Conference, MSU Public Utilities Papers, 1978, pp. 12–26.

65. The residential demand functions underlying these calculations were taken from the study by Mitchell, "Optimal Pricing of Local Telephone Service," *American Economic Review* 68 (September 1978). It is assumed that demand for telephones depends on both monthly charge and the usage charge (i.e., the charges per call). The presumption is that some people might not want even a free phone as long as it cost them, say, a dollar to make a phone call.

66. Demand for main stations in the Mitchell study was specified to depend on both the monthly charge and the charge per local call. A more general approach would also take into account the long distance charge, in which case one might expect the number of main stations to increase over the initial level due to lower long distance charges. Examples of alternative long distance rate structures with lower charges are discussed below.

    In developing the business demands underlying Tables 3-2 and 3-3, business need for phones was assumed to be independent of the level of the monthly charge or the level of usage charges per call (unlike the analysis done for residential phones). For business, the usage charge had an impact only on the number of calls made.

67. Elasticity is a measure of the responsiveness of demand to changes in price. It is defined as the ratio of the percentage change in quantity demanded to the percentage change in price, and since a reduction in price is generally needed to stimulate an increase in demand, this ratio is usually negative.

68. See J. H. Alleman, "The Pricing of Local Telephone Service," U.S. Department of Commerce. Office of Telecommunications, Special Publication, April 1977; B. E. Davis et al., "An Econometric Planning Model for American Telephone and Telegraph Company," *The Bell Journal of Economics and Management Science* 4 (Spring 1973): 29–56; A. Dobell et al., "Telephone Communications in Canada: Demand, Production and Investment Decisions," *The Bell Journal of Economics and Management Science* 3 (Spring 1972): 175–219; S. C. Littlechild and J. J. Rousseau, "The Pricing Policy of a U.S. Telephone Company," *The Journal of Public Economics* 4 (1975): 35–56; and Leonard Waverman, "The Pricing of Telephone Services in Great Britain: Quasi-Optimality Considered," *I. O. Review* 5 (1977): 1–10.

69. In addition, since the local rate structures presented above result in decreases in local calling, the increased long distance calling should not on average require additional local switching capacity.

70. See the studies by Mitchell, "Optimal Pricing of Local Telephone Service," *American Economic Review* 68 (September 1978): 517–537, and by Littlechild, "Peak Load Pricing of Telephone Calls," *Bell Journal of Economics* 6 (Autumn 1975): 191–210.

71. Littlechild's prices are much lower than the alternatives presented here because he assumes that the only costs that have to be covered are those of the best available technology rather than historical costs. In short, he does not incorporate historical continuity as a goal. Also he does not consider two-part pricing of local service (i.e., a monthly charge along with a user charge), but focuses only on the usage charge. Mitchell uses the same cost data as Littlechild but examines local residential service only. He does, though, consider two-part charges, that is, monthly charges and a usage charge, but does not consider peakload pricing (whereas Littlechild does). Mitchell also seeks to cover initial revenue, but at 1970 levels so his prices are lower than in the alternatives developed here.

72. Although the ratios for Structure C–2 are much lower than the Littlechild and Mitchell results, the desirability of Structure C–2 should not be ruled out arbitrarily since these other studies have not attempted to optimize over all three types of charges, that is, the monthly charge, the local usage charge, and the long distance charge.

Chapter 4

Natural Monopoly and Economies of Scale

A central issue in the formulation of public policy as it relates to the telephone industry is whether or not that industry, or important segments of it, exhibit the characteristics of natural monopoly. An industry is said to be a natural monopoly when the provision of goods or services in that industry can be done at least cost by one firm. In the absence of competition, a natural monopoly must, according to conventional economic and legal rationales, be regulated by a government agency.

This chapter addresses the question of natural monopoly both in general and in the particular case of the telephone industry. With regard to general considerations, the second section of this chapter explains the weakness of the argument that natural monopoly ipso facto requires regulation. Several (at least four) separate questions must be addressed before it can be concluded that natural monopoly requires regulation or that society can benefit from it.

The third and fourth sections of this chapter, specifically focusing on the telephone industry, discuss the econometric, engineering, and simulation evidence on economies of scale in telecommunications. This discussion considers the telephone system as a whole, various links of long distance telecommunications networks, and industry subsegments relevant to public policymaking, such as customer premise equipment and intercity telecommunications service.

Any economies of scale, potentially lost through permitting more competitive entry into the industry, could be at least partially offset, of course, by the costs of lessened regulation itself. A tradeoff exists, in short, between potential loss of scale economies due to competitive entry and the costs created by regulation. These issues are considered in the fifth section of this chapter. The last section is a summary of the chapter.

THE NATURAL MONOPOLY ISSUE[1]

In economic jargon, a natural monopoly is said to exist when economies of scale[2] prevail for the market area in question.[3] When such scale economies are found, it also generally follows that one producer — say, the financially strongest or the best managed — should be able to drive out all others and, having once achieved this monopolistic market position, be in a position to charge rates above those that would prevail otherwise. Specifically, the single producer should be able to achieve monopoly "rents" with prices well above actual costs. By contrast, in the static model of a purely competitive market with many producers, prices would be driven down to a level just sufficient to cover costs, including the costs of attracting the necessary capital.

The standard textbook chart for juxtaposing demand and cost curves can be used, as in Figure 4–1, to display the potential benefits of regulation in a natural monopoly situation. As diagrammed, the demand curve $P(Q)$ is relatively inelastic in the relevant ranges (i.e., in the vicinity of A), and the average cost curve, $AC(Q)$, declines fairly substantially and continuously with increases in scale. Without regulation, the private producer would maximize profits by equating marginal revenue, $MR(Q)$, to marginal cost, $MC(Q)$, at the output, Q_m, and charging the price, P_m. The profit for the monopolist at this level of output would be represented by the rectangle, indicated by dashed lines, formed by the points, P_m, A, E (the intersection of the average cost curve with the output level, Q_m) and F.

Regulators, it is presumed, would force the price down to a level that just covered average costs, specifically P^*. This would be sufficient to keep forthcoming the level of output, Q^*, associated with the average cost since, by presumption or definition, average cost, as diagrammed in Figure 4–1, incorporates a "fair return" on capital as well as allowances for all other costs. Thus, the regulators in circumstances diagrammed in Figure 4–1 would set price at P^* and output would expand to Q^*. The benefits of this regulation could then be measured by the size of the trapezoid *ABCD*. This area represents the increase in the total value, including consumer surplus, of the additional output represented by the move from Q_m to Q^* minus the increase in cost incurred because of that expansion in output.[4]

The usual justification for government regulation in circumstances of natural monopoly is thus to prevent the realization of monopoly rents. More explicitly, the justification is to achieve a pricing structure more nearly in line with costs. Regulation is justified under natural monopoly because the market cannot be relied upon to bring about the normal competitive result of prices just sufficient (more or less) to cover costs and maintain supply. It is therefore commonly alleged in discussions of regulatory issues that if it can be established that a nat-

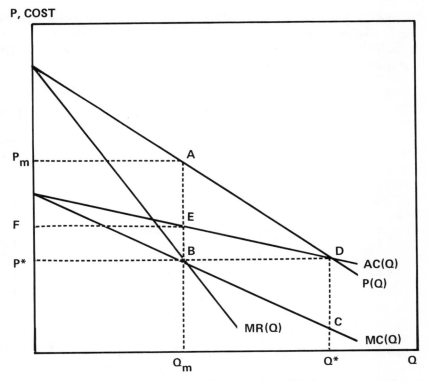

Figure 4–1. Costs and Benefits of Perfect Regulation.
Source: Adapted from Richard Schmalensee, "Estimating the Costs and Benefits of Utility Regulation." *Quarterly Review of Economics and Business* 14, No. 2 (Summer 1974): 51–64.

ural monopoly exists, regulation is ipso facto, or almost automatically, justified.

However, a leap from establishing that economies of scale might exist to justifying regulation is premature. Such a leap glosses over several important factual issues. At least four basic questions need to be answered before it can be concluded that natural monopoly justifies the imposition of regulation:

1. Are the scale economies that create the natural monopoly large, especially in comparison to the costs of regulation?
2. Is the market power of the potential natural monopolist substantial?
3. If the market power is substantial, will regulation limit its exploitation?
4. Is regulation more effective than other remedies?

These four questions place the decision about whether or not regulation is desirable in the context of a benefit/cost analysis—that is, do the potential or realizable benefits of regulation exceed the costs? The four questions are considered in turn below.

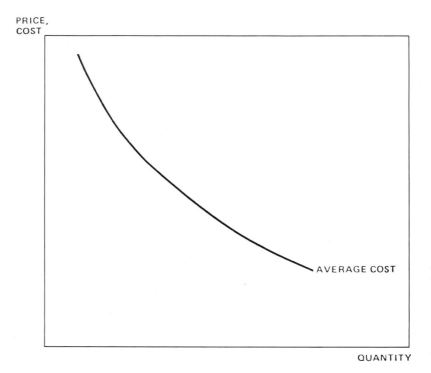

Figure 4–2. Strong Economies of Scale.

Extent of Scale Economies and Costs of Regulation

Before it is concluded that natural monopoly justifies regulation it must be determined whether or not the scale economies that create the natural monopoly are large, particularly in comparison to the costs of regulation. Scale economies come in many different forms. Figures 4–2, 4–3, and 4–4 illustrate some of the possibilities. In Figure 4–2 the average cost curve declines steadily throughout the whole range of relevant output. By contrast, Figure 4–3 represents a situation in which average costs decline very slowly over the entire range of output. Still another possibility is shown in Figure 4–4, where average costs drop rapidly with increases in output at low levels, but then stabilize. Decreases in unit costs occur very slowly or not at all after a basic threshold level of output, necessary for efficiency, has been achieved. In Figure 4–4 it has been hypothesized that the market is so large that two firms, each having 50 percent of the market, could participate without sacrificing scale economies.[5]

The different cost regimes illustrated in Figures 4–2, 4–3, and 4–4 can have very different implications for regulatory policy. The poten-

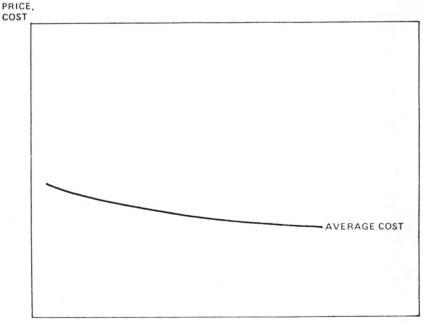

Figure 4–3.   Slight Economies of Scale.

tial cost economies from natural monopoly are likely to be relatively slight if most scale economies are exhausted at a fairly low level of output — as in Figure 4–3 and to an even greater extent in Figure 4–4. The likelihood of this result is increased if the market is large enough to support two or three or more firms of a scale equal to or greater than the threshold level at which most scale economies are exhausted. Furthermore, competition, even if only between two firms in the market, may well discipline managements to be more efficient and innovative and thereby more than compensate for any sacrifices in scale economies.[6]

On the other hand, if the cost structure of an industry is better illustrated by Figure 4–2 than by Figures 4–3 or 4–4, entry potentially implies a much greater loss of scale economies. Under such circumstances, offsets in forms of efficiency stimuli or product innovation would have to be much stronger to justify entry of more than one participant into the market.[7]

The implication of scale economies for regulatory policy is thus an empirical question. The empirical importance of scale economies also depends on how large the relevant market is or is expected to become. If an industry was in an actual situation where the cost curve behaved

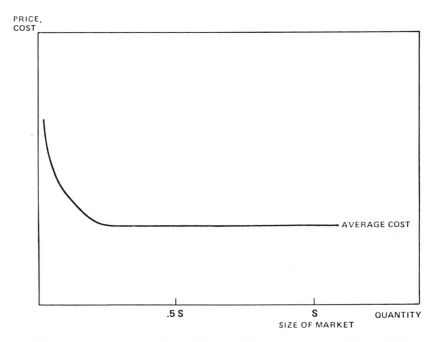

Figure 4–4. Economies of Scale Exhausted at Less Than Half the Size of the Market.

as in Figure 4–4 and the current market was just sufficient to permit one firm to achieve the threshold level that exhausts most economies of scale (i.e., equal to .5S in Figure 4–4), the decision on whether or not to establish a legal monopoly and prohibit entry would depend in great measure upon the market outlook for the industry.

For example, if considerable growth was expected, entry could be viewed as not necessarily too costly in terms of loss of scale economies. On the other hand, if the industry was characterized by stagnation or decline, the argument for creation of a legal monopoly through regulatory action would be strengthened.

The cost of regulation itself must also be considered in deciding whether or not regulation is desirable. One indirect cost of regulation, as already suggested, is the possibility that regulation may reduce managerial incentives for efficiency or innovation. If regulation keeps the firm from capturing the returns from innovative improvements (say, as "unjustifiable" profits), the regulated monopolist does not even have as strong incentives to improve as the unregulated monopolist. Regulation also involves direct costs, not only for government but also for the private firms under regulation. Hearings, legal proceedings, and other formalities are inherent in regulation and require the expenditure of resources. Government must create commissions and pay the cost of staff, paperwork, and administration. The regulated private firms must hire lawyers and other staff to meet the demands for additional records and accounts and for written and oral justifications for actions proposed or taken. Although these costs may not be too substantial relative to the total resources required in a public utility, especially initially, such costs have a tendency to escalate over time and to accumulate in unexpected and unaccountable ways.[8]

Another not so obvious cost of regulation is its inertia. Regulation, even at its best, is usually characterized by considerable lag between changes in circumstances and regulatory action to adapt to these changes. In moderation, this lag can sometimes be beneficial. However, much of the history of regulation has been characterized by attempts to indefinitely postpone or even prevent the inevitable. These tendencies arise especially when the regulatory process is confronted with technological or market changes that create political difficulties. Indeed, a major reason for inertia in regulatory proceedings is that regulation creates its own vested interests. These interests include the practitioners who earn a living by participating in the regulatory process. In addition, the industry's customers, of necessity, may have made long-term investment or consumption decisions on the basis of prevailing regulated rates and services.[9] If regulation has systematically provided poor market signals for investors and consumers, the problem of adjusting regulation to accommodate new realities, as these emerge, is greatly complicated. Furthermore, any long-postponed adjustment from regulatory abnormalities may involve transition costs, not only within the regulated industry itself, but among its customers as well.

In sum, it is a gross oversimplification to argue that the mere existence of scale economies justifies the creation of legal monopoly through the imposition of regulation. There is the question of just how large any scale economies might be. In addition, there is the issue of whether any loss in scale economies is not offset, or even more than offset, by the various direct and indirect costs of regulation. It must always be remembered, moreover, that the textbook gains from regulating a natural monopoly are based on a static analysis. Further

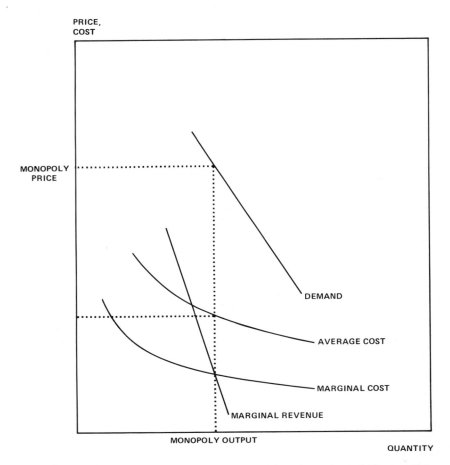

Figure 4–5. Monopoly Pricing with Low Elasticity of Demand.

analysis might show these gains to be overwhelmed by offsetting losses in a more dynamic and behavioral context.

Extent of Market Power

The second basic issue that must be considered before it can be concluded that natural monopoly justifies regulation is the extent of any market power accruing to the potential natural monopolist. Just as there can be substantial empirical differences in the scale economies associated with differant alleged natural monopolies, there can also be substantial differences in their demand curves or market power. In economic jargon, the question is whether the demand curve for the product sold by the potential natural monopolist has a relatively low elasticity with respect to price, as in Figure 4–5, or a relatively high elasticity, as in Figure 4–6.[10] If the demand for the potential monopolist's output is characterized by a relatively low elasticity, as in Figure 4–5, the argument for regulation is strengthened because the potential for extracting monopoly rents from the market is substantial. This argument is reversed if a relatively more elastic demand curve, as in Figure 4–6, prevails.[11]

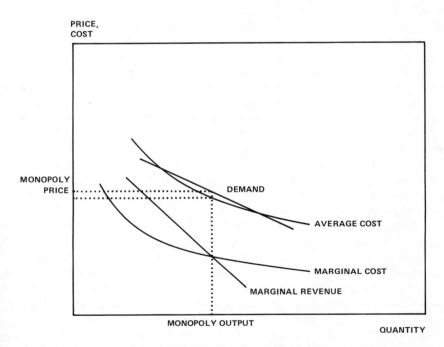

Figure 4–6. Monopoly Pricing with High Elasticity of Demand.

This empirical question of the extent of market power interacts with a quantitative determination of the extent of scale economies. For example, even if an industry was characterized by substantial scale economies, as in Figure 4–2, it might face a relatively more elastic demand curve, as in Figure 4–6. There then might be little point in imposing regulation since the empirical reality of limited market power could well discipline the producer to hold prices more or less close to cost. Any such conclusion would be fortified if the direct and indirect costs of regulation were also substantial. On the other hand, if the potential natural monopolist faced a curve of relatively low elasticity, as in Figure 4–5 (i.e., possessed considerable market power), regulation might be deemed advisable even if scale economies were relatively moderate, as in Figure 4–3.

The advisability of imposing regulation on a potential natural monopolist will also depend on the scale and importance of the product involved. Few would seriously advocate a regulatory solution to correct the potential abuses of a natural monopoly in supplying some good of very little necessity or low consumption. (For instance, little public concern might be expected if it were to be proven that economies of scale created a natural monopoly in the supply of silver toothpicks, even if the demand for such toothpicks proved to be of relatively very low elasticity.)[12] Regulation as a solution has much more appeal in situations where a good or a service is a necessity and very widely consumed.

Even if there are strong economies of scale in the production of a particular good or service, the supplier's market power may still be quite limited. Goods or services for which there are readily available and closely acceptable substitutes are not as likely to be deemed worthy of regulation as those for which there are few or only more distant substitutes. Indeed, one of the more general phenomena that may increase the elasticity of demand for almost any commodity or service is the presence of interindustry competition. Even though there may be room for only one supplier within a particular industry, there may be other sources of supply from outside the industry of goods that are almost as suitable for the same purpose. In these circumstances, a substantial rise in the price of the product in question will induce many consumers to purchase the substitute instead. Consequently, the demand for the product is relatively more elastic, as illustrated in Figure 4–6, and market power is quite limited. A prime illustration of this phenomenon often cited in the world of regulated activities is that of natural gas. Many observers would argue that even though there are scale economies in the distribution of natural gas, regulation is unnecessary because of the existence of so many close substitutes (e.g., bottled gas, electricity, or oil in home heating, and all these plus coal or a total energy package in many industrial and commercial applications).

Market power may also be limited by the existence of alternative, somewhat higher cost technologies. Such a circumstance is illustrated in Figure 4–7, which shows a case where, if the sole supplier who enjoys scale economies attempted to charge the monopoly price, entry by suppliers using the alternative technology would drive prices down to a level close to the initial sole supplier's costs.

An illustration of such a possibility might be a railroad carrying coal from a relatively remote mine site to an electric utility plant. Under such circumstances it may make sense to have only one railroad serve the market because of scale economies, and the railroad

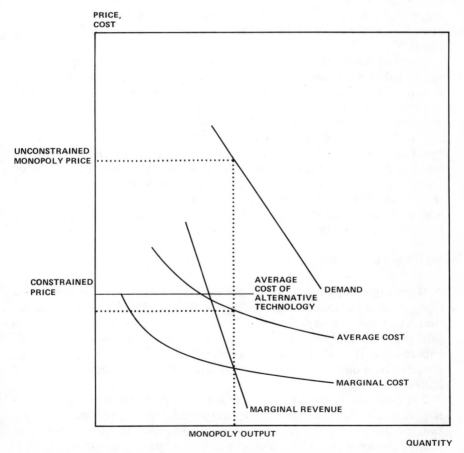

PRICE, COST

UNCONSTRAINED MONOPOLY PRICE

CONSTRAINED PRICE

AVERAGE COST OF ALTERNATIVE TECHNOLOGY

DEMAND

AVERAGE COST

MARGINAL COST

MARGINAL REVENUE

MONOPOLY OUTPUT

QUANTITY

Figure 4–7.   Monopoly Pricing Constrained by Alternative Technology.

may in fact be the lowest-cost way of moving the coal from the mine to the utility. Yet, if the railroad were to charge rates much above costs, it might well be confronted by competition from other sources of coal or by the introduction of a coal slurry pipeline or mine-mouth generation of the electricity.[13]

The threat of potential entry is something that is subject to change as technologies and markets change. In a world of steadily increasing knowledge, moreover, the trend over time normally will be toward the threat of potential entry becoming more rather than less real. Put in the vernacular, the number of ways to "skin a cat" are likely to increase rather than decrease as time passes. Furthermore, the rate at which new technology or potential substitutes appear is also likely to be a function of the extent to which a monopoly exploits its advantages — high prices and high markups will attract more search for alternatives than will lower prices and markups, all else being equal.[14]

In short, then, the question of whether a natural monopolist possesses substantial market power, to the point of requiring regulatory restraint, is at least as complicated an empirical question as determining the scope and importance of any scale economies. The extent and substance of market power is, first of all, a question of the importance and universality of the product or service produced. It is secondly a question of the elasticity of demand for the particular product or service, which, in turn, depends upon the availability of substitutes, either from other industries or from potential new ways of achieving the same ends. Finally, the juxtaposition or combination of market power and scale economies is what ultimately matters, rather than either factor taken in isolation.

Effectiveness of Regulation

A third question that must be addressed before it can be concluded that natural monopoly justifies regulation is whether or not regulation will limit the exploitation of market power. Even if it can be established that both market power and scale economies are substantial, it still does not necessarily follow that regulation will prove to be a desirable solution.[15] Much will depend upon the specific historical context.

For example, if market power in a strongly established natural monopoly situation has already been exploited, the imposition of regulation might simply serve to consecrate or preserve the status quo. Many students of U.S. transportation argue, for example, that effective regulation came to the railroad industry only after the industry had eliminated most of its early competitive manifestations and was well on its way to cartelization. In this view, regulation developed not as a cure for problems of the consumer but rather as a means for producers to achieve government help in completing the cartelization of their activities.[16]

Timing may therefore be crucial in determining the true effectiveness of regulation in achieving the goals of public policy. Regulation that appears early in an industry characterized by pronounced scale economies should have a somewhat better chance of keeping prices in line with costs than if it appears late. Not only is there a danger of cartelization and monopolistic exploitation being more fully developed in a more mature industry, but also late-arriving regulation, in attempting to undo or roll back existing market relationships, may create its own traumas and transition problems. For example, various consumers or industrial users of the monopolized good may have made location or other investment decisions on the basis of the high markups imposed by the monopolist, and they may consequently find a regulatory change in the situation extremely disorienting. Real estate or securities markets, for example, may well have capitalized any monopoly rents and incorporated them into market valuations; buyers who purchased at these values understandably could resent a change in government policy that reduced these capitalizations. Thus, the temptation may be strong for regulators to accept the established ways rather than attempt to undo a monopoly position. Given that an established monopolist is also likely to have considerable latent or real political power, as well as a superior command of the facts (since, after all, it is the monopolist's business), the difficulties of achieving a regulatory rollback in such a situation may prove substantial.[17]

It is also possible, although not probable, that a strong natural monopoly position may have existed but was never exploited. Such an innocent situation would hardly seem to justify a drastic regulatory reaction. Indeed, if a demand for regulatory action were expressed in such circumstances, the obvious question would be why. If such a request emanated from the industry itself, one might suspect that regulation was sought to achieve what the market itself would not permit. On the other hand, if a specific consumer group were seeking such action, the suspicion might be that preferential treatment was sought against other users (say to gain a competitive advantage in the case of industrial consumers).

Effectiveness of Other Remedies

The final question that must be asked before it can be concluded that natural monopoly justifies regulation is whether or not regulation is the most effective remedy to the problems of natural monopoly. The adoption of a regulatory solution to these problems must be weighed against other means of achieving the same ends. Specifically, even if it can be established that large-scale economies exist, that substantial market power resides in the natural monopolist, and that regulation could limit existing or potential exploitation of these possibilities, it still does not automatically follow that regulation is necessarily desirable. The regulatory solution should be weighed against other alternatives.

The most obvious alternative to regulation is allowing entry of new firms. An entry solution will be particularly attractive if scale economies are not too pronounced in the industry (e.g., behave more or less as in Figures 4–3 or 4–4) and if growth rates in the industry are quite substantial. In such circumstances it may be quite possible to introduce new suppliers into the industry without any great sacrifice in scale economies, especially in the longer run.[18] A determination on the desirability of entry as a solution will also depend on the extent to which it is felt that entry can improve managerial discipline for achieving efficiency, innovation, and other such goals.

Another alternative to regulation of a natural monopolist would be technological change that undermined either the cost advantages or the market power of the monopolist. With the benefit of hindsight it can be seen that a monopoly in intercity rail passenger traffic in the late 1920s should not have been viewed as very alarming from the standpoint of public policy; in the offing were car, bus, and air alternatives that eventually made rail passenger travel a little-used and generally inferior mode for most intercity travel in the United States. Similarly, in intercity freight today, as noted above, the mere threat of coal slurry pipelines or long distance transmission of electricity may be quite as effective as regulation in restricting any monopolistic tendencies of railroad coal carriers.[19]

It must not be overlooked, moreover, that regulation might not only be inferior to technological change as a means of curbing monopolistic excesses, but may also actually impede such change. For example, regulation can be used as a means of forestalling an impending technological change that would undermine or alter market relationships.[20] This would be particularly true if there is continuing validity in the conventional historical observation that regulation has a tendency to perpetuate the status quo.

Technological change can also inhibit the use of market power by modifying cost structures. In particular, a technological change that converted an industry's cost characteristics from those in Figure 4–2 to cost characteristics more like those in Figures 4–3 or 4–4 would enhance the possibility of using new entry as an alternative to regulation. Furthermore, if technological change were working to lower the threshold level at which most scale economies are realized in a market already characterized by considerable growth, the possibility of using entry as a corrective would be further enhanced.

Once again, then, the empirical facts of the situation will have much to do with deciding the issue of whether or not natural monopoly regulation constitutes good public policy. Technological trends that favor a reduction in the threshold at which scale economies are exhausted would work against regulation, all else being equal; on the other hand, technological trends working toward more pronounced scale economies would have an opposite effect. Growth of a market would tend to make alternatives to regulation potentially more acceptable and implementable, all else being equal. Rapid technological change also tends to reduce the attraction of regulation, often subverting the premise on which the regulation is based.

The desirability of natural monopoly regulation, in short, is a matter of weighing costs against benefits. It does not follow automatically or easily from the existence of scale economies that regulation is justified. Not only does no such conclusion follow automatically, but also a sensible determination on the advisability of regulation depends on a very careful evaluation of all the empirical facts, not only as they relate to current costs and demand structures but to prospective changes in them as well.

ECONOMETRIC EVIDENCE ON ECONOMIES OF SCALE

Econometric evidence on economies of scale in telephone systems comes primarily from studies in which aggregate production functions are estimated using historical data on telephone system operations. This evidence suggests that economies of scale are probably present in telephone system operations, but that these economies are not overpowering in size.[21] The evidence is not conclusive because of numerous practical problems that beset this type of research.

In econometric studies an attempt is normally made to determine whether economies of scale are present by estimating the parameters describing relationships among economic measures of a company's operations, as based on historical observations. An econometric study thus proceeds from the assumption that there are fundamental relationships generating the observed historical data. Included among the fundamental relationships are basic assumptions regarding the economic motivations and behavior of firms. Simplifying assumptions characterizing the relevant technology are also included. The latter assumptions can be crucial because economies of scale are, in essence, commonly embedded in technology.

The most basic assumption regarding the technology, or production process, is that quantities of inputs are used to produce quantities of outputs. The problem facing the econometrician is to choose a specification of the production process, in mathematical form, that can be estimated by statistical techniques. The chosen specification is necessarily an abstraction from the complex realities of the production process; it is designed to capture the crucial features of that process. The adequacy of the specification can be tested statistically when its parameters are estimated using historical data. The specification can also be tested by comparing forecasts of economic variables made by using the estimated parameters with actual or realized values of these variables experienced in the real world.

The production process can be formalized in several ways: as a production function in which a single output is produced by one or more inputs; as a cost function in which the total cost of producing a single output varies systematically with the level of output and the price levels of inputs; as a transformation function that is a multioutput analogue to the single output production function; or as a multioutput analogue to the single output cost function.

Most of the econometric evidence on economies of scale in telecommunications comes from studies in which a single output production function was estimated for all telephone system operations. Knowledge about economies of scale in toll service — or in particular segments of toll service — would be of greater value for many policy decisions regarding competition in telecommunications, but the additional conceptual and data problems involved in estimating economies of scale for components of telephone systems have deterred most econometricians from exploring the issue. Specifically, most estimates of scale economies have been made for the U.S. Bell System viewed as an entity with a few additional estimates from studies of the Bell Canada System.

Econometricians have generally estimated economies of scale after attempting to control for the independent effects of technological change, often employing heroically simplifying assumptions in the process. Nevertheless, it is important that an attempt be made to control for technological change. Studies where comparisons are possible suggest that about one-third of the economies of scale that would be estimated if the impacts of technological change were not taken into account could be in fact due to technological change.[22]

After formulating a general relationship from which a measure of economies of scale can be derived, the second step in econometric research on economies of scale is to translate the theoretical relationship into an empirical analogue in terms of available data and functions of the data that can be estimated using statistical techniques. The data used for this purpose theoretically can be either time-series or cross-section data. Time-series data consist of observations on a variable for different time periods over a period of time. Cross-section data consist of observations from various similar units at the same point in time. (Examples of such units would be Bell operating companies, independent telephone companies, state totals, or national totals in an international comparison.) In using time-series data, variations from one time period to another are regarded as clues to the nature of the production process, whereas in cross-section data, differences across the units of observation are regarded as indicative.

Econometric estimates of system economies of scale in telecommunications are essentially taken from studies using time-series data,[23] specifically annual data for a series of years in the period since World War II. For econometricians estimating economies of scale for the U.S. Bell System, this period is generally 1946 or 1947 through the early to mid-1970s. For studies of Bell Canada operations, the initial year of the data series is usually 1951 or 1952.

A general representation of a production function, the theoretical relationship that underlies most econometric studies of economies of scale, would be as follows:

$$Output = F(labor, capital, materials, land/technology)_t$$

This expression indicates that output over a specified time period, t (say a year like 1950 or 1960), is the result of the combination of various inputs such as labor, capital, materials, and land. Output in the production function is defined as the maximum amount of output that can be produced by combining given amounts of the various inputs. That is, implicit in the concept of the production function is an assumption that there is no waste or slack, and therefore no expenditures are made that do not have a payoff in terms of increased output.

The vertical line separating technology from the inputs in the above representation of the production function indicates that inputs are combined according to a particular state of technology. A state of technology consists of a particular application of knowledge regarding equipment design, system design, and management concepts and techniques. To some extent it is reflected in particular types and styles or mixes of equipment. Examples for telephone systems would be electronic, crossbar, or manual switches and coaxial cable, microwave, satellite, or optic fiber transmission.

As time changes, the historical data from the various years identify combinations of levels of outputs and inputs. Generally, output has grown over time, as have the levels of some or all inputs, and it is possible to identify by statistical techniques the parameters of the relationship between these changing levels of inputs and outputs. These parameters can in turn be used to estimate the extent of scale economies.

Evaluating whether or not economies of scale are present involves determining if equiproportional increases in all inputs yield a greater than proportional increase in output, always assuming that an unchanging state of technology exists or is adequately approximated by statistical controls. The number derived from an estimated production function used to measure economies of scale is called the scale elasticity. Scale elasticity is defined as the percentage change in output resulting from a 1-percent change in all inputs. Such a scale elasticity, *E,* can fall in three ranges:

E > 1, denoting economies of scale
E < 1, representing diseconomies of scale
E = 1, indicating constant returns to scale

If the scale elasticity, *E,* is greater than one, a 1-percent increase in all inputs yields a greater than 1-percent increase in output, and it is said that scale economies are present. If the scale elasticity is less than one, a 1-percent increase in all inputs yields an increase in output of less than 1 percent, and decreasing returns to scale are indicated. Finally, if the scale elasticity equals one, a 1-percent increase in all inputs yields an equal 1-percent increase in output, indicating the absence of either economies or diseconomies of scale, that is, constant returns to scale.

Econometric estimates in terms of the scale elasticity can also be easily translated into terms describing the behavior of costs as output increases (if increases in output are assumed to be of a scale insufficient to change the prices charged for the inputs). When the scale elasticity is greater than one, the percentage increase in output is greater than the percentage increase in inputs, and thus average cost per unit declines with an increase in output (again, always assuming no offsetting change in input prices). Similarly, when the scale elasticity is less than one, the percentage increase in output is less than the percentage increase in costs, and average cost increases. Finally, when there are constant returns to scale, output and costs increase the same percentage and average cost is unchanged or constant. In terms of typical U-shaped cost curves, scale elasticities considerably larger than one indicate strong economies of scale and rapidly falling average cost, as illustrated in Figure 4–2. Scale elasticities only slightly greater than one, on the other hand, indicate slight economies of scale and slowly falling average cost, as in Figure 4–3. The absence of either economies or diseconomies of scale (i.e., constant returns to scale) is consistent with a flat cost curve as in the right-hand, higher-volume range of Figure 4–4.

In Table 4–1 a summary is presented of the available estimates of scale elasticity in telecommunications systems obtained from econometric studies. The results for the U.S. Bell System are from four studies, one of which was done by Leroy Mantell at the Office of Telecommunications Policy[24] and three of which were done at Bell Labs.[25] The results for Bell Canada are from two studies, both of which were done by groups of researchers at the University of Toronto.[26]

The first line in Table 4–1 gives the range of point estimates of scale elasticity reported in the studies of the U.S. Bell System, irrespective of how reasonable the point estimates are on grounds of statistical measures of their accuracy or the reasonableness of particular details of the methodology employed. The estimates in the first line of Table 4–1, ranging as they do from less than one to more than two, would be consistent with increasing, constant, and decreasing returns to scale.

Table 4–1. Point Estimates of Scale Elasticity in Telecommunications Systems from Econometric Studies

Telecommunications System	Scale Elasticities
U.S. Bell System[a]	0.74–2.08
U.S. Bell System (excluding one study)[b]	0.98–1.24
Bell Canada[c]	0.85–1.11

[a]American Telephone & Telegraph Company, "An Econometric Study of Returns to Scale in the Bell System," Bell Exhibit 60, FCC Docket 20003 (Fifth Supplemental Response), August 20, 1976; L. Mantell, "An Econometric Study of Returns to Scale in the Bell System," Staff Research Paper, Office of Telecommunications Policy, Executive Office of the President, Washington, D.C., February 1974. (Also appears as Bell Exhibit 40 in Docket 20003.); H. D. Vinod, "Application of New Ridge Regression Methods to a Study of Bell System Scale Economies," Bell Exhibit 42, FCC Docket 20003, April 21, 1975; and H. D. Vinod, "Bell System Scale Economies and Estimation of Joint Production Functions," Bell Exhibit 59, FCC Docket 20003 (Fifth Supplemental Response), August 20, 1976.

[b]Excluding American Telephone and Telegraph Company, "An Econometric Study of Returns to Scale in the Bell System," Bell Exhibit 60, FCC Docket 20003 (Fifth Supplemental Response), August 20, 1976.

[c] A. Rodney Dobell et al., "Telephone Communications in Canada: Demand, Production, and Investment Decisions," *Bell Journal of Economics and Management Science 3* (Spring 1972): 175–219; and Melvyn Fuss and Leonard Waverman, "Multiproduct Multi-input Cost Functions for a Regulated Utility: The Case of Telecommunications in Canada." Paper presented at the National Bureau of Economic Research Conference on Public Regulation, Washington, D.C., December 15–17, 1977.

The second line in Table 4–1 gives the range of estimates for the Bell System scale elasticity when one study is excluded.[27] The excluded study is an exploratory study done at Bell Labs that is not directly comparable to the others.[28] Its exclusion reduces the range of estimates of scale elasticity at both extremes and narrows the range of possibilities to the general vicinity of constant returns. This modified set of Bell results is not too dissimilar to the range of estimates for Bell Canada, which include generally lower point estimates than the range for the U.S. Bell System.

Unfortunately, most of the studies summarized in Table 4–1 do not report measures of statistical significance or other information from which it would be possible to make an evaluation of the uncertainties or probable range attached to the point estimates. Estimates from the study by Mantell do, however, come with associated measures of statistical significance, specifically 95-percent confidence intervals;[29] these estimates are presented in Table 4–2. Fortunately, given the paucity of point estimates of scale elasticity for which confidence limits are reported, the two estimates in Table 4–2 seem to be quite representative of the range of estimates for the U.S. Bell System presented in the second line of Table 4–1. For Estimate 1, for example, there is a .95 probability that the unknown true scale elasticity falls between 0.91 and 1.17.

It is also possible to determine the probability that the unknown true scale elasticity is greater than one, that is, economies of scale are present. For Estimate 1, the probability is slightly greater than .75.

Table 4–2. Statistical Significance of Mantell's Scale Elasticity Estimates

Item	Estimate 1	Estimate 2
Point estimate of scale elasticity	1.04	1.16
95% confidence interval	0.91–1.17	1.03–1.29

Source: Both point estimates and the confidence interval for Estimate 2 from Leroy H. Mantell, "An Econometric Study of Returns to Scale in the Bell System," Staff Research Paper, Office of Telecommunications Policy, Executive Office of the President, Washington, D.C., February 1974. Confidence interval for estimate 1 was calculated, from the data used in Mantell's study, by H. D. Vinod, "Application of New Ridge Regression Methods to a Study of Bell System Scale Economies," Bell Exhibit 42, FCC Docket 20003, April 21, 1975.

Estimate 2 in Table 4–2 is a higher point estimate of scale elasticity and comes from a slightly different functional specification than does Estimate 1. Specifically, an alternative definition of the input of labor into the production process is utilized. The 95-percent confidence interval for Estimate 2 indicates that there is a .95 probability that the interval between 1.03 and 1.29 covers the unknown true scale elasticity. The probability that the true value is greater than one, that is, economies of scale are present, is .99.

If one were forced to reach a judgment about the existence of economies of scale based solely on these two estimates for which there are associated measures of statistical significance, the judgment would vary depending on which of the two estimates was used. Both estimates suggest that some economies of scale are probably present. However, Estimate 2 is more favorable to the case for economies of scale than is Estimate 1. Estimate 2 is not only larger, but the associated probability that the unknown true scale elasticity lies above one is also larger.

Grounds for preferring one of these two estimates over the other are tenuous, however. The estimates come from functional forms that differ only in the way the measure of labor input is derived, and there are no strong theoretical reasons to prefer one of these derivations over the other.

In general, based on the results of these econometric studies alone, a reasonable observer forced to make a judgment on whether or not economies of scale are present in telephone system operations would bet that some economies of scale are present. Nevertheless, because of the practical problems involved in econometric research and the resulting uncertainty about individual point estimates, reasonable and informed people, relying on econometric evidence, can come to quite different conclusions on these issues. Indeed this happened in the course of FCC Docket 20003. In this docket, the FCC, using a survey prepared by T+E[30] of most of these same econometric studies as are reviewed here, concluded that the case for economies of scale had not been proved. At the same time, Stanford Research Institute (SRI), employed by AT&T to review T+E's review, concluded that there probably were economies of scale in the range of 1.1 to 1.25.[31] Although the results shown in Tables 4–1 and 4–2 would suggest that the range of estimates selected by SRI is on the high side of the reasonable range of point estimates, the SRI conclusion is certainly not unreasonable.

Because the scale elasticity issue has obvious policy implications, it is important to understand certain underlying practical problems that can create uncertainty about the reliability of the econometric estimates. A major practical problem encountered in the econometric studies of telecommunication costs is the divergence between the available data on labor, capital (or plant), and technological change and the conceptually correct data that the econometrician would like to have. In these studies, for example, output is generally measured by total operating revenue less the cost of purchased materials and services — a quantity called "value added." Capital and land are always combined and measured by data on plant; some econometricians use gross plant and others use plant net of depreciation.

Econometricians would prefer, however, to have measures of these inputs and outputs in physical units. That way, in evaluating economies of scale, it would be absolutely clear, for example, that all inputs were increasing in the same proportion and exactly what the increase in output had been. Nevertheless, even if the data were available for, say, numbers of phone messages by distance and time-of-day categories and for the number of units of particular types of capital by age, and so on, using all these variables separately would result in an intractable estimation problem. Instead, dollars are used as a common denominator and different types of output or inputs are summed up to totals using their prices as weights.

Using dollar values of total output and inputs does, however, create practical problems because apparent changes in levels of output or inputs can occur simply because prices change — and if this occurs, economies of scale cannot be properly measured.[32] To control for this possibility, the dollar measures of revenue and plant are typically deflated by indices of price changes in econometric studies. The deflation procedures, however, are only as good as the indices used. If the indices are rarely updated or if they do not adequately reflect the changing mix of the components of output or the various inputs, it is likely that the value of output or inputs will be incorrectly measured.

Another very important practical problem is that the correct capital input is really capital services, not capital stock. But only data on capital stock in place is usually available. To convert stocks to measures of usage, it would be very helpful to have a measure of capacity utilization but none is available for telecommunications.

Unlike output and capital, physical measures of labor input are available, namely, the number of hours worked. But hours of employees cannot be added together meaningfully because there are differences in skill across categories of employees. Thus it is often deemed desirable, if labor hours are used, that they should be weighted by wage rates, for example, an hour of time of an employee earning $6 per hour would count double that of an employee earning $3 per hour. In addition, hours should include overtime hours as well as regularly scheduled hours. Unfortunately, data on the U.S. Bell System that could be used to take these effects into account are not available from public sources and the measures used for labor input in various econometric studies have one or the other or both of these two flaws; that is, either they do not weight different types of labor differently or exclude overtime hours.

Econometricians have generally tried to allow for technological change in the telephone industry by including the percent of long distance calls that are directly dialed by customers as a variable in the relationship being estimated. This way of allowing for technological change implicitly assumes that technological change has the same proportional impact on the productivity of capital and labor and the same proportional impact on the productivity of pieces of equipment of all ages. However, a number of people have argued that technological change in telecommunications probably increases the productivity of capital proportionately more than the productivity of labor, although this issue is not settled.

Even if there were no data or measurement problems, econometric cost and production studies can be confounded by technical estimation difficulties. In essence, statistical estimation is fitting a curve through a scatter of observations on historical data; there are usually a variety

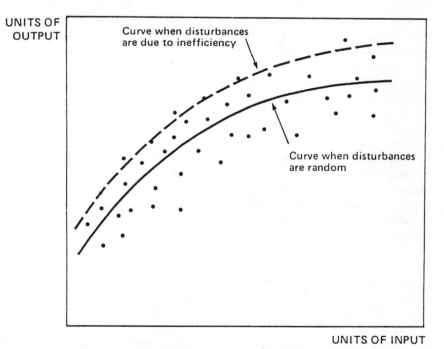

Figure 4–8. Illustrative Examples of Curves Fit by Statistical Methods on Different Assumptions about the Process Generating the Observations.

of legitimate ways this can be done. To illustrate, a scatter of points and two curves through these points are shown in Figure 4–8.[33] Which of the curves drawn in Figure 4–8 is deemed best will depend on an econometrician's judgments or beliefs and can make a difference in the results and how they are interpreted.

The curve drawn as a solid line represents basically what is done in most econometric studies of economies of scale, including studies of telephone systems. This solid curve fits through the center of the scatter of points and indicates that the econometrician believes that the practical problems, such as lack of control over all historical influences, do not operate in any systematic way so as to bias the observations on the inputs in any particular direction. Consequently, the points in the scatter are treated as if they deviate from the true relationship only because of unimportant random events.

Illustrative of other possibilities (i.e., that the points differ from the true relationships for other than random and unimportant reasons) is the top curve, drawn as a dashed line. The dashed line in this case might represent, for example, an alternative assumption that the observations lying at a distance below the dashed line are primarily due to inefficiency. It might be argued that a situation represented by the dashed line could arise in a regulated industry because of the absence of competitive discipline.[34]

In sum, because of practical problems with data or selection of a curve-fitting method or of the proper functional representation of the underlying theoretical concepts, there easily can be disagreement on the econometric evidence concerning the existence or extent of economies of scale. These problems of interpretation are not unique to econometric studies of telephone systems. Their existence, moreover, does not necessarily mean that econometric results cannot be used for making policy evaluations. Problems in interpreting econometric studies do suggest, however, that a careful evaluation of the potential impact of scale on costs should incorporate evidence from other types of studies as well, such as the engineering and simulation studies explored in the next section.

ENGINEERING AND SIMULATION STUDIES OF SCALE ECONOMIES IN LONG DISTANCE TELECOMMUNICATIONS

The engineering approach to measurement of scale economies uses cost data on current best practice technology, usually provided by design engineers in the industry under study.[35] Most engineering discussion of potential scale economies in long distance telecommunications has focused on the behavior of investment costs for new terrestrial transmission systems, both for particular links and for networks. By contrast, the econometric evidence discussed in the preceding section is derived from historical data on operating costs and the book value of capital for existing systems taken as a whole, including local facilities.

When evaluating engineering evidence on economies of scale in investment costs for long distance transmission, close attention must be paid to the components of cost that are addressed in a particular study. The three basic components of investment costs in such transmission are: (1) the cost of land and structures, including cable rights of way, buildings, land, access roads, and so on; (2) "outside" equipment such as cable, microwave towers, and antennae;[36] and (3) radio equipment, including transmitters, receivers, repeaters, and backup equipment. These three components — land and buildings, outside plant, and radio — will henceforth be referred to as "basic transmission" costs; they are almost invariably included in reported studies of long distance terrestrial transmission investment costs. Various studies do, however, differ in whether or not two other cost components, multiplex equipment and switching equipment, are included.

Multiplex equipment is needed to combine many individual channels into one group for transmission.[37] It is generally agreed that scale economies in multiplexing are minor.[38] Multiplex equipment costs can be a substantial portion of total investment cost; thus, their inclusion tends to lower estimated scale economies below what they would be for an investigation of basic transmission facilities alone. Since some studies include multiplex costs and others exclude them, this variable can be an important difference.

Switching equipment, a fifth component of investment costs for long distance transmission, establishes transmission paths between pairs of terminals within a telecommunications network; it is not needed on individual links. Switching equipment costs have generally been excluded from studies of investment costs for long distance tranmission facilities; these studies either focus only on individual links or abstract from switching costs in networks. Little evidence is available on how switching equipment costs respond as scale increases, but it is generally agreed that, as in multiplexing, scale economies for new digital electronic switching equipment are relatively minor.[39] Thus, inclusion of switching equipment costs in total investment costs will lower estimated scale elasticities below what they would be for only basic transmission plus multiplex facilities.

Figure 4–9 provides an overview of the relationship between investment costs per circuit mile for basic transmission facilities (i.e., excluding multiplex and switching) and scale for technologies in use or anticipated for long distance terrestrial telecommunications as of the mid-1970s. As shown in this figure, different types of terrestrial transmission systems display an overall downward trend in investment cost per circuit mile as scale increases. The heavy dashed lines in Figure 4–9 represent different trends that might be drawn through the systems shown. The upper and more steeply pitched line has a slope of −0.6 on a log-log graph. This implies that the elasticity of average cost with respect to scale is −0.6; in other words, a 1-percent increase

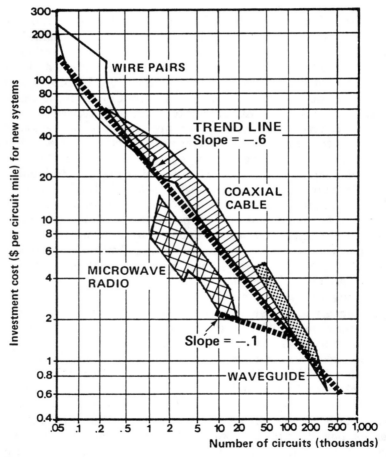

Figure 4–9. Investment Cost for New Terrestrial Transmission Systems. *Source:* Redrawn from Eugene F. O'Neill, "Radio and Long-Haul Transmission," *Bell Laboratories Record,* January 1975, p. 55. Trend lines added by Charles River Associates.

in scale results in approximately a 0.6-percent decrease in average unit cost. This steeply pitched trend line is generally presumed by AT&T to be the best description of future development possibilities in basic transmission costs. The lower or less steeply pitched trend line, by contrast, has only a -0.1 cost elasticity and would seem to hold only if future technologies provide lower unit costs at lower investment levels or if larger scales, as yet unexplored, prove less susceptible to scale economies in basic transmission costs than now expected.

To place these cost elasticities in perspective, Table 4–3 shows conversions between elasticities of average cost with respect to scale, as generally used in engineering discussions of investment costs, and the

Table 4–3. Relation Between Scale Elasticity and Average Cost Elasticity[a]

Elasticity of Average Cost with Respect to Scale	Elasticity of Output with Respect to Inputs
0	1.00
-0.10	1.11
-0.25	1.33
-0.30	1.43
-0.50	2.00
-0.60	2.50

[a]
$$\varepsilon_s = \frac{1}{1 + \varepsilon_{ac}}$$

where ε_s = scale elasticity, and
ε_{ac} = average cost elasticity

Derivation: Under the assumption that the proportions and relative prices of inputs do not change with expansion, total cost will accurately represent total inputs

Let C = total cost = total inputs
Q = output (scale)

Then: $\varepsilon_s = \dfrac{d \ln Q}{d \ln C}$

$$\varepsilon_{ac} = \frac{d \ln (C/Q)}{d \ln Q} = \frac{d \ln C}{d \ln Q} - 1$$

Therefore: $\varepsilon_s = \dfrac{1}{1 + \varepsilon_{ac}}$

production function scale elasticities relating outputs to inputs, the measure common to econometric analyses.

As the slope of the average cost line increases, a larger and larger scale elasticity is implied. If the slope is only -0.1, a scale elasticity of 1.11 is implied; if the average cost line is as steep as -0.6, as in the upper trend line in Figure 4–9, a large scale elasticity of 2.5 is indicated.

In interpreting engineering cost elasticities, it must again be stressed that some studies, purportedly examining the cost behavior of actual or simulated telecommunications networks, have in fact looked only at basic transmission costs. For example, scale elasticities have been examined by AT&T for a network model with a great number of links and nodes, but only for basic transmission facilities, specifically excluding multiplex and switching equipment from investment costs.[40] In doing this study, costs were assumed to follow the upper or more steeply pitched trend line from Figure 4–9. Thus, the principal difference between a simple extrapolation of costs based on Figure 4–9 and AT&T's more sophisticated network simulation is that the simulation allows for varying demand growth on different links in the system (e.g., Boston to Chicago growing by a different percentage than New York to Atlanta). When system links are assumed to grow at different rates, a lower elasticity normally will result. Thus in the network study, the average cost elasticity is about -0.55, compared to the single link average cost elasticity of -0.6 derived from the upper trend line in Figure 4–9.

Basic transmission costs would seem to represent approximately one-half of total long distance investment requirements. If economies of scale are lower in other components of the long distance system, the total overall scale elasticity in long distance should be lower than that found in basic transmission. Table 4–4 shows the book value of Long Lines plant in 1975, with a percentage breakdown by type in the right-hand column. The categories comprising basic transmission would be "radio," "cable," and (most of) "other plant and equipment," thus adding up to about 50 percent of the total. In actuality, basic transmission probably would be slightly below 50 percent for the entire long distance network because operating companies that own long distance facilities would generally report a higher percentage of costs in multiplexing and switching.

The categories of "switching" and "circuit" (which include multiplexing equipment and other electronics) in Table 4–4 also represent approximately 50 percent of total plant value. As already mentioned, the available evidence indicates that multiplexing and switching equipment investment costs exhibit few economies of scale. Therefore, as the proportion of multiplex and switching investment cost to total transmission investment cost increases, scale economies will decrease. Generally, the multiplex proportion will rise as the number of circuits involved increases (for both terrestrial and satellite transmission). In addition, multiplexing equipment is not needed for each repeating station but only at both ends of the transmission links (unless the configuration of channels and groups is rearranged). Therefore, the proportion of multiplex equipment cost in total terrestrial transmission cost will vary inversely with the length of each specific link; that is, the proportion will be higher for shorter links and lower for longer links.

In a study in which multiplex equipment was specifically included, based on engineering data as well as data from actual systems,[41] it was found that the elasticity of average cost with respect to scale was about −0.3. This result was, moreover, consistent with actual AT&T costs up to scales of approximately 6,000 circuits. With multiplexing included, the effect of a 1-percent increase in scale was thus closer to a 0.3 percent-decrease in unit cost rather than the 0.6-percent decrease indicated by the upper trend line in Figure 4–9.

Switching, as noted, also seems characterized by small potential economies of scale.[42] To gain perspective on how these different cost categories might be expected to interact or sum up, it can be hypothesized that Bell's historical average cost elasticity of −0.6 held on basic transmission while other categories experienced constant returns. Av-

Table 4–4. Book Cost of Long Lines Plant and Equipment, 1975

Category	Book Cost (millions of dollars)	Percent of Total
Central office equipment		
Circuit[a]	$2,210	34.4%
Radio	1,108	17.2
Switching	962	15.0
Outside plant		
Cable	1,093	17.0
Other outside plant	57	0.9
Other plant and equipment[b]	999	15.5
Total	$6,428	100.0%

Source: Book cost data taken from American Telephone and Telegraph Co., *Long Lines Statistics, 1950–1975,* April 1976.

[a]Circuit includes multiplexing equipment.
[b]Largely land and buildings.

erage cost would then decline about 3 percent with a 10-percent increase in output for a scale elasticity for the whole system of about 1.43.[43] That is, even assuming an extremely high production function scale elasticity of 2.5 in basic transmission, lack of pronounced economies of scale elsewhere would considerably lower economies of scale in the long distance system as a whole. If the basic transmission cost elasticity is lower, say only -0.3, and zero elasticities still hold elsewhere, the overall long distance scale elasticity is only about 1.17.

In sum, the engineering evidence suggests that economies of scale in long distance may not be too much greater then the range found in econometric studies of the whole telephone system; that is, the long distance scale elasticity (including multiplexing and switching) appears to be between 1.1 and 1.5 at the outside. Although good information on operating costs is not available, the best guess would be that their inclusion would further lower the scale elasticity; however, the evidence is too scanty to assert this with any confidence.[44]

The discussion up to this point deals only with terrestrial transmission. Long distance communications can, of course, also be routed through satellites, which present sharply diferent operating characteristics.[45] Satellite communications can make entry economically attractive on small scales between particular points and on particular routes under certain operating conditions and system configurations. The competitive capability of satellite communications is also expected to expand over the years as satellite costs per circuit fall with technological progress in microelectronics, satellites, earth stations, and launching vehicles.

The transmission capacity of existing individual satellites is relatively small in comparison to terrestrial systems.[46] Although satellite capacity is expected to expand over the years, it will nevertheless remain very limited for individual satellites when compared to the domestic intercity communications market as a whole (for all routes and all customers). Therefore, any economies of scale[47] in satellite systems are not likely to be significant when compared to the whole volume of intercity transmission in the United States. In short, satellite systems could well permit multiple entry into at least some segments of long distance communications without major sacrifices in scale economies on these particular segments. Such entry, however, could sacrifice some overall system economies if the entire long distance network experiences significant economies of scale and if the satellites divert traffic from the AT&T's Long Lines System.

This issue of possible scale economies in the long distance system (analyzed as a larger entity) was the central focus of AT&T's Multiple Supplier Network study (MSN).[48]The MSN study examined the cost of providing a nationwide private line network by using a network model with 178 links and 87 nodes. As such, the model is a simplified version of the nationwide telephone network. The study simulates facility construction schedules and costs over a range of time in the future for specified demand forecasts, allowing for varying demand growth on different links in the system. The study focuses solely on basic transmission facilities, excluding specifically multiplex and switching equipment, and appears to use the −0.6 average cost elasticity corresponding to the steeper trend line in Figure 4–9. The study also assumes implicitly that competitors will pursue the full private line business and build facilities accordingly when appropriate. Finally, the MSN study includes "related" operating costs and property taxes.

The MSN study provides two cost comparisons between providing the private line network by another common carrier (OCC) system and by the Bell System. The first comparison discounts the value of the cost differences back to 1971, while the second comparison discounts back to 1976. The difference between these two is that a good deal of OCC plant was in existence by 1976 but not in 1971. The discounted value for 1971 thus indicates what the cost would be of starting an OCC more or less from scratch, while the 1976 estimate treats the then existing OCC investment as sunk cost. The 1976 comparison, the results of which are summarized in Table 4–5, seems the fairer since most of Bell's costs are essentially considered "sunk" as well.[49] The 1971 comparison places the OCC costs at four times the costs of

Table 4–5. AT&T Multiple Supplier Network Study: Cost of Providing Interstate Private Line Demand (Excluding OCC Investment Cost Prior to 1977)

	Cost of Service on OCC Network[a]	Cost of Service as Part of Bell Network[a]	Difference
	(A)	(B)	(A) — (B)
Construction cost	$420	$344	$ 76
Income tax	104	107	−3
Operating costs	403	129	274
Total cost	$927	$580	$347

Source: AT&T, "Multiple Supplier Network Study: The Cost of Multiple Intercity Networks Compared to a Single Integrated Network," Bell Exhibit 57, FCC Docket 20003, p. 9.

[a]Cost data represent discounted present value (millions of dollars) in 1976 of estimated costs incurred from 1977 through 1991.

the Bell System,[50] while the 1976 analysis results in a cost about 60 percent higher ($347 million in 1976 dollars) for providing the private line network on an OCC system rather than by Bell. Presumably these cost differentials would be reduced further if later base years (e.g., 1979) were used when even more OCC investment could be deemed "sunk."

Two important objections emerge from analyzing the MSN study, however, that relate back to and help with interpreting the engineering evidence on scale economies. First, the MSN study assumes that the OCCs will serve the entire nationwide private line network. Specifically, the study assumes that the OCCs can use the Bell System until demand on a particular link increases sufficiently to justify investment. Although this assumption favors the OCCs in the cost comparison, the levels at which investment is assumed to occur are still low compared to the potentially relatively flat region of cost identified in the previous engineering discussion.[51] Thus, the MSN study may impose unrealistically low densities on the OCCs. Lower-density routes probably would be left unchallenged to the telephone companies because in these cases demand would not exceed the threshold needed to achieve relatively flat unit costs. If so, the actual cost penalty for OCCs could be considerably less than in the MSN study.

A second point about the MSN study is that it appears to assume a trend line of -0.6 for the average costs of basic transmission over the next 20 years. For reasons explained above, this would seem to be an upper-limit estimate. There are many uncertainties about how future technologies will affect these scale relationships. If a lower trend slope is used, a more favorable cost comparison results for the new entrants versus the Bell System.[52]

Furthermore, even in the presence of transmission cost penalties and prices in line with costs, competitors may still be able to enter the terrestrial long distance service business if they offer truly innovative services or use new cost-saving techniques. For example, "value-added" carriers like Telenet and Tymnet, by incorporating special signal processing techniques,[53] apparently use transmission facilities more efficiently than do the established carriers from which they lease their transmission lines.

It should also be noted in passing that there is no evidence that customer premise telephone equipment exhibits any economies of scale in the relevant ranges. This absence of scale economies is consistent with the general pattern observed in the production and distribution of other consumer durable goods, such as stereo equipment and toasters, where many firms compete in the marketplace.

In summary, engineering studies that conclude that large economies of scale are present in long distance terrestrial transmission investment costs typically focus on only one component of such costs (basic transmission), excluding all related equipment such as multiplexing (and switching when needed) and related operating costs. When these other cost elements are included, overall economies of scale appear lower. Whatever scale economies there may be, moreover, do not preclude the offering by value-added common carriers of nonconventional and innovative long distance services if cost-saving techniques or valuable new features are made available to customers. Moreover, entry into long distance may be feasible on many higher-density routes without incurring unduly large cost penalties, especially with the newly emerging satellite technologies.

THE IMPACT OF COMPETITION ON FUTURE TELECOMMUNICATIONS COSTS IN COMPARISON WITH THE DIRECT COSTS OF REGULATION

An essential question for public policy is how much competition might raise costs to consumers relative to the costs of suppressing competition through government regulation. More precisely, as the scale of the market increases, telephone system average costs should decline if scale economies exist, with or without competition. However, if competition reduces the growth rate of established telephone companies, these future economies may be less with competition than without, as the MSN study suggests.

The difference in costs to society with and without competition will depend on a number of assumptions, for example, about growth rates and the average costs of competitors versus the Bell System and the scale elasticities that are deemed to be applicable. To illustrate the possibilities, it might be assumed that revenues for all telephone services grow at 10 percent per year from 1980 through 1985 (which is roughly the historical growth rate during the 1970s). In addition, it optimistically might be assumed that in 1985 competitors or new entrants will have 2 percent of these revenues. This liberal hypothesis allows twice the usual upper-limit estimate of 1 percent of 1985 revenues conceded to specialized common carriers and would mean that these new entrants would have about $1.6 billion in revenues by 1985.[54] By way of calibration, the specialized common carriers as of 1977 (including the satellite companies) had revenues of about $140 million; thus $1.6 billion of business in 1985 implies a compound growth rate of about 36 percent a year for the specialized common carrier market between 1977 and 1985 — a most ambitious achievement.

Little information is available as to what any new competitors' average costs might be as compared with Bell. A range of assumptions about the possible relationship between established telephone companies' costs and costs of their competitors is shown in Table 4–6. The new competitors' hypothesized average costs in 1985 range from 15 percent lower than costs of established companies (a ratio of 0.85 in the left-hand column)[55] to 50 percent higher (1.5 in the left-hand column). Assumptions are also needed about the telephone system's scale elasticities; in Table 4–6, they have been placed over the range of plausible possibilities discussed earlier, 1.0 to 1.5. This range is consistent with econometric evidence on economies of scale in telecommunications systems and with the implications of available engineering evidence for system scale elasticities, as discussed in the two preceding sections of this chapter.

As can be seen from Table 4-6, under the most extreme of assumptions, telephone system costs might be a little more than 1 percent higher in 1985 if new entrants are allowed. However, it is also possible that entry by firms with cost advantages over the existing telephone companies could lead to a decrease in telephone system costs. Furthermore, the examples presented in Table 4–6 assume no offset from competition improving efficiency or inducing additional market growth (e.g., through developing new services and products). In short, technological change, and almost everything else, is hypothesized to remain the same with or without competitors.

It is very difficult to translate this analysis into a meaningful dollar estimate of what the social costs in 1985 of permitting competition might be. The competitors are hypothesized to have about $1.6 billion

Table 4–6. Percent Increase in Telephone System Total Costs in 1985 under Various Circumstances[a]

Ratio of Competitors' Average Cost to Established Telephone Companies' Average Cost	Percent Increase in Telephone System Cost with Established Telephone System Scale Elasticities			
	1.0	1.1	1.25	1.5
0.85	−.30%	−.20%	−0.07%	0.06%
1.0	0.00	0.10	0.21	0.34
1.25	0.51	0.59	0.73	0.85
1.50	0.98	1.08	1.25	1.36

Source: Calculated by Charles River Associates.

[a]Estimates are based on the assumption that competitors have 2 percent of the total telecommunications market in 1985.

in revenue in 1985, and when this amount is juxtaposed against the higher estimates for growth in telephone industry revenues reported in Chapter 2, total annual revenues of about $50 billion in 1985 (in mid-1970 value dollars) is indicated. If the 1-percent cost penalty applies systemwide, a $500-million annual cost might be incurred because of competition.[56] Alternatively, if the competitors' costs are roughly equal to their revenues of $1.6 billion and their average costs are 1.5 times that of the established telephone companies, the established companies' costs for doing the same chore would be about $1.0 to $1.1 billion and the cost penalty is again about $500 million.

These estimates, however, are almost surely on the high side. They assume pronounced scale economies within the telephone system and new entrants having 2 percent of total telephone industry revenues in 1985. They also do not allow for any possible offsets from more rapid market growth created by competition. Nevertheless, the possibility exists that for every dollar of revenue that new competitors might have in 1985, 33 percent might represent increased costs from a social viewpoint. The question for public policy is whether this risk looms so large as to justify inhibiting market entry, with all of its potential benefits. One obvious question is how new entrants can achieve such marked growth (36 percent per year) without offering better services, especially if their costs are 50 percent higher. Only very foolish and protective regulatory policy could seemingly induce such a result. Again, too, this analysis assumes that the established telephone companies will proceed down their cost curves and achieve scale economies just as fast without competition as with it.

The potential cost penalties of competition from loss of economies of scale must also be compared to the costs of regulation in assessing whether natural monopoly justifies regulation in the telecommunications industry. Among the more obvious direct costs[57] of regulating telecommunications are the salaries and expenses of the FCC commissioners and their administrative and legal staffs. The FCC's budget for fiscal 1979 is $67 million, of which one-third to one-half might reasonably be allocable to telecommunications regulation.[58] The 1979 authorization, in comparison to an actual outlay of $38 million in fiscal 1974,[59] suggests growth in federal telecommunications regulation expenditures at a rate of about 12 percent per year, at least as high as the growth rate of telephone revenues. Thus, although the federal component of direct costs of telecommunication regulation is not large in relation to industry revenues, it has been exhibiting a common tendency of regulatory expenditures to grow relative to industry scale.

Direct costs of regulation also include expenditures by the state commissioners and their staffs.[60] Most important, however, are the expenditures made by the firms in the industry on hearings and other proceedings as well as meeting routine record-keeping and reporting requirements. Only an order of magnitude can be suggested — direct costs of regulation borne by private firms could be $100 to $500 million per year. A study of the Federal Power Commission (FPC), whose proceedings were also adjudicatory, estimated total direct expenses of regulation at 3.26 times the FPC budget.[61] Application of this multiple to 50 percent of the 1979 FCC budget (on the assumption that up to 50 percent of FCC expenditures support telecommunications regulation) yields an estimate of total direct costs of regulation of roughly $100 million. And this estimate may well be low. AT&T's General Department alone spends upwards of $200 million annually on general (i.e., nontraffic-related, noncommercial) office salaries and expenses.[62] Furthermore, a recent study by Chase Manhattan Bank pegs compliance costs of regulation of cable TV and related industries at $1 billion in 1977,[63] at least half of which might be allocable to telecommunications regulation.

Thus, the direct costs of regulation of the telecommunications industry may be as large as the potential costs to consumers of loss of scale economies under competition. Moreover, any comparison of the cost implications of loss of scale economies with the cost of regulation is, to review the discussion in the first section of this chapter, only the first of several basic questions that must be answered before it can be concluded that natural monopoly justifies regulation. Specifically, important questions remain about the ability of regulation to "deliver" the benefits claimed or desired.[64]

SUMMARY

Several unanswered questions separate any determination that economies of scale might exist in telecommunications from a firm conclusion that a regulated natural monopoly should be sustained or created. Specifically, the natural monopoly argument is not sufficient to justify regulation if (1) the scale economies that create the natural monopoly are largely exhausted at volumes that allow for more than one firm; (2) there are enough substitutes for the output being produced that the monopoly firm will not have significant market power; (3) regulation is unable to limit the exploitation of market power; and (4) there are other remedies that are more effective than regulation.[65] With regard to the latter point in particular, the existence of potential or actual entrants may very well provide a discipline superior to regulation.

The prospects for growth in telecommunications thus argue for caution in accepting a regulated natural monopoly solution to the problems of the industry. The historical rate of growth in long distance communication, as documented in Chapter 2, has been at a robust rate of 10 percent or more a year, in an economy in which the real annual growth rate has averaged no more than 4 percent. Furthermore, new markets and technologies, as analyzed in Chapter 2, suggest that, as high as this high historical growth rate has been, the prospects for the next two decades or so are at least as bullish.

There are also many intangibles and uncertainties about the future of the telecommunications industry and how this future might be influenced by the presence or absence of regulation. Some of those issues are explored in greater detail in the next chapter. Although it is difficult to be conclusive when intangibles are involved, the weight of the evidence seems to suggest that the accumulation of new knowledge and technology in telecommunications, given added incentive by high markups in certain portions of the telecommunications market, will erode any monopoly market power or position in the industry as time passes. Indeed, as is shown in the next chapter, newly emerging technologies could conceivably even diminish the inviolability of the most long-standing of all natural monopoly positions in telecommunications, that of the local loops.

Overall, the empirical evidence suggests the existence of some economies of scale in telecommunications. However, even if the system as a whole shows economies of scale, these may very well stem from economies present only in one area and not in other areas of telecommunications service. For example, there seems to be no evidence of economies of scale in the production and distribution of customer premise equipment. On the other hand, evidence from engineering and simulation studies of long distance telecommunications links indicates, on balance, higher scale economies than are found in econometric analyses of the telephone system as a whole.

By assuming a generous penetration of the market by telephone companies' competitors, pronounced scale economies within the telephone system, and no incremental revenue generation effects from competition, an estimate of the maximum possible cost increases imposed upon the telephone system by competition can be obtained. This estimate approximates 1 percent of the telephone companies' estimated 1985 costs. These potential cost penalties because of competition must be balanced against the direct costs of regulation, which currently may approximate $100 to $500 million per year. It is difficult to make direct dollar-for-dollar comparisons of the direct costs of regulation versus potential costs (losses of efficiency) from competition, but the available evidence suggests that the figures may not be very dissimilar. If competition induced only minor increases in operating efficiency for the telephone system, any losses in scale economies created by competitive diversion might then be largely offset while society enjoys the benefits of reduced regulatory costs.

Finally, future technological changes might somewhat erode economies of scale in telecommunications, making them less pronounced in the future than in the past. When coupled with the very high probability of continued high growth rates in long distance telecommunications, along with the prospect of technological changes diminishing other positions of market power in the industry, natural monopoly as a justification for extensive and detailed regulation of the industry could well be something of the past. Certainly, continuing natural monopoly of a scale and strength sufficient to justify perpetuation of the present scope of regulation in the industry seems doubtful.

NOTES

1. In general, the historical development of the natural monopoly concept has been in terms of a single-product industry. However, the argument can be extended to a multiproduct industry (W. J. Baumol, "On the 'Proper' Cost Tests for Natural Monopoly in a Multi-Product Industry," *American Economic Review* 67 [December 1977]: 809-822) in which case the empirical determinations of the existence of scale economies become even more complex than in the single-product case. As the discussion in this chapter makes clear, determining the existence or nonexistence of scale economies in even a single product line is quite complex and often less than unambiguous or clearly determinable. This is especially true in industries marked by growth and rapid technological change such as telecommunications.

2. Economies of scale are said to exist when equiproportional increases in all inputs yield a greater than proportional increase in output.

3. Strictly speaking, an upturn in average costs could occur before the market was fully served and a single-supplier natural monopoly might still be efficient; in this case, the test would be whether two efficient producers (i.e., at a scale equal to minimum average costs) could be "fitted" into the market.

4. Geometrically, the increase in total value from this increase in output is represented by the total area under the demand curve, that is, by the area ADQ^*Q_m. This area represents both the increase in what people actually must pay (the rectangle BDQ^*Q_m), plus the consumer surplus (value generated by the expansion in output but not extracted from consumers) represented by the triangle ABD. The increase in the cost of the additional output is represented by the area under the marginal cost curve, that is, by the area BCQ^*Q_m. The difference between these two, the increase in total value generated by the expansion of output less the increase in total costs so generated is, by subtraction, the area $ABCD$.

5. That is, by hypothesis, most economies of scale are exhausted at a level of output equal to $.5S$ where S is the scale of market.

6. The relationship between market structure and technological innovation is discussed in Chapter 5.

7. In a multiproduct production situation these determinations become even more complex and involved with what have come to be known in the economic literature as sustainability issues. Briefly, sustainability refers to "an industry to which entrants are not 'naturally' attracted, and are incapable of survival even in the ab-

sence of 'predatory' measures by the monopolist" (as defined in W.J. Baumol, "On the 'Proper' Cost Tests for Natural Monopoly in a Multi-Product Industry," pp. 809-811). Much of the relevant literature has addressed the theoretical conditions under which a regulated natural monopoly may be vulnerable to entry by uninnovative competitors — that is, to "cream skimming."

Sustainability issues are noted in this chapter only as they affect the inferences that can be drawn between the presence of economies of scale and the justification of regulation. For example, sustainability is more likely if economies of scale in the production of individual products are not too pronounced and if the economies gained by producing multiple goods in one firm are relatively large. The relevance of sustainability to the issues of freedom of entry and predatory pricing in the telecommunications industry is discussed in Chapter 6.

8. Murray Weidenbaum, in his book *Business, Government, and the Public* (Englewood Cliffs, N.J.: Prentice-Hall, 1977), says that in 1974 businesses and private individuals spent 130 million person-hours in filling out 5,146 different federal forms — excluding those for taxes and banking. Furthermore, a recent study done for The Business Roundtable by Arthur Andersen and Company ("Cost of Government Regulation Study," Executive Summary, March 1979) states that "1977 incremental costs for 48 companies to comply with six federal agencies were $2,621,593,000." This $2.6 billion amounted to 15.7 percent of the total net income after taxes of these 48 companies.

9. Firms that create branch plants to avoid exceptionally high freight markups or commuters who locate in the suburbs expecting to receive good rail commuter service between their suburb and place of employment are two, among many, examples of this phenomenon.

10. "Relatively" as used here refers to the comparison of price elasticity of demand at monopoly price between demand curves in Figures 4–5 and 4–6. At monopoly price and output, the price elasticity of demand in Figure 4-5 is −1.3 and in Figure 4–6 is −1.9.

11. In joint production multiproduct situations the cross-elasticities of demand (with respect to price) can also be important, especially in assessing sustainability issues. For example, higher (positive) cross-elasticities increase the danger of nonsustainability; that is, the natural monopoly is more likely to be vulnerable to entry by uninnovative competitors "skimming the cream" of the best markets.

12. A strong international cartel or monopoly in diamonds is perhaps a more realistic example of an instance of market power in a luxury good not eliciting a regulatory solution.

13. See, for example, Paul W. MacAvoy and James Sloss, *Regulation of Transport Innovation: The ICC and Unit Coal Trains to the East Coast* (New York: Random House, 1967).

14. High markups are also more likely to induce cream-skimming behavior by uninnovative competitors, that is, to increase the likelihood that a regulated natural monopoly is not "sustainable." Longer-run effects of high markups are less predictable. It is possible that, if the incentives of high markups evoke substantial technological improvements over time, the stand-alone provision of individual services by separate firms may replace joint production by a natural monopolist as the least-cost way of providing the complete set of services.

15. For additional evidence, pro and con, on this question, see George J. Stigler and Claire Friedland, "What Can Regulation Regulate?: The Case of Electricity," *The Journal of Law and Economics* 5 (October 1962): 1-16; Richard A. Posner, "Natural Monopoly and Its Regulation," *Stanford Law Review* 21 (February 1969): 548-643; and William S. Comanor, "Should Natural Monopolies Be Regulated?" *Stanford Law Review* 22 (February 1970): 510-518.

16. For a more detailed discussion, see George W. Hilton, "The Consistency of the Interstate Commerce Act," *Journal of Law and Economics* 9 (October 1966): 87-113; George W. Hilton, *The Transportation Act of 1958: A Decade of Experience* (Bloomington: Indiana University Press, 1969); M. Bernstein, *Regulating Business by Independent Commission* (Princeton: Princeton University Press, 1955).

17. The problem is much like that encountered in taxation, which has led to the observation that "the only good tax is an old tax." Specifically, with the passage of time in a market economy any tax or regulation becomes embedded in prices and capital values.

18. This desired result is more likely to the extent that entry is undertaken by firms with new technologies and distinctive strategies and is perhaps less likely to the extent that entry is by firms with a technology and philosophy similar to the monopolist. In the latter case, a tacit oligopoly market situation is more likely to result, possibly to the detriment of consumers.

19. On this point, also see Martin B. Zimmerman, "Long-Run Mineral Supply: The Case of Coal in the United States," Ph. D. dissertation, Massachusetts Institute of Technology, 1975.

20. See, for example, studies in William M. Capron, ed., *Technological Change in Regulated Industries* (Washington, D.C.: The Brookings Institution, 1971).

21. Appendix D is a more detailed discussion of individual studies of economies of scale in telephone systems.

22. These results are implicit in the comparisons of functional forms discussed by Leroy H. Mantell in "An Econometric Study of Returns to Scale in the Bell System," Staff Research Paper, Office of Telecommunications Policy, Executive Office of the President (Washington, D.C., February 1974). There is a question of whether or not the effects of technological change operate independently of the scale of operation. If technological change is partially induced by increases in scale, part of the estimated impact of technological change is attributable to increases in scale. As explained in the next chapter, there is no substantial evidence that technological change is related to scale, at least insofar as enhanced scale is achieved through monopoly or regulation.

23. For telecommunications, time-series estimates seem to be the only practical way to measure system economies of scale. This is so because there is only one integrated — local and long distance — telephone system in the United States. International cross-section estimates would be an additional possibility but data limitations have thus far prevented this approach. Cross-section estimates are also possible using operating company data, but this would reflect principally local scale economies or diseconomies, not overall system effects. Chapter 5 reports the results of cost function estimation using operating company cross-section data. In addition, A. Rodney Dobell et al., "Telephone Communications in Canada: Demand, Production, and Investment Decisions," *Bell Journal of Economics and Management Science* 3 (Spring 1972): 175–219; L. Mantell, "An Econometric Study of Returns to Scale"; and L. Waverman, "The Regulation of Intercity Telecommunications" in *Promoting Competition in Regulated Markets,* edited by A. Phillips (Washington, D.C.: The Brookings Institution, 1975) report cross-section estimates. Dobell et al. and Mantell use operating company data and find scale elasticities closer to one than in their time-series estimation. Waverman uses cross-section data for private microwave systems and finds economies of scale.

24. L. Mantell, "An Econometric Study of Returns to Scale" (also appears as Bell Exhibit 40 in FCC Docket 20003).

25. The three Bell Lab studies are:
American Telephone & Telegraph Company, "An Econometric Study of Returns to Scale in the Bell System," Bell Exhibit 60, FCC Docket 20003 (Fifth Supplemental

Response), August 20, 1976; H. D. Vinod, "Application of New Ridge Regression Methods to a Study of Bell System Scale Economies," Bell Exhibit 42, FCC Docket 20003, April 21, 1975; and H. D. Vinod, "Bell System Scale Economies and Estimation of Joint Production Functions," Bell Exhibit 59, FCC Docket 20003 (Fifth Supplemental Response), August 20, 1976.

26. These studies are:
    A. Rodney Dobell et al., "Telephone Communications in Canada: Demand, Production, and Investment Decisions"; and Melvyn Fuss and Leonard Waverman, "Multiproduct Multi-input Cost Functions for a Regulated Utility: The Case of Telecommunications in Canada." Paper presented at the National Bureau of Economic Research Conference on Public Regulation, Washington, D.C., December 15-17, 1977.

27. The excluded study is AT&T, "An Econometric Study of Returns to Scale in the Bell System," Bell Exhibit 60. This study is not directly comparable because it utilizes an innovative, but as yet not widely tested, estimation method and definition of the technological progress variable. Particular problems with the estimation method are discussed in Appendix D.

28. The other studies to which Bell Exhibit 60 is not directly comparable are Mantell, "An Econometric Study of Economies of Scale"; Vinod, "Application of New Ridge Regression Methods"; and Vinod, "Bell System Scale Economies."

29. The 95-percent confidence interval indicates that there is a .95 probability that the reported interval covers the unknown true scale elasticity.

30. T+E, "A Project to Analyze Responses to Docket 20003," Final Report, Deliverable B. Prepared for the Federal Communications Commission, 24 September 1976.

31. Stanford Research Institute, "Analysis of Issues and Findings in FCC Docket 20003," Part II, Section D-2, April 1977. (Also appears in FCC Docket 20003 as Bell Exhibit 65-A.)

32. This general problem is complicated in the case of telecommunications by the flat rate structure in general use for local exchange service. Under a flat rate structure there could be changes in usage — both number of calls and duration of calls — without any change in the local service component of total telephone revenues.

33. Figure 4-8 is not intended to represent data from the econometric studies of telephone system economies of scale or the production functions estimated in those studies; rather, it is intended to illustrate the point that the econometrician's judgment about the underlying processes generating the historical observations can influence the resulting estimates.

34. John R. Meyer and Gerald Kraft, "The Evaluation of Statistical Costing Techniques as Applied in the Transportation Industry," *American Economic Review* 51 (May 1961): 313-334.

35. In discussing the engineering approach, Scherer notes that limitations include "the tendency of some engineers to underemphasize the sensitivity of plant-scale decisions to changes in input prices, and its applicability only to plants constructed using current, not past, technology. But no one disadvantage is serious, and carefully executed engineering estimates probably afford the best single source of information on the cost-scale question." F. M. Scherer, *Industrial Market Structure and Economic Performance* (Chicago: Rand McNally, 1970), p. 83.

36. For microwave systems in particular, the above costs can vary considerably depending on whether the facilities are shared with other users. For a good discussion of the cost components for microwave systems, see Leonard Waverman, "The Regulation of Intercity Telecommunications," in *Promoting Competition in Regulated Markets,* edited by Almarin Phillips (Washington, D.C.: The Brookings Institution, 1975), pp. 201-240.

37. Total multiplexing costs depend on the number of individual channels. Using frequency division multiplex, for instance, the *total* cost of multiplexing 2,400 voice

channels would be greater than the total cost for two television channels, each of which occupies a bandwidth of roughly 1,200 voice channels. This should not be confused with economies of scale, which relate to the behavior of costs *per channel* as scale changes, when the individual circuit bandwidth is held constant.

38. For evidence that multiplex scale economies are minor see Leonard Waverman, "The Regulation of Intercity Telecommunications," pp. 210, 213, 218. Discussions with major independent telephone equipment manufacturers provided further support to this point of view.

39. From discussions with major independent telephone equipment manufacturers.

40. Bernard Yaged, "Economies of Scale, Networks, and Network Cost Elasticity," Bell Exhibit 14, FCC Docket No. 20003. The same pattern holds true for another study, AT&T, "Multiple Supplier Network Study: The Cost of Multiple Intercity Networks Compared to a Single Integrated Network," Bell Exhibit 57, FCC Docket 20003.

41. In addition to the engineering evidence, Waverman fit an econometric cost function to cost data from a large number of private microwave systems. These systems varied in number of channels, from 120 to 960, and in length, from 15 to 2,262 miles; they also varied in age and in location. See Waverman, "The Regulation of Intercity Telecommunications," in particular, Equation (1) of Table 7-3, Figure 7-3, and the accompanying textual explanation. It is not clear whether some of the systems studied by Waverman also included switching equipment.

42. Switching equipment is needed for terrestrial networks, but not for private line service. Some satellite systems, presently in operation or under consideration, make use of terrestrial switching facilities while others do not; some satellites under consideration for the future may have some kind of on-board switching facilities.

43. For the conversion formula between average cost elasticity and scale elasticity see Table 4-3.

44. Waverman, in "The Regulation of Intercity Telecommunications," reaches conclusions similar to those stated above using more extensive cost data on smaller scale systems in operation in the early 1960s.

45. For a general overview of these issues, see Norman Abramson and Eugene R. Cacciamani, Jr., "Satellites: Not Just a Big Cable in the Sky," *IEEE Spectrum* 22 (September 1975): 36-40; and Burton I. Edelson, "Global Satellite Communications," *Scientific American* 236, No. 2 (February 1977): 58-73.

46. The capacity of any individual commercial communications satellite presently in use is in the same range as that of the largest operating single route terrestrial microwave system.

47. "Economies of fill" will exist for any satellite as utilization approaches the design capacity of the satellite. Economies of *scale*, however, will occur only if average cost is lower for larger satellites when the cost comparison is made at the levels of utilization for which the satellites are designed.

48. AT&T, "Multiple Supplier Network Study: The Cost of Multiple Intercity Networks Compared to a Single Integrated Network," Bell Exhibit 57, FCC Docket 20003.

49. The simulation essentially looks at only the marginal cost of providing the network incrementally on the existing Bell System.

50. The 1971 comparison appears in Table 1-2.

51. See Figure 4-9 and the related discussion.

52. Unfortunately, there is no way to test for sensitivities, as Bell has not released sufficient details to do a replication of the MSN study.

53. It may be noted that packet communications technology was developed outside of AT&T and that AT&T has not used it yet for any large-scale nationwide transmission network. See Lawrence G. Roberts, "The Evolution of Packet Switching," *Proceedings of the IEEE* 66, No. 11 (November 1978): 1307-1313.

54. Citations and data generally consistent with these extrapolations are given in greater detail in Chapter 2.
55. A ratio of 0.85 might occur, for example, if the new competitors used only the newest technology or specialized in particular services.
56. A 1-percent systemwide cost penalty could, however, be offset if a 1-percent systemwide efficiency improvement could be generated.
57. The potential indirect costs of regulation, which would be difficult, if not impossible, to quantify, are outlined in the first section of this chapter. They include reduced efficiency and innovativeness of management and regulatory inertia. The impact of regulation on innovation is discussed further in Chapter 5.
58. This would exclude budget line items for broadcast, safety and special radio services, and cable television, and roughly 50 percent of line items for commissioners, field operations, research and planning, and support (i.e., legislative, legal, etc.). See Office of Management and Budget, *The Budget of the United States Government, Fiscal Year 1979, Appendix* (Washington, D.C.: U.S. Government Printing Office, 1978).
59. Outlays in the mid-1970s are reported in Murray L. Weidenbaum, *Business, Government, and the Public* (Englewood Cliffs, N.J.: Prentice-Hall, 1977). He also discusses the substantial paperwork imposed on firms by regulation and its impacts on internal working procedures and decision making.
60. These expenditures are difficult to quantify with any certainty. The combined annual budgets of the commissions of the states and U.S. territories reporting in the *1977 Annual Report on Utility and Carrier Regulation of the National Association of Regulatory Utility Commissioners* (Table 108, p. 705) were approximately $145 million. The share of this total attributable to telecommunications regulation is unknown. One reasonable estimate could be derived by multiplying the combined budget total by the share of one-half the FCC budget in the total of reported federal agency regulatory expenditures. (Assuming that one-half the FCC budget might reasonably be allocable to telecommunications regulation, as noted in the text.) This procedure would result in an estimate of approximately $23 million. In any case, expenditures of the state commissions would most likely continue to be necessary if competition was allowed in the telecommunications industry because competition is least likely to affect the local services franchises. (Regulatory options are addressed in Chapter 6.)
61. Stephen G. Breyer and Paul W. MacAvoy, *Energy Regulation and the Federal Power Commission* (Washington, D.C.: The Brookings Institution, 1974), Table 1-1. The authors estimate expenditures by the FPC and the regulated companies for a typical year during the 1970s.
62. Federal Communications Commission, *Statistics of Communications Common Carriers for the Year Ended December 31, 1976,* Table 16.
63. As reported in Willard C. Butcher, "$100 Billion Price Tag of Regulation," *Financier* 2 (October 1978): 32-35.
64. Several of these benefit issues are explored in depth in the next chapter.
65. For evidence on potential market size, see the third section of Chapter 2. For competing substitutes to local telephone service, see the fourth section of Chapter 5 and the second section of Chapter 6. For the effectiveness of regulation, see the second section of Chapter 5.

Chapter 5

Market Structure and Regulation

Regulation is often advanced or advocated as a means of bringing into being certain kinds of broad social benefits that would not otherwise be available or easily achievable. These other benefits go beyond the price reduction and quantity "expansion" consumerist arguments that lie at the heart of the basic economic justification for natural monopoly regulation. Instead, these other benefits relate to issues that either supersede a narrow concern with prices and costs or focus on issues that are related to or conditioned by prices and costs, but in a somewhat derivative manner. Two off-cited examples of such benefits would be the possibility that regulation might (1) expedite technological innovation; and (2) simplify the achievement of universality targets, such as serving rural and other out-of-the-way locations.

These other issues have arisen in debates on almost all types of regulation, and telecommunications is no exception. Thus, AT&T often claims that the reasonably impressive technological development of American telecommunications owes much to the existence of its large-scale, vertically integrated corporate structure and regulated monopoly status; by contrast, many critics argue that technological progress in the industry has been achieved in spite of these handicaps. Similarly, AT&T argues that the industry's regulatory status has had much to do with the industry's ability to provide telephone service, even though costly, to remote places; on the other hand, critics charge that AT&T and regulation have had remarkably little to do with extending service to the most remote locations in the nation and some also insist that the cost of serving these remote locations is not obviously that much greater than elsewhere.

In most analyses, regulation ostensibly brings forth its additional benefits by creating what might be described as a particularly advantageous market structure for the regulated industry. This advantageous market structure is usually less competitive under regulation than it would be without regulation. Perhaps the extreme example of this kind of argument is advanced by the common carrier trucking industry. Without regulation trucking would reputedly be so competitive that rates and services would become very unstable and chaotic, to the point of making planning by consumers of trucking services impossibly complex and the financing of the trucking industry extremely difficult. By reducing the otherwise strong competitive tendencies of this industry, so the argument goes, regulation shields both the industry and users of trucking services from these difficult problems.[1] Similarly, railroads long argued that they needed regulatory protection on high-value intercity freight markets between major cities in order to earn enough profits to maintain their common carrier responsibilities of profitably serving out-of-the-way points. The telephone industry variation on this argument, as noted in previous chapters, is that competition in long lines would so erode the ability to earn "contribution" to overhead that local service costs would have to rise dramatically to keep the companies solvent.

Indeed, every industry that has experienced or enjoyed some price and entry regulation, but which has within it the seeds of greater potential competition, has usually argued against deregulation on the grounds that competition would bring price, service, and financial instability. The telephone industry, then, in questioning suggestions that the telecommunications industry be subjected to less formal regulation, is behaving quite conventionally. Indeed, the telephone industry, all things considered, has probably been less prone to reject suggestions of deregulation on these grounds than most other regulated industries confronted with such "threats" in the past.

A very real question arises, moreover, of whether regulation can perform the chores of stabilization and succor often attributed to it. Competition in a market economy is usually hard to suppress. In addition, regulation often creates added incentives for competitive entry by artificially holding some rates well above costs. The tendencies for competition to break through the regulatory barriers are even greater in industries characterized by rapid growth and substantial technological change, for these factors create more opportunities to circumvent regulatory inhibitions. It is therefore not uncommon for regulators to find that the competition they think they have suppressed reappears in an unsuspected form or incarnation. For example, regulated railroads found their high-value, high-markup freight increasingly carried by truckers; regulated common carrier truckers have found themselves increasingly confronted over the years by competition from private or contract trucking;[2] and the regulated common carrier airlines have found it impossible to escape the competitive challenges of chartered airlines. In addition, suppressed price competition can often reemerge as service competition.[3]

In short, competitive forces in a market economy do not disappear easily. There is no reason to suppose that telecommunications will be any exception to this rule. Indeed, it is clear that high markups on long distance telephone service have induced a considerable competitive response — many vendors are eager to enter the field and "help" the telephone companies deliver these services to the American public.

A similar competitive response has thus far not been observable in the other major segment of telephone company activity, the provision of local services. The general presumption has been that this lack of interest reflects the inherent natural monopoly characteristics of this segment of the business. Although there is surely some truth in this contention (that local service is more like a natural monopoly than most other activities in telecommunications), it may also be true that the lack of competitive interest represents the simple fact that markups on local services have apparently been low. Accordingly, an interesting question arises of whether local service would attract competitive interest if, say, rates were marked up substantially in this sector, thereby increasing the potential rewards to entry. The answer depends, of course, on what kinds of alternatives might be available to the local loop if and as rates rose to levels that created stronger incentives for entry. The structural stability and lack of competition in local telephone service might turn out, on closer inspection, to be as much a function of regulation creating a rate structure that did not establish strong incentives for entry as a function of the inherent natural monopoly characteristics of the underlying technology.[4]

Regulation, in short, might be able to create stability only to the extent that it recognizes the market realities. That is, the stabilization properties often attributed to regulation may occur only when regulation arrives at conclusions or results not too dissimilar from those that the market itself would create.

This chapter is essentially concerned with analyzing the issues that link market structure with possible additional benefits of regulation.

In the next two sections, two of the prominent alleged benefits — technological innovation and provision of service to remote locations — are analyzed and assessed. The fourth section addresses the question of whether regulation really can create its own market structure or whether market structure is, rather, more a function of underlying economic realities than of regulation. Specifically, an investigation is made of the potentialities of maintaining a monopoly position in local telephone service if an attempt were to be made to exploit the presumed inherent "natural monopoly" characteristics of this segment of the industry more fully than is now the case.

MARKET STRUCTURE AND TECHNOLOGICAL INNOVATION

AT&T's position on the impact of market structure on technological innovation is quite clear. AT&T believes the historical record demonstrates an impressive ability on the part of the Bell System to generate technological innovation; this history of innovation is a direct result of the corporate structure of AT&T and its regulated, monopoly status; policy changes leading to a more competitive market structure are likely to result in a slower rate of technological innovation. In a presentation made to the FCC, AT&T summarized its case thus:[5]

> The innovation record . . . results from the commitment of the Bell System to research and development and from the innovation incentives provided by the regulated monopoly market structure. However, even with these incentives, this innovative record simply could not have been compiled without the close coupling in the Bell System between research and the application of the fruits of that research to the provision of telecommunication services.

Bell has further claimed that:[6]

> . . . multiple supply carries with it a potential for a reduction in just the types of innovation that have made the U.S. telephone system the best in the world.

Without doubt, AT&T and the nation are much indebted to the program of research and development undertaken by the scientists at Bell Labs. The important question, nonetheless, is whether or not a change in the present regulated market structure in the telecommunications industry is likely, on either a priori or empirical grounds, to lead to a slowing in the rate of technological innovation. To answer this question, reviews of the following topics are required: (1) the history of innovation in the telecommunications industry prior to the time that AT&T was regulated; (2) the theoretical economic literature relating market structure to technological innovation; (3) empirical tests of such relationships (including the record in FCC Docket 20003); and (4) the record of innovation in other regulated industries.

The Historical Record of Telecommunications Innovation

As indicated in Chapter 2, the telecommunications industry experienced an initial period of monopoly followed by a competitive period that was followed, in turn, by a period of steady growth under slowly increasing government regulation. The history of the telecommunications industry has thus been marked by several types of market structure. Furthermore, the very formation of the telephone industry is of some significance because it arose, not from research and development by the existing communication monopolist of the day, Western Union, but as a result of a very small group of Yankee entrepreneurs financing Professor Bell's research. Thus, the telephone industry itself arose as a competitor to the existing communication monopolist and could not feel secure in its market position until a patent suit had cleared away the threat of retaliation by Western Union.

Quality of service, it seems generally agreed,[7] was poor during the patent monopoly period of Bell. This deficiency in service was directly related to technology. According to a competitor of Bell, from 1876 to 1895:[8]

> The style and efficiency of the transmitter was the same practically throughout the monopoly, and the circuit conditions had undergone little or no change, the lines being mostly grounded circuits of iron wire and the service consequently subject to noise from earth currents, crosstalk and other inductive disturbances. . . . Because of the style of the organization, they fail to properly appreciate conditions peculiar to varying localities, as effected by the general policies.

But as a result of competition after 1895 it is reported that:[9]

> The very first effect . . . was the bettering of the service rendered by the Bell Company by more careful attention to operators' work [and] the substitution of either common return or metallic circuits for grounded lines.

In the early 1900s, New York City assigned an engineer to study the question of telephone competition so that the city government might determine whether or not a competitor for New York Bell was justified. His report indicated that Bell responded to competition elsewhere by improving services and that the installation of an automatic switching system by the independent telephone industry had resulted in independent subscribers having a more efficient service than that provided by competing Bell Companies.[10]

Another report on the subject comes from Gabel, who found that:[11]

> Prior to 1907 little or no attention was given by the Bell System to what came to be known as "fundamental research."
>
> The major developments in the art, up to this point, originated outside the Bell System. The Strowger Switch, which made possible the advent of automatic telephony, was invented by an undertaker and manufactured by several of the independent manufacturers, while the use of dial telephone service was actually resisted by Bell leadership. The loading coil was developed by Professor Pupin of Columbia University around 1905. This coil tremendously improved the quality of telephone transmission, actually making possible, for the first time, long-distance telephone systems. Perhaps the most significant technical development of the period — and another major innovation from outside the Bell System — was Lee DeForest's development of the vacuum tube in 1914.

A more current indication that regulated monopoly may not always result in technological innovation comes from an examination of technological advance in local exchanges. In virtually every state of the United States, local exchanges are held by a single franchise and are state-regulated monopolists. It is, however, precisely in local exchanges and loops that the rate of technological innovation in telecommunications has been the lowest.

The recent period of incipient competition experienced by the telecommunications industry has been one of considerable innovation, although the industry would characterize at least part of this innovation as proliferation and imitation of services rather than actual technological innovation. Even those who would make such a claim, however, admit that users of interconnect equipment have benefited from innovative measures of new entrants.[12]

In sum, the history of the telecommunications industry has exhibited several forms of market structure, both regulated and unregulated, and this history is inconsistent with the hypothesis that only regulated monopoly encourages innovation. On the contrary, periods of competition in the historical record of the telecommunications industry have exhibited considerable technological innovation.

Theoretical Analyses of Market Structure and Innovation

The traditional starting point for reviewing the economics literature on the theoretical relationship between market structure and technological innovation is Joseph Schumpeter's *Capitalism, Socialism, and Democracy.* [13] Schumpeter argued that the large-scale firm was likely to be more innovative than the small and competitive firm. Furthermore, any static efficiency losses due to the noncompetitive nature of the market would be more than offset by the increased product innovation and productivity growth of large-scale firms. According to Schumpeter, a monopoly will have a greater demand for innovation because its market power will increase the firm's ability to profit from a given innovation. Monopoly will also result in a larger supply of innovations because of economies of scale in research and development and other unspecified factors.

Fisher and Temin show that Schumpeter's hypothesis does not follow from his own assumptions.[14] As a result, Schumpeter's argument now must be regarded as mainly a speculation, at least until someone provides a clear theoretical model that fully articulates the linkages between greater monopoly profit and innovation. Furthermore, in a regulated industry in which monopoly profits would ostensibly be denied by regulation, these linkages, if they exist, would have to be more complex than in the unregulated markets that seemingly dominated Schumpeter's concerns.

Arrow, in a 1962 article,[15] argued, in direct contravention of Schumpeter's views, that the incentive to create a cost-reducing innovation for a monopolist is always less than that incentive in a competitive industry. The inventor in the competitive market should be able to sell the invention at a price equal to the value of the cost reductions of total output in the market and this total output will be greater under competition than monopoly. The monopolist is limited to the difference in monopoly profits earned prior to and after the introduction of the invention. Arrow shows that this difference will always be less than the royalties to be earned by an inventor capturing all of the cost reduction in a competitive market.

A possible difficulty with Arrow's analysis, however, is that it slights the question of appropriability[16] of the benefits of technological innovation. A monopolist, for example, might be able to achieve appropriability whereas under competition the inventor would not be able to appropriate all the available benefits (i.e., cost savings). As a consequence, greater appropriability may offset the monopolist's disincentive to invent created by his preinvention monopoly profits and lower volume of output.

Another objection that has been made to Arrow's analysis is that it does not correct, when comparing the output of competitive industries with the output of monopoly, for differences in the size of the two industries' output. Specifically, Demsetz has argued that if the usual restrictive effect on output of monopoly is removed, say by having the monopolist price discriminate, then the incentive to invent is the reverse of what Arrow concludes.[17]

Kamien and Schwartz[18] have argued, however, that Demsetz's result is crucially dependent on the construction of the demand curves facing the competitive and monopolistic firms. In particular, they assert that elasticities of demand are crucially important in determining the magnitude of incentives to invent. Specifically, it can be shown that the incentive to invent (a cost-reducing innovation) is greater as the demand elasticity is greater, all else being equal.[19] Assuming industries of equal preinvention output size and unit costs, if the demand curve facing a monopolistic and a competitive industry are equally elastic, the monopoly will have the greater invention incentive. If the competitive industry faces a demand curve that is sufficiently more elastic than the demand for the monopolist's product, the competitive industry can provide a greater incentive to create a cost-reducing invention of given magnitude. Kamien and Schwartz suggest that policy emphasis on heightening the elasticity of demand for products, rather than on the more traditional antitrust policy, might be appropriate.

The Kamien and Schwartz analysis could have important implications for telecommunications. Bell typically claims that new entrants into long distance telecommunications have not been technologically innovative and have in fact merely offered services that Bell offers but at lower rates. This process of imitation as opposed to innovation is denounced by Bell as wasteful and duplicative. However, if product diversification and imitation serve to increase the number of substitutes for Bell output and thereby make the demand curve more elastic, such activities may in fact increase Bell's incentive to invent. Indeed, Kamien and Schwartz conclude:[20]

> From the standpoint of government regulation of industry, our analysis suggests that encouragement of new product lines that cause the demand curve confronted by a monopolist to become more elastic may be at least as potent a method for stimulating invention as the more traditional antitrust policy.

In a more recent paper, Kamien and Schwartz[21] argue that the market structure that generates the optimal rate of innovation is characterized by a degree of rivalry greater than monopoly and less than perfect competition. Again, Kamien and Schwartz note that their conclusion depends upon the appropriability of benefits to the innovator.

A general or pervasive problem with the theoretical literature on the relationship of market structure to innovation is that none of it deals adequately with uncertainty. All of the literature is based on models that assume that the benefits and timing of innovation, as well as innovation's phasing and entry, are known in advance by all firms in the marketplace. This abstraction is clearly somewhat limiting.

Even more important for present purposes, none of the standard articles on the relation of market structure to innovation deals with regulated monopoly as opposed to unregulated monopoly. Ever since Schumpeter, most authors have been concerned with contrasting the expected behavior of a monopoly free to exploit its position with competitors operating in an unfettered market. The question of technological innovation under regulation has only been treated, and then to a limited degree, as a by-product of the discussion emanating from the Averch-Johnson hypothesis that rate of return regulation induces inefficient behavior by regulated firms.[22]

Perhaps the most salient or suggestive contribution of this type is Baumol's[23] proposal to use regulatory lag as a policy instrument to spur adoption of cost-saving techniques. He would set the allowed rate of return at the cost of capital but would then allow a period, say three years, to elapse before reviewing the firm's earnings. Thus the firm would be free for a specified period to use research, management skills, and so forth, to escape the rate-of-return constraint. Regulatory lag could thus become beneficial by stimulating the firm to invest quickly in technological innovations that lower costs.[24]

This view of regulatory lag as being potentially beneficial is not shared by everyone.[25] If innovation requires a highly durable investment, one whose economic life would be greater than the lag, the danger arises that regulators might consider "windfall profits" in the first lag period as an argument for lower returns in subsequent periods. Baumol's policy could also create some curious, and not necessarily efficient dynamics. A tendency might arise, for example, for regulated firms to hold back on cost-reducing innovations until their periodic rate review is completed. Some very interesting possibilities might also occur for playing the "learning curve" by making a big investment just before the periodic rate review (thus increasing the rate base) but before any accompanying economies have been realized.

Inflation can also complicate the situation.[26] If technological change is insufficient to overcome cost increases due to inflation, regulated firms could be hurt by Baumol's lag policy. In such circumstances, special adjustments or cost pass throughs would be needed to garner Baumol's gains from lag. More generally, Klevorick[27] has shown that regulatory lag may affect the rate of technological innovation in a variety of ways, with the outcome depending on an elaborate specification of various demand, production, and regulatory output and reaction functions.

In summary, the theoretical literature is not sufficiently developed to justify AT&T's strong conclusion about the beneficial effects of market structure on the rate of technological innovation. Some consensus among economists may be developing that a market structure midway between monopoly and perfect competition is best for promoting innovation. However, the theoretical literature also suggests that the empirical questions involved in testing the relation between market structure and technological change are likely to be quite complex and not susceptible to easy measurement, especially when regulation is involved.

Empirical Evidence

Empirical studies of the relation between market structure and innovation almost exclusively look at unregulated industries. Nevertheless, the results may be useful in assessing the telecommunications case. Most recent empirical studies have used econometrics to examine the relation between measures of innovativeness and variables such as concentration, technological opportunity, and barriers to entry.

The innovation measures examined include productivity change, patent output, and R&D inputs such as employment or spending. In general, the results have not supported any finding of a strong positive relation between concentration and innovation. Any positive correlations have been weak, and some studies have found significant negative correlations. Other evidence suggests that the threat or actual occurrence of new entry may play an important role in promoting innovation. Barriers to entry, then, may be more important than other aspects of market structure in influencing technological change.

In a study of the relation between productivity change (as measured by unit labor requirements) and concentration for the period from 1899 to 1937, Stigler found the largest productivity increase in industries where concentration fell substantially during the period and the smallest in industries of continued high concentration.[28] However, in a study covering almost the same time period, Phillips found a positive correlation between productivity change and concentration.[29] The industries covered in the two studies were not identical, and the concentration data were from different sources. Most later studies of the relation between productivity change and concentration have found, alternatively, either positive correlations or statistically insignificant correlations.[30]

Important drawbacks to measuring innovativeness by productivity change arise from differences across industries in the capital intensity of new processes and from the fact that much innovation is undertaken by capital equipment suppliers but results in productivity increases in industries purchasing the equipment. Many studies, therefore, have looked at more direct measures of innovativeness, such as patent output or R&D resources expended.

Scherer, studying patent output, found no statistically significant effect of concentration on patents.[31] Other variables included in his specification were firm size and dummy variables representing the underlying technological opportunities of the industry. Studies that have looked solely at the relation between concentration and measures of R&D effort, such as the R&D spending to sales ratio, have generally found a positive correlation.[32]

A number of studies have attempted to examine the separate effects of technological opportunity and concentration on research effort. Although the theoretical and empirical treatments of opportunity differ among the studies, the usual result is that R&D effort is strongly correlated with the opportunity measures and the correlation of R&D effort with concentration is much weaker when opportunity variables are included than when omitted. In some studies, in fact, the correlation between concentration and R&D effort is negative when technological opportunity is also considered.

Phillips, in a study covering 11 broad industries, measured technological opportunity with a subjective index of "product changeability."[33] Scherer, in a study covering 56 industries, accounted for opportunity by dividing his sample into four product technology classes: electrical, chemical, mechanical, and traditional.[34] Comanor, in a study of 33 industries, divided the industries into two groups, reflecting his judgment about the possibility of achieving product differentiation through innovation.[35] These three studies all found a weak but still positive relation between concentration and research effort after including their measures of technological opportunity. In the Scherer and Comanor studies, moreover, an analysis within each opportunity class indicated that the relation between concentration and research effort was weakest in the higher technological opportunity classes.

Three later studies, however, have found negative correlations between concentration and research effort, at least in some opportunity classes. For example, Adams, comparing French and U.S. industries, found that for all high technology industries except instruments, the country with the larger concentration index had the smaller R&D

spending intensity; his results were not clear-cut for the lower technology industries.[36] Globerman, dividing 15 Canadian industries into two technological opportunity classes, found that the higher technology industries had a negative relation between research intensity and concentration, whereas for the lower technology industries the relation was positive but not statistically significant.[37] Finally, Wilson has attempted to measure two dimensions of technological opportunity; he found R&D spending positively and significantly correlated with the variables representing both opportunity dimensions but significantly negatively correlated with concentration.[38]

A problem bedeviling all efforts at measuring the impact of market structure on innovation empirically is that of causality. Specifically, some authors have suggested that technological opportunity gives rise to concentration. For example, high concentration in many technologically advanced fields has originated from patent barriers to entry. In addition, rapid technological change may lead to skewed distribution of firms by increasing the variance of firms' growth rates.[39]

In general, barriers to entry may be an important additional element of market structure affecting innovativeness. Thus, Comanor, controlling for concentration and product differentiation opportunity, found that industries with moderate entry barriers had much higher R&D employment relative to size than industries with higher or lower entry barriers. He suggested that with low entry barriers innovation might be discouraged because of fear of rapid imitation, whereas the insulation provided by high barriers might dull incentives.[40] In addition, Scherer cites case study evidence supporting the view that the conditions of entry affect innovation both because new entrants do the innovating and because the threat of entry spurs existing firms to innovate. One example he cites is AT&T's development of microwave relay systems when faced with the development of several potentially competing systems.[41]

An especially intensive review of the empirical relationship between market structure and technological innovation for the telecommunications industry was undertaken as part of FCC Docket 20003.[42] In an evaluation commissioned by the FCC, T+E, an independent consulting firm, surveyed the information in FCC Docket 20003 and additional FCC dockets and concluded that AT&T had failed to make its case that competition inhibited technological innovation in the interconnect market.[43] New entrants in the long distance market had in fact provided, in T+E's view, innovative services that AT&T had previously failed to supply.

AT&T, in a move parallel to that of the FCC, commissioned the Stanford Research Institute (SRI) to do a review of the same materials as those assessed by T+E.[44] The SRI report concludes that innovation increased under competition in the interconnect market. SRI also concludes that the users of interconnect equipment clearly have benefited from the presence of the new entrants in the marketplace.

With regard to competition in the long distance market, SRI concludes:[45]

> The evidence considered... is sufficient to support a finding that OCC's[46] have increased product variety and that they have produced some technical innovations. The evidence is unclear as to the net impacts of OCC's on the innovation rate in the telecommunications industry.

The SRI report thus largely fails to support any claim that innovation has suffered as a result of the introduction of competition in the telecommunications industry. SRI is markedly reticent about claiming great advances as the result of competition and is cautious, probably appropriately, about the causal nexus between competition and innovation.

AT&T also commissioned a study of the relationship between market structure and innovation by Arthur D. Little (ADL).[47] ADL produced a report that included a survey of the theoretical and empirical economic literature (much like that reported above). ADL concluded that "the question of just how much size and market concentration is ideal for innovation is unsettled."[48]

In summarizing the various presentations made to it on the innovation issue as part of Docket 20003, the FCC found that as a result of competition there had in fact been significant technological innovations and AT&T had increased its sales force and offered a more thorough and frequent analysis of customers' specific communications needs.[49] However, the FCC, in accord with almost everyone else examining this literature, found that there is simply no conclusive result about the general relationship between market structure and innovation. The FCC also notes, apparently approvingly, the emerging consensus that possibly some market structure midway between monopoly and perfect competition is most likely to secure an optimal rate of technological innovation. Most importantly, the FCC concludes that, on the basis of the record, the growth of competition has not inhibited innovation.

The empirical evidence thus provides no support for the view that regulated monopoly is the best climate for innovation. No consistently strong positive relationship between concentration and measures of innovativeness — productivity, patents, or research effort — has been found. The relation between research effort and concentration may even be inverse in industries faced with rich technological opportunities. And the limited evidence on the relation between entry barriers and innovation suggests that both low and high barriers are detrimental to innovation.

Innovation in Other Regulated Industries

The historical record in regulated industries other than telecommunications suggests that any attempt to accelerate innovation through regulatory modification of market structure will almost inevitably involve introducing new distortions into investment and developmental decisions. Specifically, the history of regulation is replete with instances where regulation itself impeded or distorted innovative activity.

A pronounced weakness of regulation, indeed perhaps its greatest weakness, lies in its static nature. Regulatory commissions typically, and usually necessarily, make decisions on the basis of past experience and only recently have begun to consider projections for the future. Regulatory decisions are thus typically based on information that is at least a year or two out of date by the time any decision is rendered. Furthermore, regulatory commissions often develop an aversion to technological change or innovation because such change can undermine the assumptions on which past and present regulatory decisions have been made.[50] Quite obviously, it is easier to establish reasonably sensible rate and entry regulations for slowly changing situations than for those marked by rapid market growth or technological innovation.[51]

To illustrate the distortions that can be created by regulation, regulation has been held by some analysts as at least partly responsible for the dependence of the northeastern United States on imported oil, specifically by preventing railroads from introducing lower trainload rates for moving coal. As MacAvoy and Sloss put it:

> It was primarily the regulation of rates on a cost saving innovation that retarded the introduction of the innovation, at least in the case of trainloads of coal to the Northeast sector of the country.[52]

The ICC was apparently opposed to lower rates for trainloads of coal because of equity and structural considerations. The ICC believed that some smaller consignees of shipments might find it difficult to arrange their affairs to accept trainload shipments, and the introduction of trainload rates also tended to throw other rates under ICC regulation into disarray. The result of allowing these other, nonefficiency considerations to weigh on the ICC decision process was to contribute an additional reason for northeastern utilities to shift from coal to imported oil.

Another historical cause célèbre in which regulation inhibited and distorted technological change is the case of the Southern Railway's attempt in the early 1960s to introduce "Big John" cars into railroad service. (Big John cars were large covered hoppers able to carry 100 tons of dry bulk cargo as compared with 70-80 tons typical in predecessor equipment.) There is more than circumstantial evidence that the Southern Railway originally wished to introduce the Big John cars as much because they offered an opportunity to circumvent traditional regulatory inhibitions on restructuring rates for carriage of dry bulk cargo, particularly feed grains, as to implement a long overdue technological change or improvement. Indeed, by some measures, the Big John cars in the circumstances of the early 1960s may not have been the best technology from the standpoint of operating efficiency; in particular, these cars may have had axle loadings that damaged the lightweight rails common at that time.[53] The ICC, at any rate, obstructed the introduction of the Big John cars for several years, apparently not because of any misgivings about the appropriateness of the new technology but rather because they feared the implications of the new and lower rates for competitors of the railroads, namely barges and trucks. A particularly piquant irony in this situation was that agricultural commodities moved by truck and barge are exempt from regulation, and thus the ICC was "protecting" the barges and trucks from railroad competition in a sector of barge and rail operations that was not subject to ICC control. Full use and introduction of the Big John cars was never really accepted by the ICC but rather was forced upon the commission by court action, which took several years to complete.[54] Today, the Big John car or its lineal descendants, as well as heavyweight rails to sustain the heavier axle loadings, are standard to the U.S. railroad industry. An interesting and unresolved question is whether this is the optimal technology or simply the best technology attainable given the workings of the regulatory and legal processes.

The airline industry also offers examples of technological change being distorted by regulatory decisions. For example, the Civil Aeronautics Board's weight and size regulations, its rate taper relating rates to distance traveled, and government subsidy policies for local service have generally created stronger incentives to develop new equipment for longer hauls than for shorter hauls. Some analysts argue that this created incentives for the airlines and aircraft manufacturers to develop and deploy jets as opposed to turbo props earlier than would otherwise have been desirable. In retrospect, the lower cost achieved by jets (lower than estimates at the time of adopting jet aircraft) may well have made this decision justifiable. However, the inhibition to invest in research in more efficient turbo props has probably impaired the development of airline travel on shorter haul routes served by local service operators receiving government subsidies. Similarly, weight and size limits for *exemption* from regulation may have prevented earlier and more extensive development of so-called third-level carriers that often seemingly can provide more frequent service to smaller towns and cities at a lower cost without subsidy than subsidized second-level local service carriers. Without so much government regulation and subsidization of shorter-haul and lower-density markets, market share competition among the major trunk lines might also have encouraged them to develop either their own or subsidiary feeder operations to a greater extent (as a means of generating traffic for their main routes or longer hauls).[55]

Regulation has also contributed its share to creating inefficient investments and dulling incentives for innovation in the petroleum extraction industry. For example, production controls aimed at stabilizing the domestic price for crude petroleum have apparently led to the underutilization of efficient wells and overutilization of high-cost and low-output wells, thereby possibly reducing the total percentage of oil recovered from underground pools. A more appropriate regulatory solution might be to unitize the oil field and allow the market to work within such a framework.[56]

More recently, government regulations have created uneconomic investments downstream in the petroleum industry by creating strong incentives to invest in uneconomically small refineries. Specifically, smaller refiners receive more government entitlements for every thousand barrels of crude they refine than larger refiners; these government entitlements are needed for a refinery to buy cheaper domestic "old" oil that is controlled to sell at a price more than $8 less than "new" domestic or imported oil (so that each entitlement is thus effectively worth over $8 per barrel). The result has been quite predictable. As one report puts it:

> The government has set off a veritable boom in small refineries. . . . The exploding refinery population consists almost entirely of inefficient, inappropriately designed plants whose main function is to enrich their owners at public expense. Subsidies to small refineries now run well over a billion dollars a year. . . . Their proliferation has helped discourage investment in bigger, more versatile refineries. . . . As a result, the U.S. may in a few years wind up with another refinery bottleneck like the one that led to shortages and price increases in the early 1970s, well before OPEC flexed its muscles.[57]

The net result of all this is, of course, that the petroleum industry is saddled with inefficiency and the consumer ultimately pays more for energy than would otherwise be the case. Of course, there may be a certain perverse rationality in all this if the higher prices created by these inefficiencies create incentives to reduce energy consumption and such conservation is deemed to be in the public interest. On the other hand, if the regulations leave more oil in the ground than would otherwise be the case (by reducing the percentage recovery in domestic oil fields), even this conservation feature loses some of its luster. Furthermore, windfall profits for a special group of entrepreneurs in society may not be widely deemed the most socially optimal means of imposing a tax to discourage energy consumption.

The telecommunications industry, as noted previously, may also not have been totally immune to regulation distorting investment and innovation decisions. The primary thrust of technological innovation in the telecommunications market has been to make the costs of long distance communication substantially lower. One of the primary policy problems in telecommunications is the sluggishness with which the telephone rate structure has adapted to these changes, thereby creating strong incentives for investment by new competitors in long distance intercity communications. The telephone companies claim that these new or outside investments are uneconomic. Critics of the telephone companies, on the other hand, have argued that telephone companies have been slow to adopt new telecommunication costs because of a rate structure protected by regulation and regulatory decisions based on rate-of-return concepts that create incentives not to use capital-saving innovations. The possibility of rate-of-return regulation distorting investment decisions is not, incidentally, limited to telecommunications and, indeed, has been widely alleged to have occurred in other regulated public utilities as well.[58]

In general, making regulatory decisions on the basis of an assessment at one point in time about a market in the throes of technological change and rapid growth is highly likely to result in mistakes. Such an approach is comparable to a critic reviewing a movie based only on still pictures from the opening scene! Even the wisest and most nobly motivated regulatory commissions simply cannot foresee all the future possibilities. Generally, the market itself has demonstrated a greater capacity for handling the information flow and the numerous decisions needed to cope with rapidly changing market and technological circumstances than regulation. When regulators are less than alert and motivated, and attempts are made to suppress technological change in order to preserve a particular pattern of industrial relationships or a specific price structure, the problem is further compounded.

Ironically, such regulatory efforts are not only inappropriate, but they are also likely to be doomed to be ineffectual, at least in the long run. Technological change in a reasonably open market economy will usually not be totally denied. When regulatory authorities attempt to channel or retard technology's progress, the usual consequence is merely to rechannel the change, perhaps distorting it and making it somewhat less efficient in the process. Thus, in telecommunications the market is clearly headed in the direction of making electronic communications costs lower and lower. Protection of existing carriers, or restricting entry into the market to preserve rates well above costs, is likely to be increasingly difficult if not impossible.[59]

Overall, then, there is no evidence to confirm AT&T's thesis that regulated monopoly is the market structure that will greatly enhance the rate of technological innovation. A consensus does seem to be slowly emerging that a moderately competitive market structure somewhere between complete monopoly and perfect competition may be better than other alternatives. Importantly, if the "desired" market structure is to be created by regulatory act, it must also be recognized that regulation has its imperfections and a history of creating its own barriers to efficient investment and innovation.

EXTENDING SERVICE TO RURAL AND OTHER LOW-DENSITY AREAS

One of the more common subsidiary arguments used to justify regulation is that it is needed to finance the rendering of costly services to remote out-of-the-way places at rates that, by implication, do not cover costs but are nevertheless not too far out of line with rates charged users in more central locations. This argument in defense of regulation has been used by the trucks and airlines, and by the telephone industry as well. Regulation's role, in this argument, is to suppress competition that might "skim the cream" off the profits of the major companies to the point where they could no longer afford to render services at high cost to out-of-the-way places. AT&T specifically contends that competition in interconnect equipment and private line service will force it to abandon rate averaging, leading to higher rates in sparsely populated regions.[60] As one consultant to AT&T has put it:

> The specialized telephone companies [the new entrants] will have one abiding advantage. The general service companies, of course, possess the assets of the network and its inherent economies of scale. But if we assume comparable technological efficiency of the two groups, the spe-

cialized telephone companies do not have the costly burden of providing telephone service on equal terms to rural and other subscribers on thinly populated routes or to residential subscribers as a class.[61]

Somewhat the same point was made, though more guardedly, by three economic advisers to AT&T:

> What this means is that nationwide rate uniformity and the entry of a substantial number of competitors are incompatible. . . The greater the degree of competition, the less the likelihood of Bell's being able to continue uniformity, supplying the sparse routes at their present prices which are relatively low in relation to cost. It will be unable to make up its losses among the denser routes because competition will render that impossible.[62]

As noted in Chapter 3, rate uniformity in other than interstate rates may be more of a mirage than a reality. Specifically, as pointed out, substantial differences exist in local connect charges, intrastate toll charges, assessed tariffs for particular kinds of equipment, and user charges or message unit charges in different states and circumstances. Furthermore, many of these differences are very difficult to explain in terms of any tangible differences in value of service rendered, costs, or other such considerations. In short, there is little empirical evidence to establish the validity of the underlying assumption that a great deal of rate uniformity does exist in several parts of the telephone tariff structure.

In addition, it is clear that AT&T is not heavily involved today in directly serving truly low-density, rural locations and, for that matter, never was. The vigorous growth in nationwide telephone service that followed the expiration of Bell's patents (see Chapter 2) is in some measure due to service being "extended for the first time to suburban and rural areas."[63] Independents and farmer cooperatives, not Bell operating companies, introduced this new service,[64] and rates for all telephone customers fell.[65]

Indeed, the number of mutual systems and other telephone organizations peaked at 60,000 in 1927. The small rural telephone companies — in which housewives often ran the switchboard in their kitchens while the husbands repaired the lines — then began to deteriorate. The operators had little knowledge of keeping accounts, maintenance, or other managerial skills. Illustrative of these deficiencies, a special technical problem arose in the 1930s as electrical lines were installed by the Rural Electrification Administration (REA); when the electric lines used the same poles as the telephone system, interference was produced on the phones. The farmer-operators did not have the resources during the Great Depression to revamp their small magneto systems to overcome this difficulty. At any rate, by 1940 there were fewer farmers with telephones than there had been in 1920,[66] some of the decrease reflecting a decline in farm population, some a drop in farm incomes, and some a deterioration in service capability.

In 1949, when Congress passed the telephone amendments to the Rural Electrification Act of 1936, only 38 percent of U.S. farms had any phone service. REA assistance subsequently increased the penetration level to 90 percent and made rural telephone service at least marginally profitable again. The main tools used by the REA in promoting rural telephone service have been low-interest loans and technical assistance.[67] The low-interest loans have been made either directly from the REA to the local entities providing the telephone service or by a Rural Telephone Bank, especially established for this purpose. In addition, the REA also guarantees loan commitments, thereby ostensibly achieving a lower interest rate for small commercial companies rendering local rural telephone service.[68]

REA aid for development of rural telephone service essentially goes to two different kinds of operating entities: cooperatives and commercial companies. Besides the low-interest loans and technical assistance rendered by REA, the cooperatives have the further advantage of having to pay corporate income taxes only on their unallocated retained earnings. In 1976, there were 633 commercial companies being aided by REA, while there were 243 cooperatives receiving such aid. The 633 commercial companies paid $109 million in taxes in 1976 as compared with only $11 million in taxes paid by the 243 cooperatives.

The approximately 870 rural cooperatives and commercial companies receiving REA aid in recent years apparently average a little over 5,000 phones per company or have approximately 4.8 million phones in total. To put the 4.8 million total number of telephones in perspective, there were 2 million farm households as of 1977 and 23.7 million rural (farm and nonfarm) households nationwide.[69]

It seems clear that the cooperatives and small commercial companies supported by the REA account for a high percentage of phone service to rural households. Some evidence suggests, moreover, that most rural telephone service not provided by REA-supported services probably originates more with smaller independent telephone companies than with AT&T itself. There are approximately 1,600 independent operating telephone companies, and many of the 750 or so of these companies not receiving REA support would also seem to be operating in relatively rural areas. By contrast, Bell, with 83.3 percent of the phones in the United States, services only 42 percent of the geographical area of the contiguous United States. If Alaska and Hawaii are included, Bell's service area falls to 35 percent. In only four states is Bell's density of phones per square mile exceeded by an independent, and nationwide, Bell's density is twice that of its nearest competitor.[70]

In short, Bell did not enter rural areas originally and is apparently not there now to any great extent. If AT&T has a presence in rural telephone service today, it is indirect rather than direct, explicitly through contributions to the revenues of local rural telephone companies in the form of toll settlements. Revenue from toll service now accounts for 55 percent of REA borrower revenues, while toll revenues account for 45 percent of Bell operating company revenues.[71] Care must be taken, however, in interpreting this difference. The variance around the average figures is large, reflecting such factors as the average size of the local calling area and, particularly for independents, the extent of tourism and resulting long distance calls from the area.[72] For instance, the very fact that rural telephone companies are very small (in the sense of having few subscribers who are directly dialable) could account for much of this difference. The smaller the unit in any kind of trade transaction, the greater the probability that external trade will be important. This is perhaps best illustrated by carrying the calculation to its limit: a phone system with one instrument would have to gain 100 percent of its revenues through toll settlements.

Most current or potential revenue problems faced by rural telephone companies apparently have little to do with competitive entry of new specialized common carriers into long distance service;[73] indeed, what problems there are seem due to the actions of existing telephone companies.[74] The REA notes that some of its rural companies or coops already face important toll revenue diversions and "at present, the most important of these adverse factors is the growth in the carriage of long distance communications (which would otherwise be toll carriage) by the large urban-based telephone companies (primarily AT&T and its associated companies) through the provision of private line services, private line network services, and other alternative service."[75]

An REA official has stated separately:

We are concerned about this [new competition] in connection with all toll carriers, whether Bell System, major Independent, or other common carriers. The REA borrower bleeds as much whether his throat is cut by Bell or MCI. And this is not theory. One of our borrowers saw his annual toll revenues cut from $360,000 to $10,000 from by far his biggest customer, the local installation of a major conglomerate which was connected to the parent's private line system. And that happened to be a Bell System-provided network.[76]

Many REA officials now believe that phased introduction of broadband facilities by REA borrowers could compensate for any reduction in toll settlements that competition in long distance service might bring. The REA analysis is still preliminary. However, it suggests that problems created for rural telephone by a more competitive industry structure may not be insurmountable. By the turn of the century, an REA broadband telephone program may be able to effectuate significant economies, cutting costs well below what they might otherwise be, "due both to lower costs of adding subscribers via broadband facilities, and the added financial strength of REA borrowers, enabling outside financing, through a combination of services."[77] A by-product cable television service (which if independently rendered is apparently not profitable in rural areas) would provide most of the additional revenues for rural telephone firms, but new facilities for health, education, and energy management might also be viable and should also improve rural life. As a consequence of these considerations and possibilities, legislation permitting REA loans for cable television service has come before Congress.[78]

Strong empirical support is also difficult to find for the proposition that rural telephone service is much more expensive than service elsewhere. To a first approximation, revenue per telephone may be a reasonable proxy for the cost of serving a particular customer. As shown in Table 5-1, in 1976 the Bell System companies averaged $244 per year in revenues per telephone. By contrast, the average revenues per telephone for all independent telephone companies was $223 and the average for all REA borrowers $187 per phone. Similar differences result from comparing operating expenses per instrument. Table 5-2 indicates that average operating expenses per telephone in the Bell System in 1976 were $157. For the larger independents, however, the average was $137, for smaller independents $148, and for REA borrowers $121 per phone.

Many reasons, of course, may explain why the smaller companies can exist on less revenue per telephone than the Bell subsidiaries. REA help and subsidization, for example, may lower the costs of the very small independent telephone companies, although this would not explain the differences between the other independent companies and the Bell System. The Bell operating companies may also own relatively more equipment dedicated to toll service and be reimbursed with additional revenue from the separations and settlements process.

This latter possibility can be quantified by examining data on Bell operating companies' interstate revenue requirements as shown in the separations process. Separations may be described as a two-part process: (1) the recovery of investment and expenses for dedicated facilities, that is, facilities that exist solely for toll usage; and (2) the allocation of other costs to toll service on the basis of usage. In 1976, the Bell operating companies received approximately $2.9 billion, or roughly $23 per phone, from interstate revenues for items of equipment and expense that appear to be dedicated to toll service.[79] Since this figure is only for interstate facilities, an adjustment that included intrastate-dedicated toll facilities would be somewhat higher. On the other hand, the independent companies own some dedicated toll equipment and a proper comparison would require adjusting the independents' costs accordingly. On balance, it would appear that the Bell companies' provision of toll equipment and services accounts for much of the difference between the Bell and independent costs per phone shown in Tables 5–1 and 5–2, but it is difficult to determine which group's costs for the remaining nontoll service is lower overall. On the average, however, the independent companies' costs per telephone appear to be roughly comparable or similar to the costs for the Bell operating companies.

Table 5–1. U.S. Telephone Industry Operating Revenues and Number of Telephones, 1976

Companies	(1) Number of Phones (thousands)	(2) Operating Revenues (thousands of $)	(3) Operating Revenue per Telephone (2) ÷ (1)
Bell Operating Companies[a]	127,407	31,085,191	$244
Independents[b]	28,209	6,300,000	223
Class A[c]	NA[e]	NA[e]	231
REA[d]	4,976	899,000	187

Sources: Row 1: *Statistics of Common Carriers,* FCC 1976, pp. 29–32; rows 2–3: Independent Telephone Association, *1977 Independent Telephone Statistics,* Volume 1, pp. 2–3; row 4: *Statistical Abstract of the U.S.,* 1978, p. 589, Table 974; all revenue data are after the distribution of settlements.

[a]The Bell Operating Companies include wholly owned or majority owned companies by AT&T, plus Southern New England and Cincinnati Bell. This figure excludes the Long Lines division of AT&T.
[b]This category contains 1,590 companies and thus includes all Class A, B, C and D companies in the United States.
[c]Telephone companies with operating revenues in excess of $250,000 per year, 649 reporting companies.
[d]939 telephone companies borrowing from the Rural Electrification Administration.
[e]Data not available.

These same issues can be investigated more formally by estimating an econometric cost function relating cost differences between companies to differences in the mix of services they provide or to the scale of the company's operations. A cost function that includes the principal products and services provided by telephone companies might be specified as follows:

$$Expenses = f(M, P, E, L, T, A/U)$$

where

Expenses	= the operating expenses of each company
M	= the number of main station telephones
P	= the number of PBX telephones
E	= the number of extension telephones
L	= the annual number of local calls
T	= the annual number of toll calls
A/U	= the number of miles of aerial wire divided by the number of miles of underground wire and cable

Table 5–2. Comparison of Operating Expenses per Telephone for Bell System and Independent Telephone Companies, 1976

Company Groupings	Number of Companies[a]	Range of Phones per Company (thousands)		Average Phones per Company (thousands)	Operating Expenses[b] per Phone (dollars)
REA borrowers	876	0—	8	5.5	$121
Medium independents[a]	22	8—	300	98.0	$148
Selected large independents[a]	19	300—	3,100	822.0	$137
Bell operating companies	24	180—	14,000	5,300.0	$157

Sources: Rural Electrification Administration, *Annual Statistical Report-Rural Telephone Borrowers,* 1976. Federal Communications Commission, *Statistics of Communications Common Carriers,* 1976.

[a]There are 750 independents in the range of 8,000 to 3.1 million telephones per company. The 41 independents in our data base were selected from Tables 16 and 17 of the *FCC Statistics of Communications Common Carriers.*

[b]It may be misleading to compare the operating expenses per phone of REA borrowers to that of other telephone companies, because of differing quality. For example, REA borrowers have lower ratios of single party telephones to total telephones and PBX telephones to total telephones. Also, REA borrowers have a lower ratio of business telephones to total telephones suggesting that REA phones might be used less intensely.

The products and services included in this specification are main station, PBX, extension telephones, and local and toll calling. The specification allows for cost differences among types of connections (main, PBX, and extension) and among types of call (local versus toll). Costs are expected to increase with increases in any of the five product and service variables. The specification also includes the ratio of aerial to underground wire as a variable to correct for cost differences related to density. This variable should become smaller the greater the proportion of urban areas in a company's territory.[80]

The observations used to estimate the cost function are the 24 Bell operating companies and 41 large independents. All data are for the year ending December 31, 1976, and all variables are expressed in units of one. The companies range in size from around 8,000 total phones to over 14 million phones.[81]

In order to examine the effect of scale on costs, the variables in the estimated equations are expressed as natural logarithms. The interpretation of the coefficient of each product variable is thus the elasticity of cost with respect to changes in the level of that product.[82] The *sum* of the coefficients of all the product variables is the elasticity of cost with respect to equiproportional changes in each product. That is, a 1-percent increase in the level of each product would lead to an X-percent increase in cost where X is the sum of the product coefficients.

As discussed in recent theoretical literature, the concept of economies of scale is difficult to analyze in the multiproduct case.[83] A closely related concept, however, is declining ray average cost, meaning that average cost declines as all outputs increase by the same proportion. For average cost to decline, total cost must increase by a smaller fraction than outputs increase, or in other words, the elasticity of cost with respect to equiproportional changes in outputs must be less than 1.0. If this elasticity is equal to 1.0, cost increases by the same proportion as outputs and average cost remains constant. If the elasticity is greater than 1.0, average cost increases as outputs increase. The sum of the product coefficients in the logarithmic specification, therefore, is a measure of the behavior of costs as scale changes and of whether larger companies have cost advantages or disadvantages over smaller ones.

Estimation results appear in Table 5–3. In order to see whether results differ for Bell versus independent companies, an equation is estimated for the total sample of 65 companies and separate equations for the Bell and independent companies. In addition, an attempt is made to see how the Bell results are affected by the inclusion of dedicated toll expenses in total operating expenses. Unfortunately, the only data available for individual Bell companies are for total interstate expenses,[84] which include both dedicated and usage-based allocations.

Table 5-3. Cross-Section Cost Estimates for Bell System and Independent Companies, 1976, Logarithm Forms^a (t-statistics in parentheses)

Dependent Variable: LN (Expenses)

Equation	Group	Constant	LN(MAIN)	LN(PBX)	LN(Extension)	LN(Local)	LN(Toll)	Aerial/under	R²	Sum of Product Coefficients
5-1	All companies N = 65	3.99 (2.90)	.75 (4.04)	.27 (4.61)	−.26 (−1.71)	.05 (.37)	.21 (2.55)	.01 (.47)	.99	1.03
5-2	Bell (total) N = 24	3.06 (1.34)	.97 (3.54)	.20 (1.49)	−.75 (−3.44)	.18 (.77)	.35 (3.33)	−.31 (−2.83)	.99	.95
5-3	Independents N = 41	4.74 (3.05)	1.18 (4.87)	.13 (1.89)	−.37 (−1.90)	−.01 (−.07)	.09 (.85)	.02 (.67)	.99	1.02
5-4	Bell (intra) N = 24	2.25 (1.52)	.94 (5.25)	.19 (2.17)	−.47 (−3.31)	.20 (1.29)	.19 (2.79)	−.13 (−1.83)	.99	1.05
5-5	All companies N = 65	5.11 (4.14)	.62 (3.60)	.26 (4.40)		−.02 (−.14)	.15 (2.00)	.02 (.61)	.99	1.01
5-6	Bell (total) N = 24	5.70 (2.09)	.46 (1.56)	.30 (1.83)		−.08 (−.27)	.29 (2.18)	−.13 (−1.05)	.98	.97
5-7	Independents N = 41	6.21 (4.46)	.97 (4.35)	.14 (1.86)		−.10 (−.65)	.01 (.06)	.03 (.88)	.98	1.02
5-8	Bell (intra) N = 24	3.89 (2.24)	.61 (3.30)	.25 (2.40)		.04 (.20)	.15 (1.79)	−.02 (−.21)	.99	1.05

^{a}LN (x) = the natural logarithm of the variable x.

Subtracting the interstate component thus corrects for more than dedicated toll expenses alone. Nevertheless, the results give some insight and are shown as Equation 5–4 in Table 5–3.

The coefficients of main and PBX telephones are positive in all equations and statistically significant except for the PBX coefficient in Equations 5–2 and 5–3. The coefficient of extension phones, however, is always negative, contrary to expectations.[85] The extension coefficient is significant at the .01 level in the Bell equations and at the .10 level in other equations. The coefficient of toll calls is always positive and is significant in the total and Bell equations. Finally, the coefficient of local calls is never significant, and the ratio of aerial to underground wire is significant only in Bell equations.

The unexpected negative sign for the extension coefficient could be due to multicollinearity. To see whether the other coefficients are sensitive to including extensions, Equations 5–5 through 5–8 were estimated with this variable excluded. Results change only slightly, with the Bell main station coefficient showing the greatest change.

As just discussed, the sum of the product coefficients gives an indication of the behavior of costs as scale changes. This *sum* is shown in the last column of Table 5–3. As can be seen the sum is generally close to 1.0. In no case is the difference from 1.0 statistically significant. Thus it would appear that average operating expenses are roughly constant as scale changes among the companies analyzed. The absence of scale effects holds for the total sample, the Bell companies with and without interstate expense, the independents, and when the perversely signed extension variable is excluded.

The high R^2s in the equations shown in Table 5–3 indicate that nearly all the variation in the logarithm of companies' expenses can be explained by variations in the logarithms of particular types of phones and calls. Since scale has little effect on cost and since the A/U variable has little effect, the principal explanation of cost differences among these companies would appear to be differences in their product mix.

A final caveat is that the companies at the low end of the size range in the above estimation have approximately 8,000 total phones. The results cannot necessarily be extended to smaller companies.[86] The companies included in the analysis do, however, account for 92 percent of all phones in the United States. In addition, companies larger than 8,000 phones but not included in the analysis would appear to account for roughly an additional 6 percent of all phones.[87] If there is a cost problem facing companies smaller than the low end of the range considered here, the problem appears to apply to roughly 2 percent of U.S. phones mainly served by companies receiving government support through the REA.

These econometric cross-section cost results seem quite consistent with the econometric time-series studies and the engineering cost evidence reviewed in Chapter 4. As noted, the econometric and engineering studies generally include different cost elements. The engineering studies tend to focus on basic transmission investment costs for long distance links to the exclusion of local service and all switching and multiplexing equipment. On the other hand, the econometric time-series studies normally relate to complete systems, including *all* components of long distance and local service costs. Generally speaking, and as observed in Chapter 4, the more cost elements incorporated into the analysis beyond basic transmission costs, the greater the tendency for scale economies to be reduced. Thus, the time-series econometric studies reviewed in the last chapter reported somewhat lower scale economies than the engineering analyses of terrestrial transmission investment costs, suggesting fewer scale economies in local than long distance service. The cross-section econometric evidence presented here further corroborates this.

Although it does not appear that average total operating expenses significantly increase or decrease with scale, it does appear that the distribution of expenses by different categories does vary with scale. This is shown by the figures in Table 5–4 that compare different categories of operating expenses for the Bell companies with those for large independents, medium-sized independents, and REA borrowers. As can be seen from Table 5–4, the smaller companies allocate a smaller percentage of operating expenses to maintenance than the bigger companies.[88] On the other hand, there do seem to be some fairly discernible economies of scale in general office expenditures, just as a priori notions would suggest. The higher percentage of operating expenses allocated to depreciation and amortization for the small companies may also be suggestive of some economies of scale. The differences in percentages allocated to traffic, commercial, and other operating expenses, for all of which the small companies seem to have some advantage, is intriguing. Among the possibilities might be less need for advertising or promotional work in rural areas and probably a somewhat lower average wage structure; in addition, much of AT&T's extra expenses in these categories might represent "institutional" efforts (such as marketing and accounting for toll calling) that benefit the entire industry.

In summary, there is little evidence to suggest that rural local phone service is receiving any disproportionate help, directly or indirectly, from AT&T. Furthermore, although the problems of rural telephone service would seem to be technologically and economically somewhat different from problems encountered elsewhere, they do not seem to result in unit costs that are substantially higher on an average than for telephone service in other areas. And REA's loan program is already in existence to aid the lowest-density rural companies that may face higher costs. Although the entry of new competition into toll service may pose some adjustments for rural telephone companies, those closest to the situation seem confident that improvements in technology and general growth in communication demands should provide an environment in which these problems can be managed. It is not clear, moreover, that these adjustment problems for rural telephone companies would be any less without new entrants into long distance service competition than with them — since Bell's competitive inroads into rural toll settlement revenues could be as great as those of any potential new entrants from outside the industry.

Table 5–4. Distribution of Total Operating Expenses for Bell Companies as Compared to Independents, 1976[a] (percentages)

Expense Category	Bell Companies	Selected Large Independents[b]	Medium Independents[b]	REA Borrowers
Maintenance	31.5	31.1	28.8	28.3
Depreciation & amortization	21.5	30.0	30.2	33.5
Traffic	11.0	9.4	12.3	4.8
Commercial	12.5	10.1	9.3	8.6
General office	8.0	10.1	10.3	14.4
Other operating expenses	15.5	9.3	9.1	10.5
Total operating expenses	100.0	100.0	100.0	100.0
Number of companies	24	19	22	876
Phones per company	5,300,000	822,000	98,000	5,500

Sources: Data from Rural Electrification Administration, *Annual Statistical Report— Rural Telephone Borrowers,* 1976; Federal Communications Commission, *Statistics of Communications Common Carriers,* 1976.

[a]All statistics on this table pertain to the year 1976.
[b]"Selected Large" independent telephone companies have from 300,000 to 3.1 million phones each. "Medium" independents fall in the range from 8,000 to 300,000 phones each.

REGULATION AND THE CREATION OF "PREFERRED" MARKET STRUCTURES: LOCAL TELEPHONE SERVICE AS A CASE EXAMPLE

The concept of "regulation creating a better market structure" relies, at its base, on the assumption that regulation has the capability to modify and create the market structure it seeks. This inevitably raises the question of whether regulation can, in the long run, really modify market structure in important ways. In industries other than telecommunications, there is considerable evidence that attempts to modify market structure through regulation have often failed.

In trucking, for example, despite much effort, regulation has not protected common carrier truckers from losing most of the bulkier truckload business to independent truckers operating on a private contract basis. Common carrier truckers have achieved protected market positions only in certain specialized niches.[89] In aviation, similarly, regulation has not provided common carriers with the protective shield expected, as charters, service competition, and other forms of rivalry have replaced the price competition suppressed by regulation.[90] Equally as importantly, *inter*modal competition, between barges, trains, trucks, and airplanes, has often undermined regulators' aspirations to achieve a "better market structure."[91] Similar experiences could be cited in other heavily regulated industries.[92]

In telecommunications the argument that regulation could create, permit, or make tolerable a preferred market structure has been deemed particularly applicable in local service. Apparently, the approximate 15-percent duplication or overlap in local services created by open competition in local telephone services around the turn of this century[93] was deemed by some to be unduly wasteful. Certainly, the creation of such overlapping services without the provision that they honor a common carrier obligation to one another was less than optimal from a service standpoint and, unless offset by innovation or other "dynamic" gains, perhaps inefficient from a societal standpoint as well. Putting aside these questions, and whether enforcement of a simple interconnecting common carrier obligation would have been sufficient, public policy apparently concurred with AT&T's judgment in the early 1900s that monopoly was a "preferred" market structure for local telephone services. Certainly, public agencies erected no major barriers to AT&T's achievement of local service monopolies in the years after Vail resumed the AT&T presidency and the Kingsbury agreement was accepted. It is, then, of particular interest whether regulation can continue to create this preferred market structure in local telephone service. Not only is local telephone service widely believed to be most efficiently rendered as a natural monopoly, but access to local service may also affect the development of competition in long distance services. However, as discussed below, the availability of alternatives to local telephone service, at least for some purposes, may well impose some limitations on the pricing discretion of local telephone companies.

Further intensifying the latent competitive pressures on local service, a number of new long distance technologies completely bypass the local loop as a means of access.[94] Other technologies not yet in place may eventually even provide a competitive alternative to traditional local service. These possibilities may in the future place a market constraint on telephone companies' ability to raise local rates in general. It is often argued, for example, that if long distance service is opened up to competition, local rates would have to increase dramatically to make up the contributions to overhead and joint costs lost from the high markups now ostensibly enjoyed on long distance services. This argument assumes that local service customers would have no choice but to pay the higher rates. Although the technologies discussed below are unlikely to be in general use as competitive substitutes for local service in the near future, they raise the possibility of an effective alternative in the more distant future — and, as such, place a limit on the market power inherent in any existing monopoly of local services.[95]

Bypassing the Local Loop

State regulatory commissions have not as yet formally encouraged new entry into local phone service, nor has any vendor been eagerly volunteering to provide any such competition to the existing telephone companies. However, recent technological and market developments have eroded the importance of the local telephone network as a link to long distance. For example, American Satellite Corporation (ASC), a subsidiary of Fairchild Industries, is already bypassing telephone company local service with its Satellite Data Exchange Service (SDX) by beaming messages directly to small two-way earth stations on site at a limited number of clients.

Other companies also have plans to bypass local exchange service. The Xerox Corporation announced in November 1978 that it is planning to introduce a new type of telecommunications service that would not use the traditional local telephone network.[96] AT&T scientists have also announced that they could design a satellite telecommunications system that would bypass the traditional local loop for both large and small clients.[97]

Satellite Data Exchange and Xerox Telecommunications Network Service are both targeted at business users that have large or specialized (mostly intercity) communications requirements that would warrant the use of such services. One would normally not expect a residential customer or a very small business to subscribe to such a service because they have little demand for high-volume intercity transmission services or for very specialized features.

It is therefore significant that the distribution of interstate call revenues is highly skewed. A few users (see Table 5-5) account for a very high percentage of total revenues in long distance telephone services. Specifically, over 75 percent of total revenues from business use of long distance are accounted for by less than 8 percent of all customers. Even more dramatically, the 25 largest business customers generate 15 percent of the total interstate business toll revenue, while the 100 largest business customers generate 20 percent of this revenue;[98] interstate revenue, in turn, represents a little over one-quarter of total telephone industry revenues.[99] Similarly, almost 50 percent of long distance residential revenues are from around 10 percent of the users.

Looking at the statistics on a per-city basis, it is apparent that calls among the 32 largest metropolitan areas account for half of total interstate MTS and WATS revenues (see Table 5-6). Thus, there is a high concentration of interstate phone service among a few metropolitan areas and a relatively limited number of large users. Clearly, these users would be, and effectively have been, the primary targets for competitive intercity services, and it could well prove difficult to

Table 5-5. Distribution of Interstate Long Distance Billing

Residential				Business			
Customers		Revenues		Customers		Revenues	
Percent	*Cumulative*	*Percent*	*Cumulative*	*Percent*	*Cumulative*	*Percent*	*Cumulative*
4.0	4.0	29.9	29.9	3.9	3.9	61.7	61.7
5.6	9.6	19.6	49.5	3.7	7.6	13.5	75.2
6.5	16.1	14.8	64.3	5.2	12.8	9.4	84.6
14.3	30.4	19.0	83.3	10.0	22.8	8.1	92.7
18.9	49.3	11.6	94.9	15.1	37.9	4.7	97.4
50.7	100.0	5.2	100.0	62.1	100.0	2.6	100.0

Source: Calculations by Charles River Associates, based on the Telecommunications Industry Task Force, *The Dilemma of Telecommunications Policy,* submitted to Congress September 21, 1977, Exhibit 8, Tables 4 and 5.

deny these users access to new long distance services because of an inability to connect through local exchange monopolies.

Future Alternatives to Traditional Local Service

Other related technological developments, now at the infancy stage, could render local loops even less necessary as hookups to competitive long distance services in the future.[100] Among these innovations are mobile land telephone services and fiber optic transmissions.

At present, mobile telephone service is used in motor vehicles (such as taxis and delivery trucks) and for other limited purposes; however, it could provide a means of directly accessing a switching facility for many new types of users. In early 1974, following Docket 18262, the FCC decided to expand somewhat the bandwidth available for land mobile communications and also approved an experiment by Illinois Bell with a new type of general public mobile telephone service, the "cellular system."[101] Still more bandwidth eventually could become available for such service, and experiments then could be made with other systems proposed by competitors to the established telephone companies. The proposed Xerox telecommunications service, for example, would use a combination of cellular radio systems, land microwave networks, and communications satellites to provide transmission of a wide variety of messages (such as voice, video, digital computer data, and business documents). One study has even estimated that with better radio spectrum management, every motor vehicle, or even every individual, in the United States could have a mobile radio transmitter (as shown in Figure 5–1, taken from that report).[102] Quite clearly, it does not take too imaginative a reorganiza-

Table 5–6. Proportion of Interstate Business MTS and WATS Revenues Generated by Largest Metropolitan Areas

Calls Among Largest Metropolitan Areas	Proportion of Total Revenue
16	1/3
32	1/2
144	7/8
400	9/10

Source: Telecommunications Industry Task Force, *The Dilemma of Telecommunications Policy,* submitted to Congress December 1, 1977, Exhibit 8, Table 2.

tion of the uses now made of the radio spectrum to enhance greatly the ability of "free" airwaves to serve local telecommunications needs. These reorganizations, however, might be as politically complicated as they would appear to be technically straightforward or simple.

Because of such complications, it is significant that optic fiber technology also offers an alternative to traditional copper and coaxial cables. If guided transmission over optic fibers becomes quite cheap, as it well might,[103] the increase in transmission capacity would greatly reduce and quite possibly eliminate economic incentives to reorganize and rationalize use of the airwaves. Given the present state of

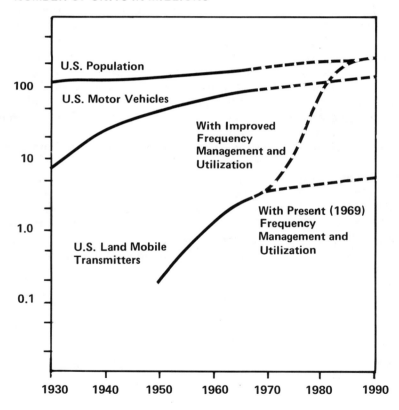

Figure 5–1. **President's Task Force Estimates on Land Mobile Radio Transmitters.** *Source:* President's Task Force on Communications Policy Staff Paper I — A Survey of Telecommunications, June 1969, Appendix F, p. 4.

technological development, most experiments in the field with fiber optics have so far focused on local trunk service between central office exchanges. This will probably continue to be the case in the coming few years, but one can envision the day when optic fibers could connect many homes as a replacement to and expansion of the traditional local loop. With a capacity equivalent to millions of times that of a copper wire, optic fibers would facilitate, for example, the emergence of "electronic households" discussed in Chapter 2.

Furthermore, as transmission costs are driven down by new technologies (electronic switching, communications satellites, optic fibers, mobile telephone service, and others) and as more clients see potential benefits in new types of services, alternative local communications networks may well expand over the years. For example, with a proper design of the networks (two-way capability, some form of switching, etc.), optic fibers could be used not only to perform the function of both the traditional local loop and coaxial cable TV, but also to provide a host of supplementary services. It would probably not be economical, however, to install an optic fiber network merely to provide traditional telephone and conventional cable TV services or, where cable television facilities have been installed, to add a network of optic fibers.

Many geographic areas, however, have not yet been wired for cable television. As can be seen in Table 5–7, there were only 400,000 subscribing households in the 25 largest U.S. cities as of mid-1978. With any introduction of new services such as those discussed in Chapter 2, the possibilities of combined service offerings that need and make economical use of substantial local loop capacity could multiply. Paradoxically, the very fact that cable TV service is practically nonexistent in the largest cities may make it easier to install additional new services in such locales (since there would be fewer vested interests to oppose such innovations).

Wherever they do exist, however, the existing coaxial cable companies are at least another source of potential competition to local telephone loops. As of June 1978, 247,000 miles of coaxial cable were used by cable TV companies in the United States connecting to 12.6 million households. A good quality coaxial cable has a frequency range of 300 megahertz, as compared to the normal local telephone company copper wire loop range of about 1 megahertz. Thus, at present, a single normal coaxial cable could provide roughly 200 or 300 times the communications capability of the ordinary copper wire loop. A large proportion of the existing coaxial network, however, can transmit information in only one direction, and thus it would not be suitable for many types of communication. The FCC has required that two-way capability be installed under certain rules, but the transmit-

Table 5–7. Cable Television Coverage in the 25 Largest U.S. Cities

Rank[a]	City	Population in the Franchised area[b]	Homes Passed	Subscribers	Year Started[c]
1	New York, New York	2,589,233	527,630	143,602	66,66,67
2	Chicago, Illinois	Not franchised			
3	Los Angeles, California	3,462,145	182,116	71,602	57,64,67
4	Philadelphia, Pennsylvania	1,950,098	34,840	11,300	72,2x
5	Detroit, Michigan	Not franchised			
6	Houston, Texas	Not franchised			
7	Baltimore, Maryland	Not franchised			
8	Dallas, Texas	Not franchised			
9	San Diego, California	2,684,122	224,893	88,868	64,64,65,71
10	San Antonio, Texas	Not franchised			
11	Washington, D.C.	Not franchised			
12	Indianapolis, Indiana	Not franchised			
13	Milwaukee, Wisconsin	Not franchised			
14	San Francisco, California	Not franchised			
15	Cleveland, Ohio	623,530	16,500	907	76
16	Memphis, Tennessee	669,005	200	162	59
17	Phoenix, Arizona	Not franchised			
18	Boston, Massachusetts	Not franchised			
19	New Orleans, Louisiana	Not franchised			

Table continued on following page.

Table 5-7 (continued)

Rank[a]	City	Population in the Franchised area[b]	Homes Passed	Subscribers	Year Started[c]
20	St. Louis, Missouri	Just franchised	0	0	1x
21	Columbus, Ohio	1,121,341	105,502	27,364	71,73,1x
22	San Jose, California	575,000	135,571	56,797	68
23	Jacksonville, Florida	Not franchised			
24	Denver, Colorado	Not franchised			
25	Seattle, Washington	503,500	57,680	14,369	51
	Total			414,971	

Source: Federal Communications Commission Cable Bureau, Data Management System, Form 325, Schedule I, Statistics List, June 26, 1978.

[a]The 25 largest U.S. cities are listed in descending order of population size, as given in the 1974 *Statistical Abstract of the Commerce Department's Bureau of Census*, Table 25, page 23. The definition of city as used here does not include the greater metropolitan area, and is consistent with the definition used by the FCC for cable TV statistics.

[b]"Population" refers to each company's franchise area. Because the franchise area may be only a portion of a city, or because franchise areas for different companies may overlap, "population" as used here may be smaller or larger than the actual city population.

[c]This column gives the year in which Cable TV companies started operations. When a city has more than one company, the start-up year of operations for each company is indicated. When franchises have been granted, but companies have not begun operations yet, it is indicated by a number and followed by an X (e.g., "2x" for Philadelphia).

ting capacity would still not be typically equal in both directions. Substantial additions of equipment, and perhaps also a different cable layout or configuration, would therefore be needed before full two-way transmission could be provided on existing cable TV systems. Morever, switching facilities might have to be added, depending on the type of communication needs to be served and technology employed.

Still another potential competitor in local loops, indeed historically the first company in such service, is Western Union. In 1977, Western Union had about 396,000 miles of underground cable that reasonably can be assumed to approximate Western Union's own "local loop" network. Unlike the cable TV companies, however, most of this Western Union mileage was probably in copper rather than coaxial cable, and therefore it represents much less communication capacity per mile than the cable TV installations. Western Union in its 1977 Annual Report mentions its intention to expand this local network to bypass the telephone company's networks and thus reduce Western Union's lease payments, which, for intercity transmission and local loops, amounted to $76 million in 1977.[104]

Conclusions

For large business users seeking access to competitive long distance services, there are a number of new technologies enabling them to bypass the local loop as a means of access. These alternative technologies to some extent will discipline telephone companies in their ability to raise local rates or to impose prohibitive access charges. As a consequence, local markets may be more subject to market discipline, and therefore in less need of regulatory supervision, than at first appears to be the case.

An important corollary is that regulation, even if it could define an optimal market structure for achieving various other goals (e.g., innovation), could encounter substantial difficulty in imposing or achieving this market structure. And, whereas it currently may be necessary to retain some regulation over local telephone service in some areas and to small-volume users, even this residual regulation may prove to be unnecessary in coming years if and as increasing competition in the marketplace offers cost reductions or supplementary services to all customers. The regulatory bodies should therefore not automatically foreclose entry or preclude new service offerings related to local telephone service. Over the long term, new offerings and offshoots of new offerings may well develop viable competitive alternatives to existing local service, while in the short term such developments will strengthen newly emerging competitive alternatives in long distance service.

SUMMARY

Regulation is commonly advanced by its adherents as being a means to improve technological innovation by creating a market structure more conducive to achieving this goal. There is, however, apparently little justification for claiming that a regulated monopoly market structure is the best structure for achieving technological change. Economists find it hard to identify any particular market structure, be it pure monopoly or perfect competition, that maximizes technological change per se. Some consensus may be developing, however, that a market structure midway between monopoly and perfect competition is best for promoting innovation. Meanwhile, regulation in other industries seems, in fact, to have often distorted technological change in ways detrimental to the overall economy.

Another conventional argument for regulation is that it is needed to extend service to out-of-the-way, rural, or uneconomic areas. It is therefore significant that the independent telephone companies and rural cooperatives were the first to extend telephone service into smaller cities and outlying areas after the expiration of AT&T's patent monopoly. Even as late as 1949, only 38 percent of U.S. farms had any phone service. It was the 1949 amendment to the Rural Electrification Act that provided low-interest telephone loans and technical assistance to rural telephone companies that assured rural area telephone service. Although problems of adequate financing for rural telephone companies do exist, they do not seem to be primarily or exclusively a function of potential new entrants entering into competition with the established telephone companies, and thus, they are not a compelling argument against allowing such competition. Furthermore, the REA currently aids the very smallest telephone companies and will continue to do so. Additionally, the claim that it costs more to provide local service to rural areas would seem to be exaggerated. Average operating expenses per telephone seem roughly the same for different sizes of operating companies.

Finally, there is the question of whether regulation can create a "target" or desired market structure in contravention to other market and technological forces. To test this hypothesis, an examination was made of the extent to which regulation can be expected to condition market structure or form in the provision of local telephone service, an industry segment that is generally viewed as being a natural monopoly. It seems, however, that the existence and increasing applicability of technological alternatives in long distance communication that bypass the local loop will serve increasingly to limit any potential ability of local telephone companies to charge monopoly prices for connection of competitive long distance networks to local service. Other technologies not yet in place may even provide a competitive alternative to traditional local service in the long run.

In sum, claims of technological innovation and extension of service to out-of-the-way areas seem to be little related to market structure as created by regulatory agencies. Even the area where telecommunication regulation would appear to be of clearest necessity, that of local exchanges, is increasingly subject to competition. It is difficult, of course, to predict the technologically efficient way of meeting communications needs in the future, local or long distance, given the wide diversity of both technologies and users. The ability of even the best-intentioned regulators to sort through similar complexities has been historically quite limited. This sorting might well be accomplished as effectively by less — rather than more — regulation and by a freer market in telecommunications, a possibility explored more fully in the next chapter.

NOTES

1. Jesse J. Friedman, "Collective Ratemaking in Trucking: The Public-Interest Rationale" (Study commissioned by Central and Southern Motor Freight Tariff Association et al., Washington, D.C., October 1977).
2. Walter Y. Oi, *Economics of Truck Transportation* (Dubuque, Iowa: C. Brown Co., 1965); D. Daryl Wyckoff and David H. Maister, *The Owner-Operator Independent Trucker* (Lexington, Mass.: Lexington Books, 1975).
3. Mahlon Straszheim, *The International Airline Industry* (Washington, D.C.: The Brookings Institution, Transport Research Program, 1969); William Jordan, *Airline Regulation in America: Effects and Imperfections* (Baltimore: Johns Hopkins Press, 1970); George W. Douglas and James Miller, *Economic Regulation of Domestic Airline Transport: Theory and Policy* (Washington, D.C.: The Brookings Institution, 1974).
4. These issues concerning the inherent national monopoly characteristics of local service are explored in a later section of this chapter.
5. First Supplemental Response of the Bell System Companies on the FCC Docket 20003, April 8, 1976, p. 15.
6. Page 2 of a cover letter accompanying First Supplemental Response of the Bell System Companies in Docket 20003, from R. E. Sageman, AT&T Assistant Vice President for Federal Regulatory Matters, to Vincent J. Mullino, Secretary, FCC, dated April 8, 1976.
7. Report of the Federal Communications Commission on the Investigation of the Telephone Industry in the United States (Washington, D.C.: U.S. Government Printing Office, 1939).
8. John H. Ainsworth and Gansey R. Johnston, *A Discussion of Telephone Competition,* report issued by ʘhe Ohio Independent Telephone Association, Columbus, Ohio, February 1908.
9. Ibid.
10. Report of Harry B. Nichols, Assistant Engineer to Bureau of Franchises, New York City, November 21, 1906, as cited in Ainsworth and Johnston, *Discussion of Telephone Competition.*

11. Richard Gabel, "The Early Competitive Era in Telephone Communication, 1893–1920," *Law and Contemporary Problems* 34 (Spring 1969): 346–347.
12. Analysis of Issues and Findings in FCC Docket 20003, April 1977, Bell Exhibit 65A.
13. Joseph Schumpeter, *Capitalism, Socialism, and Democracy,* 3rd ed. (New York: Harper and Row, 1962).
14. Franklin M. Fisher and Peter Temin, "Returns to Scale in Research and Development: What Does the Schumpeterian Hypothesis Imply?" *Journal of Political Economy* 81 (January/February 1973): 56–70.
15. Kenneth J. Arrow, "Economic Welfare and the Allocation of Resources for Invention" in *The Rate and Direction of Inventive Activity,* National Bureau of Economic Research (Princeton: Princeton University Press, 1962), pp. 609–625.
16. Appropriability simply refers to the ability of the inventor to realize the financial rewards which result from the application of his invention or innovation.
17. Harold Demsetz, "Information and Efficiency: Another Viewpoint," *Journal of Law and Economics* 1 (1969).
18. Morton I. Kamien and Nancy L. Schwartz, "Market Structure, Elasticity of Demand, and Incentive to Invent," *Journal of Law and Economics* 13 (April 1970): 241–252. As price drops, elasticity of demand will determine the size of increased revenues possible from the introduction of a cost-saving innovation.
19. Kamien and Schwartz, "Market Structure."
20. Kamien and Schwartz, "Market Structure."
21. Morton I. Kamien and Nancy L. Schwartz, "On the Degree of Rivalry for Maximum Innovation Activity," *The Quarterly Journal of Economics* 90 (May 1976).
22. See notes 11–13, Chapter 1, for citations of some of the literature.
23. William J. Baumol, "Reasonable Rules for Rate Regulation: Plausible Policies for an Imperfect World," pp. 108–123, in *Prices: Issues and Theory, Practice and Public Policy,* edited by A. Phillips and O. Williamson (Philadelphia: University of Pennsylvania Press, 1967).
24. See also William J. Baumol and Alvin K. Klevorick, "Input Choices and Rate of Return Regulation: An Overview of the Discussion," *The Bell Journal of Economics and Management Science* 1 (Autumn 1970): 162–190.
25. Elizabeth E. Bailey and Roger D. Coleman, "The Effect of Lagged Regulation in an Averch-Johnson Model," *The Bell Journal of Economics and Management Science* 2 (Spring 1971): 278–292.
26. L. Johnson, "Behavior of the Firm Under Regulatory Constraint: A Reassessment," *American Economic Review* 63 (May 1973): 90–97.
27. Alvin K. Klevorick, "The Behavior of a Firm Subject to Stochastic Regulatory Review," *The Bell Journal of Economics and Management Science* 4 (Spring 1973): 57–88.
28. George Stigler, "Industrial Organization and Economic Progress," in *The State of the Social Sciences,* edited by L. D. White (Chicago: University of Chicago Press, 1956), pp. 269–282.
29. Almarin Phillips, "Concentration, Scale and Technological Change in Selected Manufacturing Industries 1899–1939," *Journal of Industrial Economics* 4 (June 1956): 179–193.
30. Kamien and Schwartz, "Market Structure," p. 22.
31. F. M. Scherer, "Firm Size, Market Structure, Opportunity, and the Output of Patented Inventions," *American Economic Review* 55 (December 1965): 1097–1125.
32. Kamien and Schwartz, "Market Structure," p. 20.
33. Almarin Phillips, "Patents, Potential Competition, and Technical Progress," *American Economic Review* 56 (May 1966): 301–310.

34. F. M. Scherer, "Market Structure and the Employment of Scientists and Engineers," *American Economic Review* 57 (June 1967): 524–531.
35. William Comanor, "Market Structure, Product Differentiation, and Industrial Research," *Quarterly Journal of Economics* 81 (November 1967): 639–657.
36. W. J. Adams, "Firm Size and Research Activity: France and the United States," *Quarterly Journal of Economics* 84 (August 1970): 386–409.
37. Steven Globerman, "Market Structure and R&D in Canadian Manufacturing Industries," *Quarterly Review of Economics and Business* 13, No. 2 (Summer 1973): 59–67.
38. R. W. Wilson, "The Effect of Technological Environment and Product Rivalry on R&D Effort and Licensing of Inventions," *The Review of Economics and Statistics* 59, No. 2 (May 1977): 171–178.
39. F. M. Scherer, *Industrial Market Structure and Economic Performance* (New York: Rand McNally, 1970), p. 374. See also Almarin Phillips, *Technology and Market Structure–A Study of the Aircraft Industry* (Lexington, Mass.: Lexington Books, 1971).
40. Comanor, "Market Structure," pp. 652–656.
41. Scherer, *Industrial Market Structure*, p. 377.
42. In April 1974, the FCC initiated Docket 20003 to obtain information and comments from all interested parties concerning economic impacts caused by these new participants in the provision of telecommunications equipment and services. In September 1976, the FCC issued its First Report on Docket 20003. Included, as Appendix B of that report, were nine volumes of analyses of the more than 30 respondents' filings in Docket 20003 prepared by Technology and Economics Inc.(T+ E) as a contractor to the FCC. The FCC cites these T+E analyses frequently in its First Report.
43. Project to Analyze Responses to Docket 20003, September 24, 1976, see Deliverable C, pp. 22–31, Deliverable D, pp. 26–42, and reviews of Additional Submissions, pp. 56–73.
44. Analysis of Issues and Findings in FCC Docket 20003, April 1977, Bell Exhibit 65A.
45. Ibid., p. 230.
46. Other common carriers, that is, the new entrants into long distance service.
47. Arthur D. Little, "The Relationship Between Market Structure and the Innovation Process," prepared for the American Telephone and Telegraph Company, January 1976, Federal Communications Commission, Docket No. 20003, Bell Exhibit 52.
48. Ibid., p. 23.
49. "In the Matter of Economic Implications and Interrelationships Arising from Policies and Practices Relating to Customer Interconnection, Jurisdictional Separations and Rate Structures," FCC Docket 20003, First Report, adopted August 20, 1976.
50. Dr. Walter Bolter, Senior Staff Economist at the Federal Communications Commission, comments on this phenomenon: "Technology, of course, may operate immediately and at cross purposes to otherwise well conceived regulatory schemes." In, *The FCC's Selection of a 'Proper' Costing Standard after 15 years– What Can We Learn from Docket 18128,"* in *Assessing New Pricing Concepts in Public Utilities,* Proceedings of the Institute of Public Utilities Ninth Annual Conference, Michigan State University Public Utilities Papers, 1978, pp. 333–372.
51. Poor regulatory decisions can also result in investment mistakes or distortions in markets not under regulation. Regulated industries are normally quite central to the operation of an economy (e.g., transportation, power generation and distribution, and communications). Individuals, households, and businesses consuming these important services will often make locational decisions on the basis of access to and costs of these services.

52. Paul W. MacAvoy and James Sloss, *Regulation of Transport Innovation: The ICC and Unit Coal Trains to the East Coast* (New York: Random House, 1967), p. 117.
53. Aaron J. Gellman, "Freight Surface Transportation," in *Technological Change in Regulated Industries,* edited by William M. Capron (Washington, D.C.: The Brookings Institutions 1971), pp. 166–196.
54. See "Southern Railway System: The Big John Investment," available through the Intercollegiate Case Clearing House, Harvard Graduate School of Business Administration, Boston, ICH # 9–677–244.
55. Almarin Phillips, "Air Transportation in the United States," in *Technological Change in Regulated Industries,* edited by William M. Capron (Washington, D.C.: The Brookings Institution, 1971), pp. 123–165.
56. James W. McKie and Stephen L. McDonald, "Petroleum Conservation in Theory and Practice," *Quarterly Journal of Economics* 76 (February 1962): 98–121.
57. Tom Alexander, "How Little Oil Hit a Gusher on Capitol Hill," *Fortune* (August 14, 1978): 148. Courtesy of Fortune Magazine; © 1978 Time Inc.
58. On this point see Robert M. Spann, "Rate of Return Regulation and Efficiency in Production: An Empirical Test of the Averch-Johnson Thesis," *The Bell Journal of Economics and Management Science* 5 (Spring 1974): 38–52. See also, Leon Courville, "Regulation and Efficiency in the Electric Utility Industry," *The Bell Journal of Economics and Management Science* 5 (Spring 1974): 53–74.
59. An illustrative case study of what might happen when an attempt is made to proscribe or retard technological change in telecommunications is presented in the final section of this chapter.
60. Statement by AT&T Senior Vice President Crosland made in hearings before a senate subcommittee. Recorded on page 4420 of *The Industrial Organization Act,* S 1167 (Part 6 — Communications), 1974. Presented as Bell Exhibit 2 in FCC Docket No. 20003, 1976.
61. Eugene V. Rostow, "The Great Telephone Debate: Competition and the Public Interest in the Telecommunications Industry," a paper delivered at the 1977 Iowa State Regulatory Conference, Iowa State University of Science and Technology, Ames, Iowa, May 18, 1977, p. 25.
62. W. J. Baumol, Otto Eckstein, and Alfred E. Kahn, "Competition and Monopoly in Telecommunications Services," privately published by AT&T, November 23, 1970, p. 8. Presented as Bell Exhibit 46 in FCC Docket No. 20003, 1976.
63. Richard Gabel, "Early Competitive Era in Telephone Communications." See also "REA Rural Telephone Program Toasts 25," *Telephony* (September 2, 1974): 79.
64. John Brooks, *Telephone: The First Hundred Years* (New York: Harper and Row, 1976), p. 110.
65. Ibid., p. 109. (See also Gabel, "Early Competitive Era in Telephone Communications.")
66. Brooks, *Telephone: The First Hundred Years,* p. 110.
67. An October 28, 1949, amendment to the Rural Electrification Act authorized the REA to make loans to improve and extend telephone service in rural areas. In authorizing the telephone loan program, Congress directed that it be conducted to assure the availability of adequate telephone service to the widest practicable number of rural users of such service.
68. Telephone loans are made to existing companies, to cooperative nonprofit limited-dividend or mutual associations, and to public bodies. The agency also provides engineering and management assistance to its borrowers. Loans are repaid from the operating revenues of the systems REA finances.
   In the loan guarantee program, funds are obtained by borrowers from non-REA sources under 100 percent guarantee by REA. Guaranteed loans bear interest at a rate agreed upon by the borrower and the lender and may be obtained from any legally organized lending agency qualified to make, hold, and service the loan.

69. This 23.7 million represents 32 percent of total U.S. households. See the U. S. Bureau of the Census, *Statistical Abstract of the United States,* 1978, Table No. 57, "Household Characteristics, by Race and Spanish Origin: 1977," p. 44.

70. See B. A. Hart et al., *Telephone Density and Telephone Company Area Data,* Office of Telecommunications, Department of Commerce, August 1973; and *The Dilemma of Telecommunications Policy,* Telecommunications Industry Task Force, submitted to Congress on December 1, 1977.

71. Rural Electrification Administration, *Annual Statistical Report, Rural Telephone Borrowers, 1976,* Table 5, and FCC, *Statistics of Communications Common Carriers, 1976,* Table 16.

72. Some determinants of present differences in relative toll settlement receipts, such as the size of the local calling area and the number of vacation homes, may be unaffected by competition.

73. Rural telephone companies could accomplish what the present settlements or separations process does by levying a surcharge on long distance calls, if they preferred that local charges be low relative to long distance charges (and if federal communications law were changed so as to permit such surcharges). This option would not be as viable for large users, such as businesses and motels, but to the extent that the larger users require local connections there is still a potential for retaining some revenues in excess of cost. These probabilities underscore the importance of rate flexibility, as proposed in Chapter 6.

74. It should be noted that the effects of competition until now have not been entirely negative. Rural customers have benefited from a greater diversity of services and from "cost reductions brought about by the entry of new suppliers, and by innovations in design." Victor Block, "REA on the Threshold of a New Era," *Telephony* 191 (September 2, 1974): 78.

75. Rural Electrification Administration, *Management Notes on Issues Affecting the REA Telephone Program,* July 1978, p. 1, Note 2a.

76. See *Telecommunications Reports,* April 10, 1978, pp. 1–4.

77. See *Telecommunications Reports,* May 1, 1978, pp. 3–7.

78. Even if this legislation is not enacted, the REA can make loans to rural cooperative and commercial telephone companies for the provision of broadband and CATV service. On May 22, 1979, the secretary of the U.S. Department of Agriculture transferred authority for administering Sections 306 and 310B of the Farmers' Home Administration Act to the REA; these sections provide for $15 million in available loans in 1979 for rural broadband/CATV development.

79. The categories identified as dedicated to toll in this analysis are tandem dial, interexchange circuit equipment, interexchange OSP, all other plant, commercial expense, and one-half of traffic expense and accounting expense. Tandem dial ($.35 billion) includes some local equipment and therefore the total for dedicated facilities is slightly overstated. Commercial, traffic, and accounting expense are allocated to toll on the basis of work units, not usage. Data on the 1976 interstate revenue requirement for these categories from FCC Docket 20981, "Impact of Customer Provision of Terminal Equipment on Jurisdictional Separations," Appendix H, December 6, 1978, by Paul Popenoe entitled, "A Break-Even Plan for Elimination of Terminal Equipment Assignment to Toll," Table 1, Sheet 4.

80. The ratio of aerial to underground wire averages 0.49 for the Bell companies in the sample and 1.41 for the independent companies. When the variable A/U is regressed against the total number of phones for each company, the coefficient of total phones is negative in both the Bell and independent groups. The coefficient for the independents is significant at the .05 level but the Bell group coefficient is significant at only the .20 level using two-tailed tests.

81. The empirical work that follows uses information from the FCC, *Statistics of Communications Common Carriers,* 1976. Being from a single source, this information affords a good comparison of Bell and independent companies because the data items have the same definitions for both groups, and because it is easily

accessible to others who might want to check our results. In addition the FCC presents a wide variety of statistics (over 200) regarding each of the companies it lists. Variables such as "number of toll calls," "number of local calls," and "miles of wire" are difficult to find elsewhere.

Nine companies were excluded from the FCC data base due to data deficiencies. They are All America Cables and Radio, Cuban-American T&T, ITT Virgin Islands, Navajo Communications, Nemont Telephone Cooperative, Offshore Telephone Company, RCA Alaska, Virgin Islands Telephone Corp., and Puerto Rico Telephone Company. The remaining 65 companies include all 24 Bell affiliates and most of the large- and medium-sized independents. This data base represents a large proportion of the total phones in the United States. Specifically, it contains all 127 million Bell phones, and nearly 50 percent of all independent phones nationwide.

82. A/U is not considered to be a product but is included to control for this aspect of the effect of density on cost. This variable was more significant statistically than when expressed as a logarithm.

83. William J. Baumol, "On the 'Proper' Cost Tests for Natural Monopoly in a Multiproduct Industry," *American Economic Review* 67 (December 1977): 809–822.

84. NARUC presents data for the Bell operating companies that divides total operating expenses between interstate and intrastate expenses. Since the latest year these data were available was 1975, the ratio of intrastate to total operating expense for 1975 was used in Equations 5–4 and 5–8 to adjust the 1976 operating expenses from the FCC data base. See NARUC, *Economic Data of Regulation – Telephones,* July 12, 1976, revised September 29, 1976, pp. 101–107.

85. One possible explanation for the negative coefficient on extensions is that this variable is correlated with the degree of "suburbanness" of the companies and that the costs are lowest in these medium-density suburban areas. Multicollinearity may also bias individual coefficients, as discussed below.

86. Some research indicates that the result of roughly constant average costs with respect to scale may hold for REA companies but that significant economies of *density* exist for these smaller companies. The larger REA companies have more than 8,000 phones and thus overlap with the size of companies analyzed above. See John Horning and Robert W. Wilson, "Multiproduct Cost Functions for Local Telephone Service," presented at the 7th Annual Telecommunications Policy Research Conference, April 30, 1979.

87. Calculations by Charles River Associates based on data in *Annual Statistical Volumes I and II of the United States Independent Telephone Association, 1976.* The companies reporting to this source represent approximately 96 percent of all independent telephones. Since all independent companies comprise approximately 18 percent of total industry phones, the excluded companies, which are presumably very small, account for roughly 1 percent of all telephones. In addition, of the 723 reporting companies, 519 were smaller than 8,000 telephones. The total number of telephones for these smaller reporting companies was approximately 1.6 million, or an additional 1 percent of the industry total. Thus, companies smaller than 8,000 phones accounted for roughly 2 percent of industry telephones in 1976.

88. The smaller companies, being in rural areas, may have only lines on poles that are easily accessible while the larger companies, usually operating in cities, have some lines buried in difficult-to-reach subterranean installations; or the smaller companies may not maintain the same standards that AT&T is generally felt to achieve.

89. Daryl Wycoff and David H. Maister, *The Owner-Operator Independent Trucker* (Lexington, Mass.: Lexington Books, 1975).

90. See Paul W. MacAvoy, *The Economic Effects of Regulation* (Cambridge, Mass.: MIT Press, 1965) pp. 129-135; George C. Eads, *The Local Airline Experiment*

(Washington, D.C.: The Brookings Institution, 1972); Ann Friedlander and Robert Simpson, *Alternate Scenarios for Federal Transportation Policy* (Prepared by Massachusetts Institute of Technology, Center of Transportation Studies for the Department of Transportation, January 1977); W. S. Moore, ed., *Conference on Regulatory Performance, Washington, D.C. 1975* (Washington, D.C.: American Enterprise Institute for Public Policy Research, 1976); James Miller and George W. Douglas, *Economic Regulation of Domestic Airline Transport: Theory and Policy* (Washington, D.C.: The Brookings Institution, 1974).

91. John R. Meyer et al., *The Economics of Competition in the Transportation Industries* (Cambridge, Mass.: Harvard University Press, 1959), p. 11.

92. For an example in power generation, see Tom Alexander, "The Little Engine That Scares ConEd," *Fortune* 98 (December 31, 1978): 80-84. For a general description about the financial services business, see Will R. Sparks, *Financial Competition and the Public Interest* (New York: Citicorp, 1978).

93. See Chapter 2.

94. Bruce R. Carruthers and George R. Mandanis, "New Alignments in Intercity Telecommunications Services," prepared for the National Telecommunications Information Administration by Systems Application, Inc., March 26, 1978.

95. See the discussion of entry as a limit on the power of a natural monopoly in Chapters 4 and 6.

96. *Xerox Corporation Petition for Rule Making,* before the Federal Communications Commission, November 16, 1978.

97. Douglas O. Reudink and Y. S. Yeh, "A Scanning Spot-Beam Satellite System," *The Bell System Technical Journal* (October 1977); The *Wall Street Journal*, December 20, 1977, p. 8.

98. Telecommunications Industry Task Force, *The Dilemma of Telecommunications Policy,* submitted to Congress September 21, 1977, Exhibit 8.

99. Ibid., Exhibits 5 and 8.

100. This discussion focuses on local loops. Some form of switching or routing of messages would also be needed.

101. Cellular systems divide an urban area into small cells that would average between 2 and 25 miles in diameter, approximating the coverage range for typical two-way radio transmission. This would allow the same frequency to be reused many times over in noncontiguous cells.

102. Report of the U.S. President's Task Force on Communications Policy, Appendix F.

103. A major attraction of optic fibers is that the primary raw material — sand — is abundant and cheap compared to copper. As of 1978, no large-scale commercial quantities have been produced, and work is still being done on the product itself and the production methods, and also on the other components of the optic fiber system. These other components (light sources, connectors, couplers and detectors) still present problems of unsatisfactory durability and reliability and inadequate technology for field work, but a large number of companies are at work on these matters. Still, it seems unlikely that the *installed* cost (including labor, right-of-way, etc.) per *cable* mile (as opposed to *circuit* mile) of optic fibers will be much lower than that of copper wire, although on a circuit-mile basis, its unit cost could be much lower.

104. Western Union, in its public documents, does not mention any specific amount for local loops (as opposed to intercity transmission) lease payments. The distinction Western Union draws is between TWX (for which it paid $37.3 million) and non-TWX service ($38.4 million), which both use intercity and local distribution facilities. (Western Union Corporation, 10-K Report, March 1978.)

Chapter 6

PUBLIC POLICY

REGULATION AND THE CONVENTIONAL WISDOM

Government regulation of economic activity is very often justified, at least superficially, by appeal to a lengthy series of noneconomic arguments or economic arguments of spurious validity. Telecommunication regulation, as documented in preceding chapters, is no exception to this rule.[1] Basically, all such arguments should be analyzed critically and placed in perspective. The real question in most instances is the value of the ostensible benefits of regulation relative to the costs. Are the benefits really of sufficient value to justify all the costs, direct and indirect, of regulation?

Given the relatively poor record of much regulation, the burden of proof should perhaps be as much on those who want regulation as vice versa. Specifically, government regulation in many cases has led to circumstances in which neither the producers nor the consumers have been well served — with railroads, public transit, natural gas, and airlines perhaps being the most outstanding regulatory failures.[2] Add to this the consideration that market competition has generally served the U.S. economy well — in terms of resource allocation, product innovation, diversification, and so on — and it would seem that the established norm for conducting U.S. enterprise should be the market, with regulation duly deemed to be an exception to be invoked only when absolutely necessary.

This principle of parsimony in applying regulation pervades or underpins much of the discussion of public policy in this chapter. In keeping with this goal of limiting regulation to its justifiable essentials, it will be argued that both freedom of entry into the industry and freedom of the telephone companies to respond competitively should be encouraged.

ENTRY

Unquestionably the single most important condition in any industry for ensuring that market competition works and prevents monopolistic abuses is freedom of entry.[3] Therefore, if minimal regulatory involvement is sought, it automatically follows that entry, or the potential for entry, should be vigorously encouraged. Regulation should not create monopoly: if a firm is to emerge as a monopolist, it should be through market processes, not through regulatory decisions. As three advisers to AT&T have aptly put it:

> Whether economies of scale are so great as to outweigh the benefits of competition is an empirical issue. In some circumstances, full and free competition can determine whether new firms will be attracted into the industry and can achieve costs that permit them to prosper alongside the former monopoly. Such a test can, of course, be effective only if entry is not predicated on the expectation that the newcomer will be assured by regulation of a protected portion of the market if his costs turn out to be too high.[4]

The implications of a policy of freedom of entry into telecommunications are reasonably clear.[5] First, it means that the present policy of permitting more or less open entry into intercity private and specialized communications, as well as the provision of customer premise equipment, should be continued. Second, it means that there should be open interconnection between the different networks that might evolve.[6] Most important, it means that open interconnection should be required between newly emerging intercity systems and local service networks of telephone companies. As shown in the preceding chapter, such a policy of open interconnection would seem to be not only in the interest of good public policy but of the telephone companies as well.[7] Without open interconnection, those prevented from interconnecting would have strong incentives to explore other alternatives. Among the possible alternatives would be greater use of radio transmission for local service, incorporation of data or voice communication into existing cable TV systems, the development of entirely new local systems using high-capacity optic fibers or coaxial cables, relocation of data-generating activities into close proximity clusters served by a common satellite antenna, and so forth.[8]

Since the alternatives are so numerous, public policy should even consider allowing open entry into local services. Even though such entry, if it occurred, might create duplication or excess capacity in a few instances, establishing a *potential* for entry may be crucial to the development of telecommunications in an efficient and flexible pattern. Without this possibility, distortions and efficiency losses could occur, if less than optimal alternatives are sought in circumvention of any lingering restrictions on interconnection. Furthermore, if entry into new or additional local service is uneconomic, this result would emerge from market competition.[9]

The principle of freedom of entry into local services should apply, moreover, not only to potential telephone company competitors but also to the telephone companies themselves, including their potential entry into provision of cable TV and other broadband local services. Only by permitting freedom of entry in both directions — by others entering into competition with telephone companies and the telephone companies entering into competition with more specialized local services — can a true test of economies of scale or the potential of natural monopoly in local services be determined. The only limitation that public policy *might* usefully entertain on this bidirectional freedom of entry would be based on classic infant industry considerations, namely, that independent cable companies should be spared the "full weight" of telephone company competition until such time as these independents have established themselves financially and in market strength. This, however, is at best an argument for delay in permitting telephone company entry into cable TV and not for a permanent prohibition.[10] There is always danger that an infant industry can have an extremely long-lived adolescence!

The regulator's role, moreover, should not be — indeed never can be — to prevent all foolish investment. Bad investments have been and are made in most economies, whether free enterprise or socialist. Investments are undertaken on the basis of projections about the future, and the future is seldom perfectly foreseeable. In addition, regulators have already proven to be no more endowed as "seers" than are private entrepreneurs. No special public policy need seems to exist for denying entrepreneurs the opportunity to make mistakes in telecommunications, just as they can and will do in other industries. All that must be remembered is that freedom to fail is as much an integral part of a free market system as the freedom to succeed; if the market proves unfavorable to some of these endeavors, they should be permitted to disappear or to be reorganized as appropriate.

The only seemingly plausible argument for some long-term[11] continuation of entry controls in telecommunications is embodied in what has been termed the "sustainability" hypothesis in the economics literature.[12] This hypothesis refers to a situation in which a natural monopoly, even if it is the most efficient way to supply a group of products or services given the existing technological options, might not be able to maintain (i.e., "sustain") provision of all these products or services if freedom of entry or competition is allowed.[13] A particular form of this argument is the concern that competition may create difficulties for the established telephone companies if they are required to be carriers of last resort. Specifically, new entrants are said to engage in "cream skimming" (i.e., competing only in the provision of the most lucrative products and services), leaving the regulated firms with all the less lucrative business (in the particular case of telecommunications, this is widely believed to be local exchange service). It has been alleged, therefore, that entry threatens the survival of the telephone companies both as carriers of last resort and in local service.[14]

The validity of this concern depends on empirical propositions. First, for sustainability to be an issue in the multiproduct case relevant to telephone services, both the monopolist and the potential entrant must have fairly pronounced scale economies in individual services (since otherwise there would be little loss from the existence of multiple suppliers and little incentive to increase market share for the purpose of achieving lower costs). Second, consumers of different telephone industry services must view these services as substitutes to some extent. Third, if there are joint economies of production, but these economies are only modest or slight, the problems with sustaining the full set of services are increased.[15,16]

The available empirical evidence, although limited, would not suggest that sustainability should be a major public policy concern. The cost functions investigated in Chapters 4 and 5 were characterized by relatively modest economies of scale in long distance service and negligible economies of scale in local service. On the basis of this evidence, local services are likely to be sustainable since, as discussed above, the existence of multiple suppliers of long distance services will not lead to significant cost increases in local or long distance service.[17] Furthermore, as the analyses of Chapters 2 and 3 indicate, local service would seem to be sustainable without a major realignment of the rate structure; the most that would be required would be slightly more upward flexibility in local rates to keep pace with general inflation. Concern about the impact of entry on continued provision of local services, particularly residential service, thus appears misplaced.

Evidence on the cross-elasticities of demand for various telephone services, as noted in Chapter 3, is virtually nonexistent. Nothing in the available evidence, however, suggests that consumers view local service and long distance service as significant substitutes. If anything, these services are probably complementary, and thus the likelihood of local sustainability would be enhanced rather than reduced if new long lines competition expanded the demand for long distance services. On the other hand, some considerable portion of long distance message toll service (MTS) might be vulnerable to various kinds of wide area (WATS) and private line competition, and therefore the sustainability of MTS service might be jeopardized by selective entry into WATS and private line service. Whether this would entail serious economic losses, however, seems questionable because any erosion of MTS volume would occur only slowly and should be substantially offset by general growth in this rapidly expanding business.[18]

In general, accurate assessment of the extent of any threat of competition and alleged cream skimming to the continued provision of the full set of telephone services now offered depends upon knowing a great deal more about demand and cost functions than is presently known, or is likely to be known without considerably more market experience than is now available. Too much of the requisite information needed to estimate the underlying functional relationships lies outside present or historically observable experience. Furthermore, these relationships can be expected to change as technology or markets change; thus a set of telephone products and services that might not be sustainable at one point in time might become sustainable (or vice versa) as circumstances changed.

The use of entry control or prohibition as a means of meeting an unsustainability problem has its own costs and risks as well. For example, it means removing many of the incentives for cost discipline and, possibly, innovation as well. Most prospective entrants make claims to introducing new products or services; regulators are thus confronted with the task of sorting legitimate innovators from "simple" low-cost intruders when deciding whether to permit entry. Entry prohibitions can also protect an inefficient monopoly. Under certain commonly accepted pricing rules used by regulators, entry prohibitions may even allow a monopolist to usurp and dominate competitive markets in which the monopolist has no efficiency advantage.[19] Futhermore, if the monopolist is protected from entry in the basic franchised market, the monopolist may under certain conditions have incentives to move away from cost-minimizing input combinations.[20] In short, entry protection can perpetuate and extend the scope of an inefficient joint monopoly and make this monopoly even less efficient than it might otherwise be. Finally, entry can sometimes benefit consumers, at least in the short run, even if sustainability proves to be a long-run problem.[21] This will be achieved, however, at the expense of producers; and whether this is a desirable transfer from the standpoint of public policy is debatable.

Moreover, methods other than entry restriction can be identified for handling the public policy problems that might be created by nonsustainability. In essence, a lack of sustainability means that market demand is insufficient to continue production of some service or product while charging only a single price for each service. One alternative to entry restrictions to perpetuate or sustain production would be to permit more price or rate discriminations in order to fill the revenue void. Indeed, if sustainability is a potential problem, the case is generally strengthened for relaxing rate regulation. Besides the possibility of capturing needed additional revenues through price discrimination or other rate changes, flexibility in rate making will allow the monopolist to respond to threats of entry, a capability that can greatly attenuate the chances of nonsustainability.[22] Another possible solution to sustainability problems would be to use government subsidies of some form if the threatened activity were deemed worthy.

It must also be recognized that the sustainability argument is typically static in its assumptions and characterizations. It is possible that if important services were withdrawn because of sustainability problems, technology and enterprise might then find ways to fill the gap. In this connection, freedom of entry could be helpful in creating circumstances in which the necessary technological and new product explorations would be undertaken, that is, in providing some of the incentives necessary to create dynamic solutions to potential sustainability problems.

Any financial difficulty caused by a sustainability problem, moreover, is likely to appear only slowly in telecommunications (as the financial projections in Chapter 2 strongly suggest). Accordingly, if sustainability were to prove a problem in telecommunications, there would almost surely be time to consider various public policy options, including the possibility of restoring some restrictions on entry if this were deemed the best way to meet any such difficulty when and if it appeared.

The first step, then, toward a regime of minimal regulatory involvement in telecommunications is to cast away as many restraints on entry as feasible as quickly as possible. This step is the basic prerequisite to unleashing the competitive forces at work in this industry. The strength of these forces is well attested by the rapid growth and technological change of the industry — to the point of al-

ready undermining and frustrati?
a decision is made to continue f
should not be required so long
oriented to consumers rather '
regulation is the goal, with
forces, freedom of entry is a'

their operations a
markets.[24]
Applying the
unless necessa
rate regulati
service. Si
message
overhea
sector
is th
by

RATE REGULAT'

It would be comforti..
could be asserted with great confide..
communications would eliminate any fur..
in the telephone industry. Although freedom ..
such a situation in the not too distant future, the tra..
present to the future is likely to be troublesome. Regulation,
ously noted at many points, creates its own distortions and veste..
terests, and for this reason its removal or reduction almost inevitably
involves transition problems of some complexity. The solution of these
transition problems, created in large measure by the failures of past
regulation, often requires in turn (and quite ironically) some continued regulation. Local exchange service, in particular, will retain
many of the characteristics of natural monopoly for at least a few
years to come and thus will be subject in all likelihood to some continued rate regulation.[23] Allocation of the radio spectrum will also continue to be a major government concern.

In short, a combination of transition problems, a residue of some
continuing natural monopoly in telecommunications, and political
considerations will lead to a continuation of rate regulation in the industry for at least the short to medium term. The essential policy
question thus becomes how to channel this regulation to make it as
constructive as possible and, in particular, capable of flexible adaptation to new and more competitive circumstances as they emerge.

In designing any continued rate regulation for local service, the
goal should be to create a framework flexible enough to permit a restructuring of local rates along the lines suggested in Chapter 3. Specifically, local service charges should become more usage-sensitive
and vary more with the peak loadings on the system. The essential
public policy problem is how to develop experimentation with the new
rate structures while still protecting consumers from indiscriminately
large and unexpected changes in local rates. Some control on maximum rates for local service, where telephone companies have a
monopoly franchise, might also be deemed desirable to lower
the probabilities that the telephone companies inefficiently extend

...d monopoly powers into otherwise competitive

...principle of regulatory forbearance (i.e., no regulation
...ry), there would seem to be no reason for any minimum
...on in most telephone service and certainly not in local
...ce telephone companies themselves insist that local
...service is limited in its contribution to general corporate
...ds and joint costs, the point is perhaps moot. The only local
...wherein minimum rate regulation might arise as a policy issue
...e provision of cable TV and similar services if they were offered
...the telephone companies;[25] specifically, the telephone companies
...ight be able to practice predatory pricing in providing cable services.
However, as explained later in this chapter, the potential social harm
from predatory pricing can be easily overestimated. Furthermore, independent
cable TV companies would now seem to be, or shortly will
be, sufficiently well established and financed to no longer need any
infant industry protection from predatory pricing, although some exceptions
to this might be identifiable.[26]

The essential question is thus what kind of maximum rate regulation
might make sense in local service. Again the principle of forbearance
seems well advised. Since substantial restructuring and
experimentation seem desirable in local charges, maximum rate regulation
might be usefully limited to basic monthly connect charges and
off-peak usage charges. By contrast, a very strong case can be made
for elimination of virtually all maximum rate regulation on usage
charges during peak periods and especially during the normal business
hours of nine to five.[27]

The only objection to elimination of all maximum rate regulation
during business hours is the potential hardship this might inflict on
small businesses, for whom alternatives to the telephone companies'
local loops do not now exist. A few residential users might be unduly
affected by this change as well. Accordingly, some upper limitation
might be placed on the size of upward revisions in local peak period
usage charges permitted in a given time period, say one year, with
increases within this limit being automatic and not subject to regulatory
review.

Another approach to inhibiting unduly harmful upward revisions of
peak period local service charges would be to insist upon horizontal
equity. Specifically, discriminatory or differential charges in rates
among users more or less similarly placed (especially with regard to
costs of rendering service) would not be accepted by regulators if regulation
persisted or by law in the absence of regulation. In this way, the
potential for entry or the use of alternatives to the local loop by large

businesses would discipline rates charged smaller businesses and residences as well. In general, application of the horizontal equity principle is a means of avoiding predatory pricing as well as helping achieve equity goals (e.g., treating all sizes and types of consumers more or less equally).[28]

The principles of capped or time-phased maximum rate regulation outlined above could be applied to off-peak and monthly connect charges as well. If applied in these cases, however, some rollback or reduction in these rates might be well advised in terms of achieving many generally accepted policy goals, as discussed in Chapter 3. The early stages of a transition from the present structure of local telephone rates to a more desirable structure (i.e., a structure that had lower connect charges and relied more on usage charges) might, moreover, involve regulators in trading reductions in off-peak and connect charges for increases in peak period charges.

All of the preceding discussion subsumes a basic change in regulatory thinking, specifically that regulators will move away from heavy reliance on high fixed monthly charges to finance local services and toward a regime of lower fixed monthly charges coupled with increased revenues from local usage charges. The politics of any such transition may be very difficult. An economically rational approach to local service rate making depends, however, on making this change. Accordingly, its promotion and dissemination should be a principal goal of federal communications policy, always recognizing that in the final analysis rationalization of local rate structures will require the cooperation of many local and state regulatory agencies.

In long distance message toll service (MTS), somewhat less need for regulatory involvement is identifiable than in local service. Nevertheless, MTS for residential, small business, and remote users, as well as incoming WATS, may not be as susceptible to incipient competition in intercity communications as private lines and outgoing WATS services or most MTS originating with bigger businesses. With several years of open entry into intercity service, potential competition in MTS and incoming WATS should eventually be sufficient to discipline any excessive monopoly instincts on the part of telephone companies. However, because of the fears of some consumers, some time-phased maximum rate regulation might be appropriate for a few years.

In summary, rate policy should proceed along the following guidelines:

1. Minimum rate regulation should be largely or completely abolished.[29]
2. Maximum rate regulation should be greatly relaxed and made

more flexible by gearing it to a zone of reasonableness within which automatic increases in rates can be made, with a cap geared loosely to general price changes in the economy.

3. Maximum rate regulation can and should be phased out in many currently regulated sectors of telecommunications in the very near future, especially if open entry is permitted.

With the application of these principles to rate regulation in the industry, the scene would be set for a substantial restructuring of rates to better achieve the goals of efficiency, simplicity, and acceptance generally trumpeted as being the objectives of rate regulation.[30] In particular, the telephone companies would be given the tools to respond more flexibly and creatively to the new competition that is now pervading the field. If stifled in these efforts by continuation of conventional rate regulation, no one will be well served — except possibly some small segment of the population and some small companies of the future industry that could not survive without such protection.

PREDATORY PRICING AND UNECONOMIC ENTRY

The policy issues of predatory pricing and uneconomic entry are inseparably linked. Under value-of-service rate making, as discussed elsewhere in this book, rates may far exceed costs on some services, thereby attracting new entry. It is natural for the regulated firm, faced with a loss of business to new entrants, to attempt to lower its rates. It is equally natural for the new entrants to make an attempt to forestall rate reductions by the established regulated firm. Allegations of predatory pricing, especially through the vehicle of rate hearings, are one device for making such an effort.

The predatory pricing issue can arise in particularly acute form when the same firm serves different markets, some of which are not subject to competition. In such circumstances the possibility exists that "monopoly profits" from the protected markets might be used to price services rendered in competitive sectors below cost, thereby deterring entry by others into these potentially competitive sectors. Even worse, a franchised monopoly firm might choose an inefficient joint production process if this process had low incremental costs for the service in the potentially competitive market, recouping its higher costs in the franchised monopoly market by charging higher prices there.[31] In this case, the franchised monopoly's price in the competitive market would meet an incremental cost test while its overall rate of return would still withstand regulatory scrutiny. A critical assumption in this analysis is that entry by others is not allowed into the market where the franchised monopoly has unduly high prices; otherwise, the high prices (based on unnecessarily high costs) would attract entry by other firms.[32]

Conversely, it is also possible that joint production by one firm might be the lowest total-cost way to serve several different markets and that prices above the entry (stand-alone) costs of a potential competitor are needed in some of the markets in order to "sustain" the low-cost joint producer; that is, without some ability to price above stand-alone costs in some markets, the efficient single joint producer may be forced to withdraw from supplying some or all of the relevant markets.[33]

Because of these and other complications, considerable controversy exists among economists and lawyers about the proper definition of predatory pricing. The definition, however, that seems most widely accepted is that a price is potentially predatory if it is set below the firm's own short-run marginal cost.[34] Certainly, most observers would consider such a price to be predatory if it is also below average costs in most (though not necessarily all) situations;[35] in addition, some would also consider prices below short-run average variable or long-run marginal costs as predatory under certain circumstances.[36]

Whatever definition is used, a regulator attempting to prevent predatory pricing inevitably will need a cost-accounting scheme that categorizes costs according to the service or product that generated them. Such a scheme raises a number of difficult practical questions, especially when some costs may in fact be truly joint or common to a number of services or products.

The telephone industry provides a particularly striking example of the difficulties of properly attributing costs to different services and products under joint production circumstances. Under current accounting procedures, AT&T classifies about 43 percent of its costs as joint and common costs. In addition, it offers services in at least three different kinds of competitive environments. In some services, such as local exchange service, the local operating company is a regulated monopolist, and competitive entry into the provision of these services is formally prohibited. In others, such as private line services, there is regulation, but some entry is permitted. And in still others, such as customer premise equipment, entry is so easy that competition, not regulation, is (or shortly will be) the primary determinant of prices.

If costs are not accounted for with extreme care, and perhaps even if they are, it may be extremely difficult or even impossible to determine whether a given rate is below cost or not. To start, some costs may simply not be allocable to a particular activity or any economically rational basis. In other cases, costs may in fact vary with output of a service but, for reasons such as convenience or lack of analytical capability, are lumped in with overhead costs, and calculation of the marginal costs or average variable cost for this service will then be understated. It is therefore significant that the uniform system of accounts historically used by the telephone industry is inadequate to determine marginal or average variable costs, as well as not being sufficiently detailed to determine the costs of individual products or services. The current proceeding to revise this accounting system may lead to a system that remedies these deficiencies.[37] It is beyond the scope of this book to discuss how the system of accounts ought to be restructured, or even to discuss the conceptual and practical difficulties involved in properly assigning costs to individual products and services within a diversified firm engaged in joint production of some services. Additionally, though a scheme that properly attributed all attributable costs is certainly possible in principle, many costs could still be unassigned because they are in truth joint or common, and any allocation must be essentially arbitrary. In practice, furthermore, the expense of fully assigning all assignable costs may simply be prohibitive or of insufficient benefit to justify the effort.

Moreover, the benefits to consumers of preventing predatory pricing can be easily exaggerated. Predatory pricing means, in essence, that someone is providing services at a price that does not cover their costs of production. From the standpoint of a public policy oriented to consumer interests, this form of "philanthropy" would not normally be a matter of major concern, at least so long as regulation kept prices within proper bounds in the franchised monopoly that provided the base for any act of predation. From an economic standpoint, it mainly becomes a concern only if the "predator," after driving others from the field, is able to increase prices to or close to monopoly levels and thereby extort undue profits from the consuming public; or if the predator extends this monopoly into markets where the predator is not the least-cost supplier and is able, somehow, to recoup this inefficiency from charges assessed to the consuming public. The only other anxieties that might engage public policy in such circumstances would be a concern that the stockholders or employees of the predator would be unfairly limited in their factor rewards if too much of this so-called philanthropy were indulged or if the predator used the threat of unrealistically low prices to unduly restrict innovation and experimentation by new small firms.

When phrased in this way, predatory behavior that is harmful from a public policy standpoint requires an ability on the part of the predator to indulge in monopoly practices if and as the predator prevents others from participating in the market, or in other markets where the predator is protected from competition. Among other considerations, this means that firms prevented from entering the market by predatory pricing are unavailable to pose the threat of reentry after the predatory pricing has taken place. It also means that no other simple policy alternatives exist to prevent "monopoly extortion" from being practiced after the predator has driven others from the field, or in other markets where this predator has a protected position.

Simple remedies can be identified, however, that would diminish the profitability or other attractions of predatory pricing behavior. In regulated industries, unless Averch-Johnson incentives are at work, all that is normally needed is some regulatory inhibition on raising rates in other (i.e., protected) markets. Since virtually all rate regulation involves some form of maximum rate regulation, this power is widely vested in regulatory agencies concerned with so-called natural monopolies. Indeed, the possibility that predation might be used to extend a monopolist's domain beyond its efficient limits is one of the stronger arguments for retaining some authority for *maximum* rate regulation, at least until competitive entry becomes a well-established possibility.

Simple augmentation of the rules might be imagined, too, by which a regulatory commission could further prevent predatory pricing. For example, if a new rate is declared potentially predatory in character, the regulatory agency might insist that under no circumstances (except possibly substantial cost inflation) could a further increase be considered until a specified period of time has elapsed, say three to five years or so. Furthermore, even after this term of prohibition, the regulatory agency might insist that such rates be subject to full hearings and traditional regulatory review and approval before implementing any increase. This rule is not free of difficulties, for it might make the established firm overly cautious about reducing rates, especially during an inflationary period, but it is an example of the kinds of changes in the rules that might deter below-cost pricing.

Another and more general inhibition on predatory pricing exists in the body of antitrust laws. Specifically, treble damages are attainable under such laws for harm inflicted through proven predatory pricing. A major objection to the effectiveness of this antitrust remedy is the long time and high cost often required to litigate such suits. However, it is perhaps also easy to underestimate the deterrence effect of these laws; a few substantial findings of damages, when trebled, may inflict substantial penalties on even large firms.[38] Furthermore, the mere bringing of such a suit might hurt the defendant in the financial markets because stock prices often promptly reflect the uncertainties created by such suits.

Another reason for skepticism about the dangers of predatory pricing, especially as practiced by private firms, is that it will usually conflict with profit seeking.[39] It is, of course, conceivable that in modern corporations, where management and ownership are somewhat divorced, nonprofit incentives may operate[40] — perhaps even to the point of being strong enough to give rise to behavior in which certain services are deliberately rendered at prices below cost in order to achieve other goals such as maximization of gross revenues or general size of the corporation. However, it cannot automatically be concluded that such behavior should be a matter for public policy concern. The public policy question, once more, is balancing the potential advantages to consumers from these lower rates against the losses to others, for example, diminished returns to shareholders or employees or an "overconcentration" of economic power in the private sector.

Even if managerial aggrandizement or other nonprofit motives do enter into corporate decision making and thereby create possibilities for predatory pricing, a private enterprise faces clear-cut limits to how much such indulgence it can practice. Specifically, it cannot extinguish all profits or go beyond a profit level that prevents raising the capital needed to pursue the firm's expansion goals.[41] In a private enterprise economy, moreover, too much nonprofit-oriented behavior could well raise the specter of a takeover or other kinds of stockholder rebellion. Even for a firm as large and diversified in its ownership as AT&T, for which a takeover may be an unlikely threat, depressed stock prices and institutional disfavor are certainly possible. In short, pursuit of nonprofit goals in a private enterprise is limited to levels that maintain enough profitability to keep shareholders reasonably happy and contact with financial markets sufficiently fluid and viable to maintain requisite fund flows.[42]

The danger of predatory pricing in telecommunications can also be easily exaggerated because the arena for potential new competition or entry is where the established telephone companies now seem to enjoy their highest markups and profits, whereas the activities most akin to a true natural monopoly (or least vulnerable for the moment to new entry) are seemingly the least profitable. As one consultant to AT&T put it in a paper reviewing policy options in telecommunications:

> There must be something wrong, they [the potential new entrants] say, with allowing the vast telephone network to reduce prices so drastically in a competitive area when it makes so much money in areas where there is no competition. Since the field in which the general service telephone companies have a true monopoly position — that of providing residential services — is also their least profitable business, there is irony in the argument.[43]

There exist, in short, definite limits to the resources available for private telephone companies to indulge in predatory pricing activities.

This will be especially true if some rate regulation is retained over local residential services, as there is every expectation that it will be, and if freer entry is permitted and eventually becomes a technological possibility in local services, as again might be expected if current technological trends continue.[44]

The only residual public policy danger in predatory pricing in telecommunications, it would seem, is that the mere threat of predatory pricing by a very large firm, whether pursued as a conscious policy or unconsciously, might be sufficient to deter entry by very small, technologically innovative firms with finances that are miniscule in comparison. The solution to this problem may reside as much in public policies to encourage the risk of venture capital in general as in pursuing restrictive regulatory policies in a particular industry. Indeed, policies to promote the supply of general venture capital (e.g., special capital gains tax treatment, low-cost loans through government guarantees such as loans now provided by Small Business Investment Corporations or the Small Business Administration, or expansionary monetary policies),[45] should be as effective in telecommunications as in other high technology industries. While these policies clearly have their limitations,[46] as a means of promoting innovation they would seem to have fewer drawbacks than developing too much deliberate regulatory protection for small firms, although (as noted above) some infant industry justifications for regulation in the short run might be justified.

Although the potential harm done by predatory prices can easily be overstated, regulatory acts to prevent predatory pricing can often simply become purely regulatory protection of the inefficient or the unneeded. One of the great virtues of a competitive market is that it weeds out the inefficient. The history of regulation does not suggest that regulators have either the knowledge or the will to judge questions of who should fail or who should succeed as well as the market; regulatory decisions on these matters are too often made as much on political grounds as on economic realities.

In addition, it must always be remembered that regulatory restrictions placed on rate reductions by natural monopolists, frequently in the name of preventing predatory pricing, *may* result in a truly efficient or legitimate natural monopoly being unable to sustain all of its other offerings. While the sustainability argument can, like predatory pricing, be overdrawn as a public policy issue, it cannot be totally ignored. Furthermore, it can easily be shown that freedom for a monopolist to adjust rates when confronted with new competition, or its threat, is one of the best ways of enhancing sustainability (actually, whether legitimate or illegitimate).[47] In short, there are dangers in adopting either too lax or too strict a policy with regard to predatory pricing. An obvious compromise is to be permissive about price reductions, but then to use the regulatory framework, even in a residual or truncated form, to inhibit the grosser possibilities of exploiting predation.

If one assumes that the potential harm from undue concern about predatory pricing is likely to exceed the benefits in most regulated industries, it also follows that most suggestions for restructuring a regulated industry to separate so-called competitive activities from the noncompetitive are likely to have limited appeal. The Ad Hoc Committee for Competitive Telecommunications (ACCT), for example, has advocated such a separation or restructuring to achieve objectives that have been summarized as follows:[48]

1. To permit full and fair competition to exist, except in those sectors of the telecommunications market where public interest clearly requires monopoly services.
2. To reduce heavy involvement and increase the efficacy of the FCC and the operations of the competitive market.
3. To minimize the incentives and ability of the major monopoly carriers to cross-subsidize their competitive services.
4. To maximize the visibility of the costs of providing competitive services.

In connection with achieving these objectives, the ACCT has proposed two alternative strategies: (1) creation of a wholly owned telephone company subsidiary that would be essentially involved in marketing and managing the so-called interstate competitive services, such as private lines, data communications, video, and private network services; or (2) creation of a fully spunoff independent company to undertake these same services and for which the stock would be distributed on a pro rata basis to the current shareholders with the new company having a totally independent board of directors, and so on. It is clear, on even cursory reflection, that the independent new company is more likely to achieve the stated goals of ACCT policy than a wholly owned subsidiary.[49] In large measure, this whole approach can be considered a structural alternative to regulation as a solution to the so-called predatory pricing problem, the elimination of which is clearly the crux of the four stated goals of ACCT policy summarized above.

Although such structural reorganization might lessen the chances of socially unattractive predatory pricing behavior being indulged by the telephone companies, it is not clear that such structural solutions could be any more effective than the relatively minor regulatory reforms suggested above. It may also be true, as some have suggested, that such restructuring would actually be beneficial to stockholders of

AT&T or other telephone operating companies by unleashing within the companies latent marketing abilities that may have been suppressed, probably unintentionally, through the more monolithic corporate structures that now exist.[50] Since some feel that vertical integration provides incentives to undertake more basic research, a restructuring that divested research laboratories from their parent companies might also lead to a reduction in basic research in telephones and related electronics. Structural reform would also seem to be an unwieldy and politically complicated means of achieving a relatively simple end — elimination of predatory pricing, a goal with only limited public policy benefits in any event. On the other hand, if predatory pricing were really viewed as a major concern, for political if not for economic reasons, restructuring is certainly not inconsistent with achieving these goals.[51] However, this would appear to be a political judgment, not readily amenable to economic analysis and therefore beyond the scope of this book.

In sum, the public policy stance should be one of skepticism or at least limited "nervousness." As is often the case with regulatory "devils," they should be exorcised only when and if they appear, and not prematurely or in anticipation.

TOWARD A MORE RATIONAL DOMESTIC TELECOMMUNICATIONS POLICY: SUMMARY

Communication policy is currently in a state of considerable flux. FCC hearings and court cases on entry into telecommunications are common. Various bills are before Congress for modifying communications regulation. Is this legislation and litigation needed? If not, is the status quo totally acceptable? Or should public policy explore still other and far different possibilities from those thus far proposed?

The analysis in this book strongly suggests that the generally proper direction for regulation is that toward which the FCC, the courts, Congress, and other government agencies have been evolving during the past few years. Specifically, policy emphasis should be on promoting rate and entry experimentation and avoiding premature panic about the consequences. Most of the harmful effects attributed to these experiments are hypothetical, conjectural, and hotly disputed — largely because they concern future developments in technology and market demands that are uncertain. The telephone system today is, moreover, in sound financial health, and there is no indication that any prospective entry would undermine that health precipitously or without warning.[52] Accordingly, if a more openly competitive environment were to undermine the continued provision of telecommunications services or begin to cast doubt on the financial viability of the telephone companies, corrective action could be taken if and when this hazard materialized. To identify the dangers of competition in advance of the facts, especially in an industry undergoing as much technological and market change as telecommunications, would require a prescience denied to most individuals, not excluding regulators. Furthermore, experiments with more competition can create new markets and induce new demands that might leave everyone, including telephone companies, better off than they would be without competitive challenges.[53]

In short, this is an industry in a state of change, with many attendant uncertainties best resolved by proceeding with the market tests that have been started during the past decade. So far, the industry has not crumbled because of these experiments; indeed, there are many reasons to believe that it may have been strengthened by the challenge. Many ways can be identified, moreover, to remedy any adverse developments if, by chance, they do occur. In fact, if regulatory policy incorporates the basic changes suggested above — permitting free entry into the industry, granting the telephone companies freedom to meet the competition through rate reductions (when justified by costs), and freeing the participants from rigid rate regulation — then there is considerable reason to believe that the market test will end up where free markets usually do, with the most efficient producers, and therefore consumers as well, all benefiting.

NOTES

1. For example, the potential impact of competition on costs due to loss of economies of scale is discussed in Chapter 4. Potential impacts of competition on local service rates and the financial performance of AT&T are analyzed in Chapter 2. The allegations that competition could result in reduced availability of service in rural areas and a reduced rate of technological innovation are evaluated in Chapter 5.
2. See Paul W. MacAvoy, *The Economic Effects of Regulation* (Cambridge, Mass.: MIT Press, 1965) pp. 129–135; George C. Eads, *The Local Airline Experiment* (Washington, D.C.: The Brookings Institution, 1972); Ann F. Friedlaender, *The Dilemma of Freight Transport Regulation* (Washington, D.C.: The Brookings Institution, 1969); James Miller and George W. Douglas, *Economic Regulation of Domestic Airline Transport: Theory and Policy,* (Washington, D.C.: The Brookings Institution, 1974); and William Jordan, *Airline Regulation in America: Effects and Imperfections* (Baltimore: Johns Hopkins Press, 1970).
3. Joe S. Bain, *Barriers to New Competition* (Cambridge, Mass.: Harvard University Press, 1956).
4. W. J. Baumol, Otto Eckstein, and Alfred Kahn, "Competition and Monopoly in Telecommunication Services," report prepared for AT&T (reprinted by AT&T, November 23, 1970). For further discussion see Yale Brozen, "Competition and Antitrust,"

in *The Competitive Economy,* edited by Yale Brozen (N. J.: General Learning Press, 1975), pp. 6–15; and George J. Stigler, "A Theory of Oligopoly," *Journal of Political Economy* 72 (1964). This observation relates as well to the issues of predatory pricing discussed later in this chapter.

5. The principle of freedom of entry should also apply to removing present restrictions on the telephone companies engaging in various activities, including those related to but not generally classified as telecommunications, such as offering computer services.

6. This is subject, of course, to the usual safeguards for safety and "integrity of the network."

7. Open interconnection does not imply connection at a zero price. The appropriate access charge is currently a topic of both practical and theoretical interest. For a preliminary analysis of the theoretical issues, see Robert D. Willig, "A Theoretical Analysis of Network Access Pricing," paper delivered at the Tenth Annual Conference on Current Issues in Public Utility Regulation, Williamsburg, Virginia, December 11, 1978.

8. Chapter 5 contains a discussion of some of these possibilities.

9. Some external diseconomies might arise, of course, if a proliferation of local services led, for example, to too many telephone poles or dug-up streets. It would be far better to deal with these problems directly through explicit local regulations or ordinances (e.g., setting quality specifications or land-use codes) than indirectly through entry limitation.

10. It should be noted that this infant industry argument is subject to much the same limitations as arguments for protecting infant industries in international trade. For further discussion of this argument, see F. W. Taussig, "Wages and Prices in International Trade," *Quarterly Journal of Economics* 20 (August 1906): 497–522; or more recently, Richard Caves, *World Trade and Payments: An Introduction* (Boston: Little, Brown, 1973).

11. Again a *short*-term argument might be made in some sectors on infant industry grounds.

12. John C. Panzer and Robert D. Willig, "Free Entry and Sustainability in a Natural Monopoly," *Bell Journal of Economics* 8, No. 1 (Spring 1977): 1–22.

13. One conceptual complexity almost universally ignored in discussions of the sustainability problem is what would happen if two natural monopolists competed with one another in a third competitive market. If both of the natural monopolists were to prove nonsustainable without the revenues from the third market, public policy would be faced with the difficult decision of which output should be "saved" by bequeathing the third market's franchise to that monopolist. Making such a determination would proliferate the already substantial information requirements of determining sustainability.

14. A second carrier-of-last-resort issue occurs when customers using a competing service turn to the franchised telephone company for temporary service. This could occur, for instance, if a customer found that the private line supplied by a specialized common carrier was in use and therefore used the public switched network instead. If such backup use of the phone system accentuates peak demand, additional capacity could be required, the cost of which would need to be recovered in rates. Unless the temporary users can be charged for the additional capacity needed to accommodate their use, the normal customers of the carrier of last resort, it is argued, would suffer because they would have to pay for part of the additional capacity. (See, for instance, E. G. Davis and D. M. Dunn, "The Economic Difficulties of Being a Supplier of Last Resort," Bell Laboratories Economic Discussion Paper No. 52, February 1976.) The argument is subject, first, to the qualification that if scale economies are

sufficiently strong, increased usage may lower the revenue requirement per unit of capacity. The problem of providing sufficient capacity for both regular customers and "last resort" users can also be mitigated by improved rate structures such as those discussed in Chapter 3. Improved peak load pricing would to some extent discourage last resort users from accentuating peak demand. In addition, by distributing usage more evenly over the day, peak load pricing would temporarily postpone the need for expanded capacity. Bulk or volume contract rates, such as those increasingly used in transportation, might also provide some solution to this problem.

15. There are economies of joint production, for example, when production of two goods by one firm is less costly than the combined costs of production of two firms, each of which specializes in one good.

16. This result may perhaps be clarified by considering two polar cases. If there are strong economies of joint production, the firm supplying the full set of services will enjoy a strong cost advantage over any firm attempting to supply less than the full set of services and, if permitted, the joint supplier will be able to price at a level where the higher cost firms cannot profitably operate. If there are no economies of joint production, there are no advantages or disadvantages of supplying the services independently; consequently, a reduction of output of one of the services does not affect the costs of any of the other services. In the cases of both strong economies of joint production and no economies of joint production, other things equal, sustainability thus tends *not* to be a problem, albeit for different reasons. With only modest or slight economies of joint production, however, entry into a limited set of the services may still be economically feasible, but such entry will tend to increase the costs of supplying the remaining services, if the established firm is forced to reduce its output of the services into which entry has occurred. In this way, therefore, sustainability problems tend to be worsened by modest or slight economies of joint production.

17. These cost functions do not, however, shed any light on the questions of economies of joint production discussed in the previous paragraph and note. Since there is little or no evidence on the size of joint economies of production, it is difficult to evaluate any sustainability problems that joint economies might raise.

18. See Chapter 2.

19. Kenneth C. Baseman, "Open Entry and Cross-Subsidy in Regulated Markets," paper prepared for the NBER Conference on the Economics of Public Regulation, December 15–17, 1977.

20. Ibid.

21. This can happen because nonsustainability can force at least temporary provision of services at prices equal to short-run marginal costs and below long-run average costs.

22. Baseman, "Open Entry and Cross-Subsidy in Regulated Markets," op. cit.

23. Much of long distance message toll service, particularly between residences and smaller business and from and to more remote locations, may retain monopoly characteristics for at least a decade or more as well.

24. Baseman, "Open Entry and Cross-Subsidy in Regulated Markets," op. cit.

25. It was argued above that the entry of telephone companies into these services should be sanctioned in keeping with the principle of allowing freedom of entry to substitute market competition for regulation in the industry. Such entry may also permit the use of certain potentially more efficient technologies (e.g., optic fibers) sooner than might otherwise be the case by enhancing scale and density in certain applications.

26. Of course, as pointed out above, infant industry protection may be all too readily extended beyond its legitimate useful life.

27. In concentrating on the work day as the major peak period, it is not meant to imply that peak charges might not also be desirable in the early evening as well when most systems seem to be impacted by a third peak.

28. However, the danger must be recognized that if and when local markets become competitive, horizontal equity could become a subterfuge for price fixing or signaling among the competitors. See *U.S. v. General Electric Company and Westinghouse Electric Corporation,* "Plaintiff's Memorandum in Support of a Proposed Modification to the Final Judgment Entered on October 1, 1962 Against Each Defendant," Civil No. 28228 (U.S. Eastern Court for the Eastern District of Pennsylvania, December 10, 1976).

29. The only qualifications on abolition of minimum rate regulation are related to preventing so-called predatory pricing; this issue has assumed such an importance in public policy discussions that it is discussed at length in the next section of this chapter.

30. These goals are treated in greater detail in Chapter 3. An additional benefit of decreased regulation would be increased political accountability for the decisions affecting public welfare now made by the telephone industry, the FCC, and the state regulatory commissions on an ad hoc basis. These decisions include selection of rate structures that, as noted in this chapter and discussed in Appendix A, may affect income distribution. Increased competitive pressures bringing rates more in line with costs should serve to place these practices in the public forum for direct evaluation.

31. "Inefficient," as used here, means that the total costs of providing all the services are greater than necessary. Besides the assumptions stated here, the analysis depends on a number of conditions concerning the cost and demand functions. A fuller discussion can be found in Baseman, "Open Entry and Cross-Subsidy in Regulated Markets."

32. Such considerations reinforce, of course, the case for opening up entry as widely as possible, even including entry (as in the local service telephone case) into markets sometimes considered to be natural monopolies.

33. Panzar and Willig, "Free Entry and Sustainability in a Natural Monopoly," and G. Faulhaber, "Cross-Subsidization: Pricing in Public Enterprise," *American Economic Review* 65 (December 1975).

34. Philip Areeda and Donald F. Turner, "Predatory Pricing and the Sherman Anti-Trust Act," *Harvard Law Review* 88 (1975): 697–714.

35. A price below short-run marginal and average costs might not be deemed predatory, for example, if learning curve or scale economies were expected to lower costs in the near future and a lower price at the moment was needed to develop the market.

36. Frederick M. Scherer, "Predatory Pricing and the Sherman Act," *Harvard Law Review* 89 (1976): 868–890. Areeda and Turner respond to these criticisms in "Scherer on Predatory Pricing," ibid., 891–900.

37. *Revision of the Uniform System of Accounts and Financial Reporting Requirements for Telephone Companies (Parts 31, 33, 43 and 43 of the FCC's Rules),* Federal Communications Commission Docket No. 78–196, Notice of Proposed Rulemaking (1978).

38. In *U.S. v. United Fruit Corporation,* CCH 1958 Trade Cases, Para 68, 941, United Fruit agreed to relinquish a part of its banana barony by establishing and spinning off a new firm capable of handling 35 percent of all U.S. banana imports.

   A suit alleging monopolization of color film processing, *U.S. v. Eastman Kodak Company,* CCH 1954 Trade Cases, Para 67, 920; CCH 1961 Trade Cases, Para 70,100, was settled by an agreement by Eastman Kodak to reduce its share of the processing market to below 50 percent within seven years, helping new firms enter the industry by licensing its patents and conveying its know-how to them.

In another case, *U.S.* v. *Radio Corporation of America,* CCH 1958 Trade Cases, Para 68, 245; 1963 Trade Case, Para 70, 628, RCA agreed to license large numbers of patents on a royalty-free basis.

39. The Averch-Johnson effect, for example, could make it rational profit-seeking behavior for a regulated firm to provide some services below cost in order to expand the rate base and thereby justify taking even more profits out of already remunerative services. However, this effect may be largely a phenomenon of the past. Inflation, and attendant increases in capital costs, may have eliminated the conditions necessary for fulfillment of the Averch-Johnson effect. Indeed, today the effect may well have been "put in reverse," with pressures to economize dominating the incentives of regulated firms. However, the phenomenon could reappear so that some restructuring of regulatory procedures might be desirable so as to minimize this threat.

40. See Robin Marris, "A Model of the Managerial Enterprise," *Quarterly Journal of Economics* (May 1963): 195–209; Oliver Williamson, *Corporate Control and Business Behavior* (Englewood Cliffs, N.J.: Prentice-Hall, 1970); R. M. Cyert and J. G. March, *A Behavioral Theory of the Firm* (Englewood Cliffs, N.J.: Prentice-Hall, 1963), Chapter 7; H. A. Simon, *Models of Man: Social and Rational* (John Wiley and Sons, New York: 1957); J. Williamson, "Profit Growth and Sales Maximization," *Economica* 33 (February 1966): 1–16; A. A. Berle, Jr., and Gardner C. Means, *The Modern Corporation and Private Property* (New York: Harcourt Brace, 1969); and W. J. Baumol, "The Theory of the Expansion of the Firm," *American Economic Review* 52 (December 1962): 1078–1087.

41. The funds available for indulging nonprofit management goals may be more limited in a regulated environment than in many unregulated situations. Specifically, regulation usually implies some restrictions on rates charged in monopoly markets so that if the regulated firm pursues policies of "giving its services away" in competitive sectors, it may have limited ability to recoup through rate increases on its monopoly services. Any limitation on monopoly sector prices will also chill the incentives to pursue Averch-Johnson effects or for the monopolist to extend activities into markets where the monopolist is not the true low-cost supplier. If, in addition, the regulated monopoly markets face potential entry from new suppliers or technologies, the ability to generate monopoly profits to cover predatory prices may be limited by economic forces as well as by regulation.

42. These points are well recognized in the literature on the alternatives to profit maximization by stressing such concepts as "satisficing," which implies that a range of alternative goals may be pursued by management subject to meeting certain minimum constraints on such important variables as profits and basic financial ratios. See Marris, "A Model of Managerial Expertise," op. cit.; Simon, *Models of Man: Social and Rational;* Williamson, "Profit Growth, and Sales Maximization," op. cit.; Berle and Means, *The Modern Corporation and Private Property;* Baumol, "The Theory of the Expansion of the Firm," op. cit.; and Cyert and March, *A Behavioral Theory of the Firm.*

43. Eugene V. Rostow, "The Great Telephone Debate: Competition and the Public Interest in the Telecommunication Industry," *Proceedings of the Sixteenth Annual Iowa State University Regulatory Conference on Public Utility Valuation and the Rate Making Process,* May 1977, p. 183.

44. As discussed in Chapter 5.

45. One *might* also envisage certain changes in the patent laws being helpful to smaller enterprises through providing better protection for technological innovations in their early years.

46. Charles River Associates, "An Analysis of Capital Market Imperfections" (Cambridge, Mass.: CRA, April 1977).
47. Baseman, "Open Entry and Cross-Subsidy in Regulated Markets," op. cit.
48. As taken from a report by Arthur D. Little, Inc., "Analysis of Telecommunications Industry Restructuring," prepared for the Ad Hoc Committee for Competitive Telecommunications, September 1977, p. 4.
49. Ibid.
50. Bro Uttal, "Selling Is No Longer Mickey Mouse at AT&T," *Fortune* (July 17, 1978).
51. ACCT clearly envisages restructuring as a means of relaxing the regulatory grip on telecommunications (see its goal (2) as enunciated above).
52. See the analyses in Chapter 2.
53. The extrapolations reported in Chapter 2 suggest some of the possibilities for continued or accelerated growth in the demand for communications services.

Distributional Impact of an Alternative Rate Structure

As described in Chapter 3, several alternative rate structures can be constructed with characteristics that serve various public policy goals reasonably well — goals that are only partially met and less well served by the existing telephone rate structure. Rate Structure A–5, discussed in Chapter 3, can be cited as a particularly appropriate and suggestive example. In this rate structure, local calls are priced to reflect peak and off-peak usage; toll charges are lower on the average and are charged only by duration of call (rather than by distance) to reflect the distance insensitivity of transmission costs via satellites; and station connect charges are lowered (to promote the goal of universal service).[1] To understand better the incidence complications of adopting such a rate structure, or structures of similar character, this appendix explores the changes in annual telephone expenditures that might be made by families in each of 12 income classes if Rate Structure A–5 were adopted.[2]

In addition to these "household effects," businesses will pay slightly more for the telephone services they utilize under Rate Structure A–5 than under the existing rate structure. This slight increase in expenditure will reflect itself in changed prices for the goods and services that the businesses sell or in changed payments to the inputs utilized in producing these goods and services. As a result, consumers, either through their income or through their general consumption expenditures, will be affected by changes in business telephone expenditures.

For a variety of reasons, it is not possible to interpret either an increase or a decrease in per-family telephone expenditures as a straightforward decrease or increase in the well-being, or welfare, of a family. To make welfare determinations in such circumstances, various price indices must be compared.[3] One commonly used index of price change (the Laspeyre index, L) weights prices in any two periods under examination (say, before and after the adoption of a new telephone rate structure such as A–5) by the quantities consumed in the initial period. Another alternative (the Paasche index, P) would be to weight prices by quantities consumed in the final period. Still another possibility (the index of income change, E) would be to weight prices in each period by the quantities consumed in each period, respectively; this, of course, would simply be the ratio of final expenditure to initial expenditure.

For a family in a given income class, if the ordering of the various indices is such that the index of expenditure change, E, is greater than both the Paasche index, P, or the Laspeyre index, L, it can then be concluded that the family is better off after the change in prices. Conversely, when E is less than both P or L, the individual family's welfare can be said to have fallen under the new price regime. Other orderings of these three price indices are either logically contradictory or indicate that no conclusion can be drawn as to the change in a family's level of welfare.[4]

BACKGROUND DATA AND INFORMATION

In order to compute price indices before and after a change such as Rate Structure A–5, it is first necessary to estimate the various prices and quantities being consumed (before and after) in various sectors of family telephone expenditure: residential local, residential toll, and distributed business expenditures. Estimated 1975 telephone rates and revenues under the existing and proposed (A–5) rate structures are presented in Table A–1 (with the relevant telephone statistics mostly taken from Appendix C, "Alternative Rate Structures"). All telephone service rates and revenues shown in Table A–1 have been deflated to 1972–1973 values in Table A–2. Expenditures on telephone service by income class (shown in Table A–3) were taken from the 1972–1973 Consumer Expenditure Survey (CES).[5]

Table A–1. Telephone Service Rates and Revenues in 1975 and under Rate Structure A-5

Initial—1975

Type of Service	Business Rates (dollars)	Business Revenues (billions of dollars)	Residential Rates (dollars)	Residential Revenues (billions of dollars)
Local:				
Station[a]	$21.00/month	$2.86	$7.43/month	$5.95
Extension[c]	3.00/month	.30		
Call[e]	0.0187/call	1.78	0.0041/call	.40
Total		4.94		6.35
Toll:[g]	0.21/minute	5.52	0.21/minute	7.88
Total		10.46		14.23

A-5

Type of Service	Business Rates (dollars)	Business Revenues (billions of dollars)	Residential Rates (dollars)	Residential Revenues (billions of dollars)
Local:				
Access[b]	$10.00/month	$1.28	$3.00/month	$2.55
Equipment[d]	1.00/month	.21	1.00/month	.85
Call[f]	0.054/call	4.67	0.042/call	3.31
Total		6.16		6.71
Toll:[h]	0.109/minute	4.53	0.109/minute	8.98
Total		10.69		15.69

Source: Calculations by Charles River Associates, based on sources listed below.

[a]Initial station rates represent estimated average charges. Residential initial local station revenues were derived by multiplying the number of present main stations (66.7 million) by the initial monthly charge and then by 12 to give annual revenue. Business revenues represent the total of 9.4 million business main stations at $21 per month and 1.3 million PBX trunks priced at $31.50 per month. See Appendix C, Table C-12, for more complete explanation.

[b]A–5 access charges and revenues are taken from Local Rate Structure A of Appendix C, Table C-8.

[c]The business extension charge is assumed to be roughly twice the residential extension estimate of $1.50. Business extension revenue was estimated by estimates of 8.4 million business and Centrex extensions, which is the total after deducting 39.9 million residential extensions from 48.3 million total industry extensions (excluding PBX and KTS). Residential extension revenues are not included in this analysis.

Footnotes continued on following page.

Table A–1 (continued)

[d]Business and residential equipment charges are taken from Rate Structure A, Appendix C, Table C–8. Business equipment revenues under A–5 were derived by estimates of 9.4 million business main stations and 8.4 million business extensions (Appendix C, Table C–7). Annual residential equipment was estimated using 70.7 million main stations.

[e]Initial average per-call charges were derived by assuming 120 calls per household per month. This implies 97.1 billion residential local calls (67,447,000 households); from a total of 192.3 billion calls (see footnote d, Table C–8), 95.2 billion business local calls are left. Since 10 percent of residential and 46 percent of business phones have metered local service, 95.2 billion business calls x 0.46 = 43.79 billion measured business measured business local calls and 97.1 billion residential calls x 0.10 = 9.71 billion measured residential local calls. Since a total of 53.5 billion metered local calls results in an average per-call charge of $.0406 ($2.17 ÷ 53.5), this implies an average residential charge of $0.0041 per call ($0.0406 x 0.10), and $0.0187 per call ($0.0406 x 0.46) for business calls. Initial local call revenues for both business and residential are then derived by multiplying the total number of local calls by the respective price per call.

[f]Assuming the peak, shoulder, and off-peak local call distribution as explained in Appendix C, Table C–8, footnote d, the A–5 average per-call charge is derived as follows:

$$0.80(\$.06) + 0.10(\$.04) + 0.10(\$.02) = \$0.054 \text{ for business calls and}$$

$$0.35(\$.06) + 0.40(\$.04) + 0.25(\$.02) = \$0.042 \text{ for residential calls.}$$

Under a local Rate Structure A (Appendix C, Table C–8), residences are assumed to make 93 calls per month at $.04 shoulder call charge. This implies 78.9 billion residential calls per year (70.7 million main stations), and $3.31 billion call revenues per year, given average per-call charge as derived above. Assuming businesses make 12 local calls per day at $.06 peak charge, 250 business days per year yields 86.4 billion local calls per year, and local calling revenues of $4.67 billion.

[g]$.21 per minute is derived by dividing $13.4 billion of initial total toll revenue by total interstate and intrastate toll minutes of calling volume of 62.8 billion at the old rates. Total initial toll revenue and toll calling volume are taken from Appendix C, Tables C–9 and C–12.

Business and residential initial toll revenues were calculated from total revenues using a 40–60 business-residential split that is explained below. In 1972, business long distance interstate messages were 46.7 percent of total; residential was 50.1 percent. For 1973, the figures were 47.4 percent and 49.6 percent, respectively. Thus, the split for business for 1972–1973 is 47/97 = 48 percent and that for residential is 50/97 = 52 percent (data from Long Line Statistics 1950–1975.) Bell Statistical Manual shows an average for the years

Footnotes continued on following page.

228

Table A–1 (continued)

1972–1973 of 3,394,599 long distance interstate messages. Forty-eight percent of this figure is 1,629,408 total business long distance interstate messages and thus 1,765,192 residential messages. According to *Long Line Statistics*, for 1972–1973 the average residential long distance interstate message length by paid minute was 9.67, and businesses averaged 6.09 minutes in length (see page 118 of *Long Lines Statistics 1950–1975*). Therefore, 6.09 minutes x 1,629,408 messages = 9,923,095 total business minutes, which equals 37 percent of total, and 9.7 minutes x 1,765,192 = 17,122,362 total residential minutes or 63 percent of total. The distribution of messages by time of day, assuming the length of call is the same in each category, is:

	Day (percent)	Evening (percent)	Night/Weekend (percent)
Business	74.2 (1)	22.3 (.65)	3.5 (.40)
Residential	36.0 (1)	55.5 (.65)	8.6 (.40)

The numbers in parentheses indicate the proportion of the 1978 full rate taken from the 1978 Boston area phone book (New England Telephone and Telegraph). Thus, the percent of business calls paying full rate is 90.1, and that for residential is 75.5 percent. Then the share of total long distance minutes times the percent of calls paying full rate serves as an approximate measure of the share of total toll revenue that each service category earns. This implies 41.2 percent for business and 58.8 percent for residential. Estimate of total toll service revenue was taken from Appendix C.

[h]Business and residential toll per-call charge under Rate Structure A–5 was taken from Appendix C, Table C–11e.

Assume residential price elasticity for toll calling of –1.2, and business price elasticity of toll calling of –0.7. Given initial quantity demanded for business of 26.29 billion minutes ($5.52 billion ÷ $.21/min), then new quantity demanded under A–5 is $4.53 billion. Similarly, initial residential toll minutes of 37.52 billion (7.88 billion ÷ $.21/min) result in A–5 toll revenues of $8.98 billion.

Table A–2. Telephone Rates and Revenues in 1975 and under Rate Structure A–5 Deflated to 1972–1973

Type of Service	Initial — 1975 Business Rates (dollars)	Business Revenues (billions of dollars)	Residential Rates (dollars)	Residential Revenues (billions of dollars)	Type of Service	A–5 Business Rates (dollars)	Business Revenues (billions of dollars)	Residential Rates (dollars)	Residential Revenues (billions of dollars)
Local:[a]					Local:[a]				
Station	$18.14/month	$2.47	$6.42/month	$5.14	Acess	$8.64/month	$1.11	$2.59/month	$2.20
Extension	2.59/month	0.26			Equipment/phone	0.86/month	0.18	0.86/month	0.73
Call	0.0162/call	1.54	0.0035/call	0.35	Call	0.047/call	4.06	0.036/call	2.84
Total		4.27		5.49	Total		5.35		5.77
Toll:[b]	0.20/minutes	5.26	0.20/minute	7.50	Toll:[b]	0.104/minute	4.31	0.104/minute	8.55
Total		9.53		12.99	Total		9.66		14.32

Source: Calculations by Charles River Associates. 1972–1973 local index computed from local rate indices reported in American Telephone and Telegraph, *Statistical Report,* 1976. 1972–1973 toll index computed from message toll rate indices reported in National Association of Regulatory Commissioners, *Toll Rates in Effect July 12,* 1976.

[a]Average 1972–1973 local rates are 86.4 percent of those rates in 1975:

$$\frac{117.0 + 124.4}{2(139.7)} = 0.864$$

where 117.0 = 1972 local rate index
124.4 = 1973 local rate index
139.7 = 1975 local rate index
(Based on 1960 = 100)

[b]Average 1972–1973 toll rates are 95.2 percent of those in 1975:

$$\frac{100.9 + 102.6}{2(106.9)} = 0.952$$

where 100.9 = 1972 toll rate index
102.6 = 1975 toll rate index
106.9 = 1975 toll rate index
(Based on 1960 = 100)

Residential Local Expenditure

Under Rate Structure A–5, residential consumer savings from lower station connect and equipment charges amount to $2.97 per month or $35.64 per year (Table A–2) per main station.[6] The result of this decline in price is an estimated increase in the number of residential main stations, or subscribers, of approximately 4 million.[7]

The 4 million families who newly subscribe to the telephone network represent a group for whom the move to Rate Structure A–5 is clearly an improvement over the existing rate structure; presumably, these new connections will be made only if the individuals felt that their level of well-being would thereby be increased. Given the configuration of residential connect and calling charges, this group probably desires, in the main, connection to the system and does not represent a large demand for calls, either toll or local. This new group of 4 million main stations should therefore have calling and welfare patterns significantly different from those of existing subscribers to the telephone system. Specifically, existing subscribers are more likely to make large numbers of calls, at least local calls, under the present rate regime, which features very low charges for local calling. The essential or difficult welfare question, in fact, is whether the gain from reduction in connect and equipment charges for existing subscribers is or is not offset by welfare losses from increased expenditures for local calling as the price on those local calls rises significantly under Rate Structure A–5. Because existing and new subscribers are very likely to have different calling characteristics and because information on new subscribers is necessarily speculative (e.g., as to quantity of local and toll calling), this analysis will separate new subscribers from the determination of welfare changes created by the change in price structure. The families for whom welfare changes are computed therefore will represent existing subscribers to the telephone network. This procedure means that the analysis is biased toward concluding that families in particular income classes, specifically the lowest two or three income classes, will show a diminished level of well-being under Rate Structure A–5. Were new connections to the telephone network also included, many of these connections would be made by families in the lowest two or three income classes,[8] and these new connections would provide an undeniable welfare gain for these lowest income groups. In short, the deletion of new subscribers from the welfare analysis means that much of the beneficial welfare impact of moving toward even greater extensions of service to the U.S. public will be missed by this analysis.

Table A–3 indicates that the average annual telephone expenditure of the average (reporting) American family in 1972–1973 was $170.90.

Table A–3. Telephone and Consumption Expenditures by Income Class, 1972–1973

(1) Family Income Before Taxes	(2) Assumed Midpoint Income	(3) Average Annual Telephone Expenditure Per Family (dollars)	(4) Number of Families (thousands)	(5) Average Annual Consumption Expenditure Per Family (dollars)	(6) Total Consumption (billions of dollars)[a]	(7) Proportion of Total Consumption by Income Class[b]
Total reporting		$170.90	67,447	$7,883.95	531.75	1.0000
Under $3,000	$1,500	86.21	9,065	3,039.34	27.55	.0518
$3–3,999	3,500	112.68	3,991	3,999.69	15.96	.0300
$4–4,999	4,500	124.88	3,624	4,531.32	16.42	.0309
$5–5,999	5,500	128.03	3,282	5,099.71	16.74	.0315
$6–6,999	6,500	148.81	3,401	5,724.97	19.47	.0366
$7–7,999	7,500	156.85	3,251	6,147.90	19.99	.0376
$8–8,999	9,000	166.78	6,594	6,921.21	45.64	.0858
$10–11,999	11,000	178.38	6,278	7,888.79	49.53	.0931
$12–14,999	13,500	188.93	8,375	8,889.67	74.45	.1400
$15–19,999	17,500	213.42	9,996	10,639.26	106.35	.2000
$20–24,999	22,500	229.19	5,028	12,591.46	63.31	.1191
$25 and over[c]	37,500	289.25	4,560	16,737.95	76.33	.1435

Sources: Columns 1, 3, 4, and 5 were taken from U.S. Department of Labor, Bureau of Labor Statistics, Consumer Expenditure Survey Series: Interview Survey, 1972–1973, Report 455–4, 1977, pp. 4–9. Calculations resulting in columns 2, 6, and 7 by Charles River Associates are based on columns 1, 3, 4, and 5.

[a]Column 6 = column 4 × column 5.

[b]Column 5 × column 4 divided by $531.75 billion of total family consumption expenditures.

[c]Assume $50,000 as upper limit.

Under the existing rate structure, as shown in Table A–2, residences are charged $6.42 per month on the average for connecting one main station to the telephone network. Converting this charge to an annual figure of $77.04 (12 x $6.42) and subtracting this figure from total family expenditure on telephone service results in a sum of $93.86 ($170.90 − $77.04), which should represent the average family's annual expenditure for local and toll calls. Such calculations were made (see Table A–3) for each of the 12 income classes under examination.

To break down total annual expenditure for local and toll calling further into its constituent parts, the average number of annual local calls per family for each income class must be estimated.[9] Data from the Consumer Expenditure Survey[10] indicate that the average age of the head of the family in the lowest income class is 57 years as compared to an average age in the low- to mid-40s for the $8,000 to $25,000 income class. The lowest income class also has a disproportionate number of persons aged 65 or over as compared with other income classes. Conversely, the distribution of children under age 18 is such that the lowest income class has disproportionately few children. The size of the family generally increases as the level of income increases. In short, the lowest income class is heavily weighted toward single-person families who are older and retired.

If it is reasonable to suppose that calls per family increase as the size of the family increases, as the number of children (teenagers in particular) in a family increases, and as work contacts increase, per-family calling should then rise as income increases. Given the paucity of information, the safest assumption would seem to be that local calling is distributed across income classes simply in proportion to population in that class.[11] Column 1 of Table A–4 shows the initial number of calls per household per year as derived from the distribution of population across income classes in the Consumer Expenditure Survey data.

In order to compute the estimated residential local calling expenditures under Rate Structure A–5, the elasticity of calls to a change in calling charge in different income classes must be estimated. Data limitations preclude rigorously estimating "actual" demand elasticities. The elasticities used, as shown in Column 4 of Table A–4, do possess, however, various desirable characteristics. They become more inelastic with higher income;[12] they are consistent with an aggregate elasticity of approximately −0.1;[13] and there are no large jumps in elasticity between income classes.

Table A–4 illustrates some of the problems in identifying changes in levels of well-being with changes in expenditures. As a result of moving from the initial rate structure to Rate Structure A–5, for example, the lowest income class reduces its expenditures on connect

Table A–4. Average Residential Expenditures for Local Telephone Service by Income Class: Initially and under Rate Structure A-5

Income Class	(1) Initial Number of Calls per Family per Year[a]	(2) Initial Local Calling Expenditure per Family[b]	(3) Initial Total Residential Expenditure per Family[c]	(4) Elasticity of Calls to per Call Charge	(5) A-5 Estimated Number of Calls per Family[d]	(6) A-5 Local Calling Expenditure per Family[e]	(7) A-5 Local Residential Expenditure per Family[f]	(8) Change in Total Local Expenditure per Family[g]
Under $3,000	706	$2.47	$79.51	−0.3	351	$12.64	$ 54.04	−$25.47
$ 3–3,999	960	3.36	80.40	−0.3	477	17.17	58.57	− 21.83
$ 4–4,999	1,060	3.71	80.75	−0.25	592	21.31	62.71	− 18.04
$ 5–5,999	1,210	4.24	81.28	−0.20	759	27.32	68.72	− 12.56
$ 6–6,999	1,261	4.41	81.45	−0.20	791	28.48	69.88	− 11.57
$ 7–7,999	1,363	4.77	81.81	−0.15	961	34.60	76.00	− 5.81
$ 8–9,999	1,412	4.94	81.98	−0.10	1,118	40.25	81.65	− 0.33
$10–11,999	1,615	5.65	82.69	−0.10	1,279	46.04	87.44	+ 4.75
$12–14,999	1,715	6.00	83.04	−0.10	1,358	48.89	90.29	+ 7.25
$15–19,999	1,816	6.36	83.40	−0.05	1,616	58.18	99.58	+ 16.18
$20–24,999	1,917	6.71	83.75	−0.04	1,746	62.86	104.26	+ 20.51
$25,000 and over	1,917	6.71	83.75	0	1,917	69.01	110.41	+ 26.66

Source: Calculations by Charles River Associates.

[a] Derived by assuming local calling was distributed by the proportion of population in each income class and dividing the income classes' calling volume by the number of families in that income class.

[b] Column 1 × $0.0035 (estimated per-call charge for local residential calls; see Table A–2).

[c] Column 2 plus initial station connect charge of $77.04 (derived from Table A–2: $6.42 × 12 months).

[d] Estimated volume of calls under rate Structure A–5 at $0.036 per call.

[e] Column 5 × $0.036 (estimated per-call charge for local calls under A–5. See Table A–2).

[f] Column 6 plus A–5 annual access and equipment station charges of $41.40 ($3.45 × 12 months).

[g] Column 7 less column 3. A plus sign indicates an increase in expenditure under Rate Structure A–5 and a negative sign a decrease.

and local calling by approximately $25 per year. However, this reduction in expenditure is accomplished at the price of cutting the annual number of local calls per family in half. Rather than making two phone calls per day under the initial rate structure, the typical family (previously connected to the system) in the lowest income class will be making approximately one phone call per day under Rate Structure A–5 (a similar comment can be made about the next to the lowest income class as well). However, the per-call price of local calling under Rate Structure A–5 employed in Table A–4 is derived by aggregating calling patterns across various times of day, each time of day having its own cost-based price. Implicit in the application of the same aggregated per-call charge for local calls across all income classes is the assumption that each income class will distribute its calling pattern by time of day in exactly the same fashion.[14] Lower-income classes may be more sensitive to higher peak charges and could shift their calling to times of day in which the charges are lower. As a result, the effective local per-call charge for the lower-income classes when averaged across all times of day might well be lower than that for higher-income classes; the reduction in local calling by 50 percent derived for the lowest income class is thus probably an overstatement. Without knowing the nature of the shift in calling patterns for the families in the various income classes, however, it is impossible to know exactly how many additional phone calls above those shown in Table A–4 under Rate Structure A–5 would be generated by shifts in calling patterns by time of day.

Table A–4 suggests that the change in local expenditures per household moves from negative to positive under Rate Structure A–5 as income level rises. For higher-income classes with (by assumption) virtually inelastic demands, there is a significant increase in annual expenditure, presumably tapping the willingness to pay as revealed by the inelastic demands of these income classes.

Residential Toll Expenditures

Column 1 of Table A–5 displays the average family's annual expenditure on toll calls under the existing rate structure for each income class. These estimates are derived by subtracting local calling expenditure under the existing rate structure, shown in Table A–4, as well as the annual connect charge of $77.04 from the total annual family expenditure for telephone services.

By dividing average family expenditure on toll calling by the estimated existing average toll charge of $0.2 per minute (see Table A–2), it is possible to derive, as shown in Column 2 of Table A–5, estimates of the number of toll calls made by families in various income classes

Table A–5. Residential Expenditures for Toll Telephone Service by Income Class: Initially and under Rate Structure A–5

Income Class	(1) Initial Residential Toll Expenditures per Family[a]	(2) Initial Number of Minutes per Family[b]	(3) Call Elasticity	(4) A–5 Minutes per Family[c]	(5) A–5 Residential Toll Expenditures per Family[d]	(6) Residential Change in Residential Toll Expenditures per Family[e]
Under $3,000	$ 6.70	34	−1.2	75	$ 7.80	$+ 1.10
$ 3– 3,999	32.28	161	−1.2	353	36.71	+ 4.43
$ 4– 4,999	44.13	221	−1.2	484	50.34	+ 6.21
$ 5– 5,999	46.75	234	−1.2	513	53.35	+ 6.60
$ 6– 6,999	67.36	337	−1.2	739	76.86	+ 9.39
$ 7– 7,999	75.04	375	−1.2	822	85.49	+10.45
$ 8– 9,999	84.80	424	−1.2	929	96.62	+11.82
$10–11,999	95.69	478	−1.2	1048	108.99	+13.30
$12–14,999	105.89	529	−1.2	1159	120.54	+14.65
$15–19,999	130.02	650	−1.2	1425	148.20	+18.18
$20–24,999	145.44	727	−1.2	1593	165.67	+20.23
$25,000 and over	205.50	1028	−1.2	2253	234.31	+28.81

Source: Calculations by Charles River Associates.

[a]Derived by subtracting column 3 of Table A–4 from column 3 of Table A–3.

[b]Estimated volume of calls at $0.20/minute per call charge (see Table A–2: column 1 ÷ $0.20/minute.

[c]Estimated volume of calls under Rate Structure A–5 using $q_1 = q_0 \, (P_0/P_1)^E$ and call elasticity (E) in column 5.

[d]Column 4 × $0.104/minute (see Table A–2).

[e]Column 5 − column 1. A plus sign indicates an increase in expenditure under Rate Structure A–5; a minus sign indicates a decrease.

under the existing rate structure. To derive toll charges under Rate Structure A–5, an overall aggregate toll elasticity of −1.2 is assumed.[15] Applying this elasticity across all income classes — since there is no available evidence, pro or con, as to whether or not toll elasticities change across income classes — the number of toll calls under the new Rate Structure A–5 can be estimated, as well as the level of toll expenditures under the new rate structure as shown in Column 5 of Table A–5.

As a result of the assumed toll elasticity being greater than one, all families spend more money on toll calling under the new rate structure than under the old. The amount of increase is minimal for the lowest-income class, $1 per year, but is as much as $28–29 dollars per year for the highest-income class. Once again, though, these changes in expenditure probably mask significant changes in price and quantities.

Distributed Business Telephone Service Expenditures

The telephone companies often argue that a major cut in toll charges, such as embodied in A–5, would necessitate a sizeable increase in local residential rates. Under Rate Structure A–5, however, toll calling charges for both businesses and residences are reduced, but total business expenditures for telephone services are increased, thereby negating the need for dramatic price rises in residential local rates. Under A–5, business toll revenues fall by approximately $1 billion (as shown in Table A–2), but this $1 billion is offset by a $1-billion increase in business expenditures on local telephone service.[16]

Businesses, of course, represent contractual arrangements linking dollars of revenues to payments to the factor inputs that generate these revenues. What businesses pay for telephone services will ultimately be reflected in changed prices to consumers for the goods or services produced by the business or in changed payments to the factors of production (land, labor, capital, management) utilized in the production process. One way of distributing telephone service expenditures by businesses would be to consider these expenditures as a type of sales tax; that is, telephone expenditures are passed on to consumers as a constant percent of sales. In essence this implies that the U.S. economy is generally competitive, and therefore changes in telephone expenditures by businesses are paid by consumers in the long run, rather than "absorbed" by reduced payments to the factors of production. Thus, under these assumptions, consumers in a given income class would pay business telephone service expenditures in proportion to this income class's share of total consumption expenditures.

To measure the effect of changes in business telephone expenditures as they flow through to families in the various income classes, business telephone service expenditures for 1972–1973 (approximately $9.53 billion) were distributed by each class's share of total annual consumption, as shown in Column 1 of Table A–6. Total business telephone expenditures are estimated to rise to $9.66 billion under A–5 and this is distributed in Column 2 of Table A–6. Assuming that the changes in telephone expenditure among various business sectors are not so large as to change consumer prices and shift consumption patterns among classes perceptibly, both initial and A–5 business expenditures were distributed by each income class's proportion of total consumption expenditures (from Table A–3, Column 7). Since the size of the total increase in business expenditure was small, approximately $130 million, each family was faced with a very small increase in distributed business expenditures. The lowest income class faced an annual increase of less than $1 per household while the highest income class faced an increase of approximately $4.

OVERALL CHANGE IN FAMILY EXPENDITURES FOR TELEPHONE SERVICE

The estimated changes in residential local, residential toll, and distributed business expenditures created by moving to Rate Structure A–5 are summarized in Table A–7. It is clear that the upper-income families will pay significantly more for their telephone services under Rate Structure A–5 than under the existing rate structure. A family in the lowest income class, however, will pay approximately $2 per month less for its telephone service. (Recall, however, that the lowest income class also reduces its local calling by approximately one-half under Rate Structure A–5.) Although Rate Structure A–5 does result in significant expenditure increases for several higher-income classes, these increases occur because these families are assumed to make large numbers of local calls at peak hours and to more than double their toll calling under new lower toll rates, thus offsetting lower connect and equipment charges.

WELFARE CHANGE AS MEASURED BY PRICE INDICES

If the existing rate structure is changed to Rate Structure A–5, the overall impact of various price and quantity changes can be compared by computing price indices for each of the 12 income classes under examination. Table A–8 displays the results of these price index computations.[17]

As the ordering of indices displayed in Table A–8 exhibits, the movement to Rate Structure A–5 makes families in none of the income classes unmistakably worse off. Indeed, all income classes but the very lowest are made better off by the move to Rate Structure A–5.[18] For the lowest income class, the fact that E is greater than P means that the individual family in the lowest income class does not

Table A–6. The Change in Distributed Business Telephone Service Expenditures from the Existing Rate Structure to Rate Structure A–5

Income Class	(1) Initial Distribution of Total Consumption by Income Class[a]	(2) Column (1) × $130 Million[b] (millions of dollars)	(3) Number of Families[c] (thousands)	(4) Change in Expenditures per Family[d]
Total	1.0000	$130.00	67,447	$+1.93
Under $3,000	0.0518	6.73	9,065	+0.74
$ 3– 3,999	0.0300	3.90	3,991	+0.98
$ 4– 4,999	0.0309	4.02	3,624	+1.11
$ 5– 5,999	0.0315	4.10	3,282	+1.25
$ 6– 6,999	0.0366	4.76	3,401	+1.40
$ 7– 7,999	0.0376	4.89	3,251	+1.50
$ 8– 9,999	0.0858	11.15	6,594	+1.69
$10–11,999	0.0931	12.10	6,278	+1.93
$12–14,999	0.1400	18.20	8,375	+2.17
$15–19,999	0.2000	26.00	9,996	+2.60
$20–24,999	0.1191	15.48	5,028	+3.08
$25,000 and over	0.1435	18.66	4,560	+4.09

Source: Calculations by Charles River Associates, based on sources cited below.

[a]From column 7, Table A–3.

[b]$130 million is the increase in business telephone expenditures as rate structures move from the initial structure generating $9.53 billion in business expenditures to Rate Structure A–5 generating $9.66 billion in business expenditures, see Table A–2.

[c]Column 4 of Table A–3.

[d]Column 4 ÷ column 3. The positive sign indicates an increase in expenditures under Rate Structure A–5.

suffer a fall in standard of living by a move to Rate Structure A–5; however, the fact that E is not greater than L means that the new rate regime cannot be described as one in which the family's standard of living has been increased.

Two important considerations, moreover, bias the analysis of this appendix toward concluding that families in the lowest income class will be harmed by moving to Rate Structure A–5. First, the lowest income class should contain a significant portion of the 4 million new subscribers gained under Rate Structure A–5, the gains from which are ignored in this analysis. Second, the assumption that families in all income classes adjust their local calling patterns in exactly the same way to a new regime of local time of day user charges requires that the lowest income class make the same adjustments in calling patterns as the highest income class. The lowest income class seems more likely than the highest income class to shift local calling to times of day in which calling charges are lower.

Table A–7. Estimated Increase (+) or Decrease (−) in Total Annual Expenditures by a Family for Telephone Services under Rate Structure A–5, by Income Class

Family Income Before Taxes	Change in Residential Expenditures			Change in Business Expenditures	Change in Total Annual Telephone Expenditures
	Local	Toll	Total		
Under $3,000	$−25.47	+ 1.10	−24.37	$+0.74	$−23.63
$ 3– 3,999	−21.83	+ 4.43	−17.40	+0.98	−16.42
$ 4– 4,999	−18.04	+ 6.21	−11.83	+1.11	−10.72
$ 5– 5,999	−12.56	+ 6.60	− 5.96	+1.25	− 4.71
$ 6– 6,999	−11.57	+ 9.39	− 2.18	+1.40	− 0.78
$ 7– 7,999	− 5.81	+10.45	+ 4.64	+1.50	+ 6.14
$ 8– 9,999	− 0.33	+11.82	+11.49	+1.69	+13.18
$10–11,999	+ 4.75	+13.30	+18.05	+1.93	+19.98
$12–14,999	+ 7.25	+14.65	+21.90	+2.17	+24.07
$15–19,999	+16.18	+18.18	+34.36	+2.60	+36.96
$20–24,999	+20.51	+20.23	+40.74	+3.08	+43.82
$25,000 and over	+26.66	+28.81	+55.47	+4.09	+59.56

Source: Calculations by Charles River Associates.

Table A–8.   Index Numbers as Indicators of Welfare Change[a]

	Index Numbers Excluding Distributed Expenditures				
Income Class	E	L	P	Index Relationship	Welfare Change under Rate Structure A–5
Under $3,000	0.72	0.82	0.66	$L>E>P$	Inconclusive
$ 3– 3,999	0.85	0.82	0.64	$E>L>P$	Positive
$ 4– 4,999	0.91	0.82	0.64	$E>L>P$	Positive
$ 5– 5,999	0.95	0.85	0.67	$E>L>P$	Positive
$ 6– 6,999	0.99	0.82	0.65	$E>L>P$	Positive
$ 7– 7,999	1.03	0.83	0.66	$E>L>P$	Positive
$ 8– 9,999	1.07	0.82	0.67	$E>L>P$	Positive
$10–11,999	1.11	0.84	0.67	$E>L>P$	Positive
$12–14,999	1.12	0.84	0.67	$E>L>P$	Positive
$15–19,999	1.16	0.82	0.67	$E>L>P$	Positive
$20–24,999	1.18	0.81	0.67	$E>L>P$	Positive
$25,000 and over	1.19	0.75	0.65	$E>L>P$	Positive

Source: Calculations by Charles River Associates.

[a]Index numbers derived as follows:

Index of income change $E = \dfrac{\Sigma p^1 X^1}{\Sigma p^0 X^0}$

Laspeyre index $L = \dfrac{\Sigma p^1 X^0}{\Sigma p^0 X^0}$

Paasche index $P = \dfrac{\Sigma p^1 X^1}{\Sigma p^0 X^1}$

NOTES

1. See Chapter 3 for a complete examination of the attainment of the goals under existing and alternative rate structures.
2. Numerous simplifying assumptions are required to effectively model changes in family expenditures on telephone service (or any other good or service). Every attempt has been made to clearly identify these assumptions.
3. For a more complete explanation, see C. E. Ferguson, *Microeconomic Theory* (Homewood, Ill.: Richard D. Irwin, 1969), pp. 67–72.
4. Strictly speaking, an individual family's welfare change cannot be measured without examining *all* of its expenditures. This analysis will assume that telephone expenditures are sufficiently small and separable from remaining expenditures so

that changes in the former can be examined without concern for changes in the latter. In addition, data do not exist to analyze changes in all expenditures.

5. The CES data were used for determining the distribution and level of total telephone expenditures. No breakout of actual expenditures by service category within the total of telephone expenditures was available at the time of this analysis. See *Consumer Expenditure Survey Series: Interview Survey, 1972–73*, Report 455–4, Bureau of Labor Statistics, U.S. Department of Labor, 1977, pp. 8–9.

6. Residential extension phones have not been included in this analysis because of the lack of data on the distribution of extension phones across income classes. The exclusion of residential extensions reduces estimates of total initial revenues by $0.7 billion (39.9 million extensions at an estimated monthly charge of $1.50).

7. See Table 3–4.

8. See Table 7, page 51, of Bell Exhibit 21 in FCC Docket 20003. This exhibit consists of a report on the proportion of households with telephones in various sociodemographic categories, including income classes. The report was authored by Lewis J. Perl and entitled "Economic and Demographic Determinants of Telephone Availability," prepared by National Economic Research Associates, April 15, 1979.

9. One source of data for such an estimation indicates that the lowest income class (from $0 to $3,000 per year) makes approximately three times as many phone calls as the average family in the highest income class (from $30,000 up). See Bridger Mitchell, "Optimal Pricing of Local Telephone Service," *American Economic Review* 68 (September 1978): 517–537. Mitchell's source for residential calling by income level was a study by American Telephone and Telegraph entitled *Subscriber Line Usage Study, May 1972June 1973*. According to Mitchell, this study was based on 10 California exchanges which used No. 1 ESS switching equipment.

10. Consumer Expenditure Survey Series: Interview Survey 1972–1973, pp. 4–5.

11. Although not reported, the analysis of this appendix was also conducted using two alternative distributions of local calls: a California distribution via Bridger Mitchell ("Optimal Pricing of Local Telephone Service"); and a uniform distribution by proportion of families in each income class. The conclusions about the well-being of families in the various income classes before and after a change to structure A–5 seemed largely insensitive to which distribution was used.

12. For a mathematical derivation of this general result, see Bridger Mitchell, "Optimal Pricing of Local Telephone Service," p. 43.

13. This is the approximate aggregate elasticity that is reflected in the number of calls and calling charges for residential local calling in Table A–2. Such an elasticity is at the inelastic end of the range of estimated elasticities cited on p. 302 in Appendix C. A doubling of the elasticities across all income classes does not greatly modify the welfare results of the analysis.

14. Rate Structure A–5 assumes that 35 percent of all residences pay the peak charge, 40 percent of all residences pay the shoulder charge, while the remaining 25 percent of calls pay the off-peak charge.

15. This elasticity is consistent with data on p. 302, showing that the elasticity of toll calling is greater during the evening than the day and greater for residential customers than for businesses. An elasticity of −1.2 is in the midpoint of the range of estimates for the evening/night period and for residential customers.

16. This $1-billion increase is consistent with lower connect and equipment charges for business coupled with higher local usage charges, representing the demand for capacity imposed by daytime calling of businesses.

17. Table A–8 does not include the business expenditures in the welfare analysis because data are not available on the myriad price and quantity changes that would

occur as businesses flow through the increased expenditures on telephone services. Given that the change in distributed business expenditures per family is approximately $1 to $4 dollars per year, it is unlikely that the deletion of business expenditures from this analysis changes the results. If, however, the existing rate structure were adjusted to one in which business expenditures in toto were dramatically increased or decreased, a complete welfare analysis would have to include some measure of the impact of distributed business expenditures on family expenditure patterns.

18. Reworking this welfare analysis based on the distribution of residential local calling as shown in the ten California exchanges cited in Bridger Mitchell, "Optimal Pricing of Local Telephone Service," does result in the existing subscribers in the lowest income class being worse off. For reasons argued in the text, this distribution seems unrepresentative.

Sources and Methods Used in Projecting AT&T Financial Performance under Various Hypotheses

This appendix describes the sources and hypotheses used in building the AT&T financial models reported in Chapter 2. It also presents AT&T's financial statements year by year for each of the scenarios, as well as a comparison of these results to actual AT&T results, as an indication of the validity of the general structure of the models.

DESCRIPTION OF HISTORICAL FINANCIAL RATIOS

It is common practice, when doing financial projections, to tie operating and income statement ratios to some other item of the income statement. For this book, it was decided to project revenues and costs independently to see what effect they had on the financial integrity of the Bell System. The growth rates used to project revenues and costs are essentially the same rates reported in Chapter 2.

In order to keep the modeling exercise within manageable proportions and still obtain useful and meaningful results, it was necessary to reduce the number of items appearing on AT&T financial statements. Two different sources of financial information from AT&T, the annual reports and the embedded direct cost (EDC) studies, were used. To reconcile these different sources, some financial items had to be extrapolated and the definitions of others had to be altered. These modifications are summarized below and in Tables B–1 and B–2. The structural relationships established between elements of the financial statements and used as hypotheses for doing the projections appear on the right-hand side of the tables; they are also included on the financial projection tables that appear later.

Income Statement Ratios (Table B–1)

Revenues (Lines 1, 2, and 3) A comparison of annual revenues as taken from the annual reports to revenues mentioned in the Bell EDC studies indicates that the EDC revenues are consistently lower than the annual report revenues by about 5 percent; 94.9 percent was used as a correction factor to convert extrapolations of annual report revenue estimates into EDC equivalents and vice versa.

Operating Costs (Line 4) The definitions used for operating costs for the projections reported here are different from those used in either the EDC studies or the annual reports. To summarize briefly, the following definitions were used for operating costs:

EDC revenues
(minus) Interest charges (from annual reports)
(minus) Federal income tax charges (from
annual reports)
(minus) Net income (from annual reports)
(equals) Operating costs

This definition permits combining many elements of AT&T's income statement into only one item, while not creating any systematic biases in the reporting of interest charges, federal taxes, and net income. Operating costs, as thus defined, appear on line 4 of Table B–1. These operating costs are further broken down into four categories in a later section of this appendix.

Interest (Lines 5, 6 and 7) The effective embedded interest rate for the projections was defined as the ratio of interest payments charged to the income statement during the year to the beginning-of-

Table B–1. Financial Ratios Used in AT&T Pro Forma Income Statements, 1971–1977

Item	Units	1971	1972	1973	1974	1975	1976	1977	Financial Ratios Selected
					Income Statistics				
1. Annual report revenues	$ millions	18,442	20,904	23,527	26,174	28,957	32,816	36,495	
2. Embedded direct cost study revenues	$ millions	NA	NA	22,250	24,830	27,470	NA	NA	
3. Percent of annual report (2 ÷ 1)	Percentage	NA	NA	94.6	94.9	94.9	NA	NA	94.9%
4. Operating costs	$ millions	12,578	14,141	15,560	17,480	19,747	22,225	24,648	
5. Interest deductions	$ billions	1.29	1.50	1.73	2.06	2.30	2.40	2.44	
6. Beginning-of-year debt	$ billions	—	21.23	24.14	26.64	29.54	31.79	32.52	
7. Effective interest rate [5 ÷ 6]	Percentage	—	7.1	7.2	7.7	7.8	7.5	7.5	7.5%+
8. Profit before tax (PBT)	$ millions	3,635	4,202	4,956	5,293	5,322	6,515	7,542	
9. Federal tax: charges	$ millions	1,433	1,670	1,963	2,123	2,174	2,686	2,998	
10. paid	$ millions	NA	NA	932	678	129	582	617	
Federal tax/PBT									
11. charges [9 ÷ 8]	**Percentage**	39.4	39.7	39.6	40.1	40.9	41.2	39.8	40.0%
12. paid [10 ÷ 8]	**Percentage**	NA	NA	18.8	12.8	2.4	8.9	8.2	
13. Net income	$ millions	2,202	2,532	2,993	3,170	3,148	3,829	4,544	
14. Dividends	$ millions	2,504	1,634	1,783	2,040	2,177	2,490	2,823	
15. Dividend/net income	Percentage	67.5	64.5	59.6	64.3	68.8	65.0	62.1	65.0%

Source: Calculations by Charles River Associates, based on information provided in AT&T's annual reports and embedded direct cost studies.

year debt. It was assumed that this ratio would grow continuously for the projection years up to a specified percentage, as indicated in the specific projections. The interest and debt figures were taken directly from the annual reports for all these calculations.

Federal Tax (Lines 9 to 12)   A distinction must be made between federal income tax charges as they appear in the AT&T annual reports and the tax amount actually paid to the IRS. The difference between these two amounts (lines 9 and 10) creates deferred credit items on the balance sheet, as explained below.

In the pro forma income statements, the federal tax charges were assumed to be at the same level as in the last few years, about 40 percent of profits before tax. The statutory federal tax rate is 48 percent, but for a number of reasons explained in AT&T annual reports the actual rate was lower by 8 percentage points on the average between 1971 and 1977. Profit before tax is defined as the sum of net income (line 13) and federal tax charges (line 9), both of which are taken directly from AT&T annual reports. All other types of taxes (state, local, and property) are included in operating costs.

Dividends (Lines 13, 14, and 15)   AT&T's dividend payment as a percentage of net income has varied between 60 and 69 percent over the last few years (line 15). This percentage was set equal to 65 percent of net income for the projection years.

Balance Sheet Ratios (Table B–2)

Plant and Equipment Investment (Lines 1 to 11)   AT&T's investment in plant and equipment consists of many diverse items such as local loops, switching equipment, terminal equipment, and transmission facilities. Cost and physical count estimates were not available for each type of equipment for the base period of 1971 through 1977. For this reason, the projections for plant and equipment were linked to the increases in total number of phones in service in the Bell System. Of course, this should not be construed as meaning that investment in plant and equipment is due directly or exclusively to net phone installation. This relationship was used only because of its availability and usefulness in the projections. The total number of phones is assumed to grow for the projection years at the rate experienced in the 1973–1975 period (3.6 percent) for the three historical scenarios and in the 1971–1977 period (4.2 percent) for all the others (line 3).[1]

Three alternatives were considered for estimating the average plant investment per phone. One option would be to take the embed-

Table B–2. Financial Ratios Used in AT&T Pro Forma Balance Sheets, 1971–1977

Item	Units	Balance Sheet Statistics							Selected Financial Ratios
		1971	1972	1973	1974	1975	1976	1977	
1. Number of phones (end of year)	Millions	100.3	105.3	110.3	114.5	118.5	123.1	128.5	
2. Change from previous year [Δ1]	Millions	NA	5.0	5.0	4.2	4.0	4.6	5.4	
3. Change from previous year	Percentage	NA	5.0	4.8	3.7	3.5	3.9	4.4	3.6% & 4.2%
4. Plant: gross in service at end of year	$ millions	57,616	63,920	70,170	77,689	84,619	91,318	98,717	
5. Change in plant/change in number of phones	$	NA	1,255	1,347	1,690	1,737	1,444	1,370	$1,400
6. Plant addition and interest charged to construction	$ millions	7,411	8,290	9,301	10,058	9,595	9,964	11,521	
7. Plant addition per phone [6 ÷ 2]	$	NA	1,658	1,860	2,395	2,399	2,166	2,134	
8. Depreciation expense	$ millions	2,764	3,042	3,332	3,690	4,088	4,484	5,045	
9. Depreciation rate [8 ÷ 4]	Percentage	4.8	4.8	4.7	4.8	4.8	4.9	5.1	5.0%
10. Net plant	$ millions	47,197	52,600	58,568	64,935	70,442	75,922	82,398	
11. Percent of gross in service [10 ÷ 4]	Percentage	81.9	82.3	82.7	83.6	83.2	83.1	83.5	83.0%
12. Operating costs	$ millions	12,578	14,141	15,560	17,480	19,747	22,225	24,648	
13. Investments	$ millions	3,260.5	3,227	3,307	3,566	3,555	3,663	3,801	

Table continued on following page.

Table B-2 (continued)

Item	Units	1971	1972	1973	1974	1975	1976	1977	Financial Ratios Selected
14. Percent of operating costs [13 ÷ 12]	Percentage	25.9	22.8	21.3	20.4	18.0	16.5	15.4	18.0%
15. Current assets and deferred charges	$ millions	4,062	4,788	5,176	5,546	6,159	7,132	7,773	
16. Percent of operating costs [15 ÷ 12]	Percentage	32.3	33.9	33.3	31.7	31.2	32.1	31.5	32%
17. Current liabilities	$ millions	4,986	5,855	5,956	7,203	6,856	7,859	9,062	
18. Percent of operating costs [17 ÷ 12]	Percentage	39.6	41.4	38.3	41.2	34.7	35.4	36.8	37%
19. Deferred credits	$ millions	1,055	1,738	3,229	4,714	6,876	9,114	11,513	
20. Percent of operating costs [19 ÷ 12]	Percentage	8.4	12.3	20.8	27.0	34.8	41.0	46.7	
21. Change in deferred credits/change in plant [Δ19 ÷ Δ4]	Percentage	NA	11	24	20	31	33	32	30%

Source: Calculations by Charles River Associates, based on information provided in AT&T's annual reports.

ded cost of plant per phone (line 4 divided by line 1). This would result in an investment per phone below $800, which does not give a realistic indication of more recent investment costs. This option was therefore discarded.

A second option would be to use the increment in gross plant in service for every year and divide this figure by the yearly increase in number of phones (line 5). This method provides a better approximation of marginal costs, and it was the option actually used in the projections. Specifically, a figure of $1,400 average gross new investment per phone was used because four of the six years from 1972–1977 averaged close to this number.

A third option would have led to even higher estimates. From the statements entitled "changes in financial position" in AT&T's annual reports, each year's aggregate outlays on new construction (line 6) can be obtained. When divided by the net increase in the number of phones, this calculation yields an average of more than $2,100 investment per phone in the most recent years (line 7).

As can be seen, the three options result in widely diverging estimates of investment per phone ($800, $1,400, and $2,100). As mentioned, the first option was eliminated because it represents an average investment figure calculated over many decades of historical experience and would seem to have little relevance to current costs at the margin. The third option was discarded because the second option was better suited than the third for present projection needs. The concern here is with the net increase in number of phones, taking into account not only installation of new phones but also termination and transfer of service, and the second estimate takes this netting into account better than the third. Choice of the second alternative also avoided any need to reconcile "changes in financial position" (CFP) statements with the rest of the structure used for the projections; in particular, figures taken from the CFP statement could not be related directly to the balance sheet accounts without further analysis and adaptation. The second option also made it easier to estimate the net plant in service because the ratio of net to gross plant in service has been quite stable (at 0.83) over the years (line 10). It should also be noted that by using this approach some items of the balance sheet can be combined (essentially plant under construction and plant held for future use).

Using a ratio of 83 percent for net-to-gross plant implicitly assumes that the historical depreciation policies (plant life, depreciation methods, etc.) will continue as in the past. The implicit depreciation rates used for the 1971–1977 period are shown on line 9; this item is included in the table as a memorandum, for it does not figure explicitly in the income statement or balance sheet projections. In the income statement, depreciation is assumed to be part of operating costs as defined in the EDC study. In one set of pro formas, however, the depreciation rates were modified in the projection years, effectively doubling from 5 to 10 percent.

Investments (Lines 12, 13, and 14) This item represents mainly the equity investment in Western Electric and is taken directly from the AT&T annual reports. For the projection years, this figure was tied to the operating costs (line 12), as they were defined for Table B–1. As can be seen from line 14, the ratio of investments to operating costs has declined continuously over the years. Since "investments" is only a minor item of the balance sheet, it was assumed that from 1976 and afterward investments would be held at a constant percentage of operating costs, probably resulting in a slight overestimation that should not have a material effect on the projections.

Current Assets and Deferred Charges (Lines 15 and 16) This total, which combines a couple of entries from AT&T's annual report, is fairly stable over the years at around 32 percent of operating costs (line 16). It includes mainly cash and accounts receivable.

Current Liabilities (Lines 17 and 18) This item appears as such in AT&T's annual report and combines entries such as accounts payable, taxes accrued, and debt maturing within one year. It varies over the years as a percentage of operating costs (line 18), but without any discernible trend.

Deferred Credits (Lines 19, 20, and 21) This item, which appears as such in the AT&T annual reports, consists of accounts entitled "accumulated deferred income taxes" and "unamortized investment tax credits." These accounts arise mainly because of the differences in the accounting methods as used for reporting to shareholders and to the IRS. These deferred credits have grown rapidly over the last few years and can be expected to continue to do so in coming years, barring a major change in the U.S. tax code. For this reason, deferred credits cannot sensibly be projected as a fixed percentage of operating costs (line 20). Instead, they have been tied to the item that largely generates them, investment in new plant and equipment; specifically, the increase in deferred credits for the projection years was assumed to be a fixed percentage of the change in plant and equipment account (line 21).

Modifications to AT&T's Embedded Direct Cost Study: Removing the Implicit Return on Capital

In the EDC studies AT&T included an implicit return on capital, without explaining how it was estimated. For the purposes of the projections, it was necessary to take out this return on capital from the cost projections. To do so, it was assumed that the EDC "return on capital" included interest charges, federal income taxes, and net income. These items, which also could be called earnings before interest and taxes (EBIT), were attributed to the four familiar cost categories (local, joint, common, and others) in proportion to the gross plant and equipment investment for each of these categories. This method was consistent with AT&T's statement that when EDC expenses could not be assigned specifically and directly to any one cost category, they were attributed on the basis of a proxy, such as relative investment.

Gross plant and equipment (P&E) classifications as published by AT&T do not match directly with the cost classifications used in the EDC study. However, using estimates published by Probe Research Incorporated, we obtained reasonable estimates of P&E categories.

This distribution was used in Table B–3 to remove the implicit return on capital from the cost categories as they appeared in the EDC study. The revised costs appear in the right-hand column of Table B–3.

These revised costs were the basis for the financial projections, as reported in the following section. They serve as base year (1975) numbers to which specific growth rates were applied for the projection years (1976–1985).

PRO FORMA FINANCIAL STATEMENTS, 1975–1985

Each projection of the financial statements over the 1976–1985 period was based on one specific set of assumptions about a possible future. Each such set will be called a scenario and should *not* be viewed as a forecast but rather as an analysis of what might happen under certain assumptions about the future.

General Description of Table Formats

The financial projections consist of a set of two tables, an income statement, and a balance sheet, for each scenario. On each set of two tables all financial items are included for the full 1975–1985 period, as well as a special column that summarizes briefly the hypotheses used in the projection. A general description of each item in these tables is given below.

Income Statements (Table B–4) Revenues for 1975 are the same as appear in the Bell EDC study in the two categories "other" and "local." Some scenarios include a line called "potential for diversion"; this line is generally an estimate of the revenue that could be diverted by competition without jeopardizing unduly AT&T's financial integrity. During the projection years, revenues are assumed to grow at the rate indicated in the column entitled "hypotheses."

Costs of 1975 are tied directly to costs appearing in the Bell EDC study, but they have been reduced proportionately so that the following line, earnings before interest and taxes (EBIT), can be the sum of its components as they appear in the 1975 AT&T Annual Report. This adjustment to costs had to be made because the EDC study assumed an implicit return on capital. Costs are thus defined as being equal to revenues minus EBIT for the year 1975. For the following years, cost

Table B–3. Modifications to AT&T's Embedded Direct Cost (EDC) Study for 1975

EDC Category	Initial EDC Costs (billions of $)	Distribution of Plant and Equipment (Percent)	Adjustments to Initial Costs (billions of $)	Revised EDC Cost (billions of $)
Local telephone service	3.92	21.1	(1.61)	2.31
Joint (access) costs	6.99	27.3	(2.08)	4.91
Common (overhead) costs	5.02	12.8	(0.98)	4.04
All other OTC costs	11.54	38.8	(2.96)	8.58
Total	27.47	100.0	(7.62)[a]	19.85

Source: Calculations by Charles River Associates, based on information provided in AT&T's annual reports, AT&T's Embedded Direct Cost Study for 1975, and "The Future of AT&T" (a report by Probe Research Incorporated, circa Fall 1976).

[a]This total is equal to the return on capital for the year 1975, defined as being the sum of interest charges, provisions for federal income taxes, and net income, as they appear in AT&T's 1975 Annual Report.

Table B–4. Sample AT&T Pro Forma Income Statement, 1975–1985 (billions of $, unless otherwise noted)

	Item	Hypothesis (percent)	1975	1976	1977	1978	1979	1980	1981	1982	1983	1984	1985
I	Revenues												
	Other												
	Surplus for diversion												
	Other net												
	Local												
	Subtotal												
II	Costs												
	Other												
	Common												
	Joint												
	Local												
	Subtotal												
	Net Subtotal												
III	Earnings Before Interest and Taxes (EBIT)												
	Extra Depreciation												
	EBIT (Revised)												
IV	Interest												
V	Profit before taxes												
VI	Federal tax												
VII	Net income												
VIII	Dividends												
IX	Retained earnings												
X	Net income/beginning-of-year equity												
XI	EBIT/interest												

categories are assumed to grow as indicated in the column "hypothesis."

All categories of costs are added in the "subtotal" line to form the equivalent of "operating costs" that appear on line 4 of Table B–1 and line 12 of Table B–2. As already explained, this definition of costs combines many miscellaneous items and cannot be compared directly with the item called "total operating expenses" that appears in AT&T's reported financial statements.

Earnings before interest and taxes (EBIT) for 1975 is taken from the 1975 Bell Annual Report as the sum of net income plus interest charges plus federal income tax charges. Other items in the 1975 Bell Annual Report are combined as positive or negative adjustments in the costs category (e.g., property taxes, Western Electric profits). For the projection years, EBIT is obtained by subtracting the "costs subtotal" from the "revenues subtotal."

In one scenario, an item called "extra annual depreciation" is deducted from an "initial" EBIT to give a "revised" EBIT. This extra depreciation results from an assumption in this scenario that the depreciation rate gradually doubles over the years. This extra item was not included as an element of the costs subtotal so as not to modify the ratios used for estimating some items on the balance sheet.

Interest for 1975 is the item named "interest deductions" as it appears in the 1975 Bell Annual Report. For the following years interest payments are computed and calculated as being the product of the interest rate (growing as indicated in the "hypothesis" column) and the total debt at the beginning of the specific year (as appearing on the balance sheet).

The interest rate is assumed to increase to 10 percent in the final year for the historical scenarios, to 8 percent for the *No-Inflation Scenario* and to 9 percent for all nominal scenarios. This combines the effect of a long-term trend toward higher interest rates and rollover of existing debt.

Profit before tax (PBT) does not appear as such in the AT&T annual reports; it is here defined as being the difference between EBIT and interest charges. This item is needed to compute federal tax charges.

The federal income tax charges for 1975 are the sum of the items "federal income, current and deferred" and "investment tax credits" as they appear in the 1975 AT&T Annual Report. For the following years, federal tax is assumed to be equal to 40 percent of PBT. As explained elsewhere, AT&T uses a different set of financial statements for reporting to IRS, and its actual payments to the federal treasury are lower than these estimates, which is taken into account in estimating the deferred credits item of the balance sheet.

Net income for 1975 is the same as appears in the AT&T annual report. For the following years, it is taken as being the difference between PBT and federal taxes. In some scenarios, it will be assumed that net income is constrained never to exceed 12 percent of equity. When that is the case, it is so indicated in the hypothesis column.

Dividends for 1975 include both common and preferred dividends as they appear in AT&T's annual report. For the following years, they are assumed to be 65 percent of net income.

Retained earnings for 1975 do not appear as such in the AT&T annual report; they are defined to be equal to the difference between net income and dividends. This item serves to increase the funds available to AT&T and is thus added to the equity account of the balance sheet in the projections.

The item "net income/beginning-of-year equity" is not part of the income statement but is a financial ratio that measures the return on equity. It is one of three measures of financial integrity used, the others being the interest coverage ratio and the debt-to-capitalization ratio. Return on equity is thus defined as being equal to the ratio of net income to the equity account at the beginning of that year (as it appears on the balance sheet). Thus it may differ slightly from other definitions that may weight the equity account for any variations occuring during the year, or may distinguish between common and preferred shares.

The EBIT/interest item is used as a measure of interest coverage. It is an indicator commonly used by institutional lenders. Other definitions of interest coverage, such as cash flow to interest, are also used by lenders but are not included here.

Balance Sheets (Table B–5) The number of phones represents the total number of phones in service at the end of each year. The 1975 figure is taken from AT&T's annual report, and the number of phones in the ten following years grows at the indicated rate. The line below the number of phones shows the variation in number of phones that occurred during the year, as derived by subtracting the previous line.

Plant and equipment represents the annual investment in plant and equipment. The 1975 figure is taken from AT&T's annual report, and the following years are obtained by multiplying the annual variation in the total number of telephones by the average investment per phone, which starts at $1,400 in 1975 and is assumed to grow every year by the percentage rate indicated in the hypothesis column.

The 1975 gross plant in service, before depreciation, is taken directly from AT&T's annual report. It represents a cumulative investment and increases each year by the amount computed above as plant and equipment investment.

The net plant and equipment (P&E net) for 1975 is taken directly

Table B–5. Sample AT&T Pro Forma Balance Sheet, 1975–1985 (billions of $, unless otherwise noted)

Item	Hypothesis (percent)	1975	1976	1977	1978	1979	1980	1981	1982	1983	1984	1985
I # Phones (millions)												
II △ Phones (millions)												
III △ Plant and equipment (P + E)												
IV Gross Plant In Service (GPIS)												
V P + E net												
Extra depreciation												
Annual (percent) (billions of $)												
Cumulative (percent)												
Revised P + E net												
VI Investments												
VII Current assets and deferred charges												
VIII Total assets												
IX Current liabilities												
X Deferred credits												
XI Equity: Beginning-Of-Year (BOY)												
Retained earnings												
Net issue												
End-Of-Year (EOY)												
XII Debt: BOY												
Net issue												
EOY												
XIII Total liabilities												
XIV Debt/capitalization (EOY)												

from AT&T's annual report; it is net of depreciation and adjustment for plant in construction, and so on. For the projections, this number is assumed to be a fixed percentage of the gross P&E subtotal as indicated in the hypothesis column.

In one scenario, it is assumed that the depreciation rate is effectively doubled over a period of years. Four lines are added for that scenario. The first indicates how many percentage points are added to the depreciation rate, the second computes the extra annual depreciation expense, which is subtotaled in the third line and netted out in the fourth line to give a "revised" P&E net.

The figure for investments for 1975 is taken directly from AT&T's annual report and consists mostly of Bell's equity investment in Western Electric. For the projection years, this item is assumed to be a fixed percentage of operating costs as they appear in the income statement. The percentage used for this calculation is indicated in the hypothesis column.

The figure for current assets and deferred charges for 1975 is taken from AT&T's annual report and consists mostly of cash and accounts receivable. For the projection years, this item is assumed to be a fixed percentage (as indicated in the hypothesis column) of operating costs for that year, as they appear in the income statement under the label "costs subtotal."

The total assets figure for 1975 is taken directly from AT&T's annual report. For the following years, it consists of the sum of three previous items: P&E net (or P&E net revised, if applicable), investments, and current assets and deferred charges.

The current liabilities figure for 1975 is taken directly from AT&T's annual report and consists mostly of accounts payable and of the current portion of long-term debt. For the following years this item is assumed to be a fixed percentage (as indicated in the hypothesis column) of operating costs for that year, which appear in the income statement as "subtotal costs."

Deferred credits for 1975 come directly from AT&T's annual report and consist of accounts entitled "accumulated deferred income taxes" and "unamortized investment tax credits." For the following years, in all scenarios but one, the variation in deferred credits is set equal to 30 percent of the variation in gross plant in service during the year. As explained before, this item arises because of the difference in financial reports presented to the shareholders and to IRS. For the scenario in which accelerated depreciation is used, deferred credits are calculated separately.

The beginning-of-year (BOY) equity account for 1975 is the same as in the annual report. For the projection years, this item is exactly the same as the end-of-year (EOY) equity account for the immediately preceding year. This account includes all components of equity: common and preferred shares, cumulative retained earnings, and so on.

The retained earnings account is an annual amount that is taken from the income statement. The figure for 1975 is derived from AT&T's annual report.

The net issue item includes all stock issues (both common and preferred) and any other adjustments to the capital account. The figure for 1975 is derived from AT&T's annual report. For the projected years it is determined in conjunction with the item labeled "debt: net issue." Together, these two net issue items must fill any gap (positive or negative) that may arise between total assets and total liabilities. They are residual items that are calculated every year.

End-of-year equity consists of the sum of the three items just above it: equity BOY, retained earnings, and net issues. The figure for 1975 is the same as in AT&T's annual report.

The beginning-of-year (BOY) debt account for 1975 is the same as in AT&T's annual report. For the projection years this item is exactly the same as the end-of-year (EOY) debt account for the immediately preceding year.

The "debt: net issue" item includes all new debt issued, net of redemptions of outstanding debt. The figure for 1975 is derived from AT&T's annual report. As explained above, it is determined in conjunction with the item labeled "equity: net issue" for the projection years.

It is also assumed that AT&T issues debt, rather than stock, to fill any financial gap, as long as the debt/capitalization ratio does not exceed a predetermined percentage (45 percent in the projections).

The end-of-year debt for 1975 is the same as in AT&T's annual report. This item consists of the sum of debt BOY and debt net issue for all the projection years.

The total liabilities figure for 1975 is taken directly from AT&T's annual report. For the projection years this item is, by definition, set equal to total assets. It is also equal to the sum of current liabilities, deferred credits, equity EOY, and debt EOY.

The debt/capitalization ratio, the third measure used for financial integrity, is obtained by dividing debt EOY by the sum of debt EOY and equity EOY. For the projection years it is assumed that this ratio should decline continuously and never again exceed a predetermined level of 45 percent. If outside financing needs were too large to be accommodated within this limit, equity would be issued in order to maintain the 45-percent debt ratio. Of course, equity issues have the effect of diluting the return on equity so that equity issues drive down the first measure of financial integrity, return on equity.

SCENARIOS AND RESULTS FOR PROJECTION YEARS

The financial statements for each scenario for the ten-year projection period appear on the following pages. The graphs and summary tables included in Chapter 2 are not reproduced here but are useful reference.

Historical A Scenario (Tables B–6 and B–7)

This scenario suggests what AT&T's financial condition might be if the 1973–1975 experienced growth trends were extrapolated to 1985. The 1973–1975 period was chosen because, owing in part to difficulties created by the external environment, AT&T's financial condition was under increasing pressure. We should note that in this period common (overhead) costs, as defined in AT&T's embedded direct cost (EDC) studies, were growing at 16.5 percent a year, which was much faster than the overall growth rate for operating costs. It is not clear what proportion of this increase was due to higher incurred costs or merely to a reclassification among cost categories.

The growth rates are taken directly from Chapter 2, where they were estimated on the basis of AT&T's EDC studies.

It is further assumed that the investment cost per net new phone will increase by 10 percent annually over the projection years. This assumption is quite generous for AT&T since there is no such observable trend in the 1971–1977 period, and technological progress should also help keep down the costs of most transmission equipment.

Finally, it was assumed, as in all other scenarios, that the shareholders would receive 65 percent of net income each year as dividends and that the debt-to-capitalization ratio would be constrained not to exceed 45 percent in the projection years.

Given these assumptions, the financial condition of AT&T improves gradually over the early years, and then it edges down slowly at the end of the period. This slight degradation of financial indicators is mainly because of the unconstrained growth of overhead costs, which clearly need not happen in reality. The *Historical B Scenario* takes this factor into account.

Nevertheless, the financial situation is better for each of the projection years than was the case in 1975. As can be seen from Tables B–6 and B–7, return on equity closes at 12 percent, interest coverage ratio stands at 3.4, and the debt ratio is at its constrained maximum of 45 percent.

Historical B Scenario (Tables B–8 and B–9)

The *Historical A Scenario* is modified in the *Historical B Scenario* by constraining common (overhead) costs to remain at the same proportion of total operating costs as they were in 1975. This implicitly assumes that AT&T would be able to control the growth of its overhead costs. All other hypotheses are strictly the same as in the *Historical A Scenario*.

The results of this scenario, as they appear in Tables B–8 and B–9, show a constant improvement over the years, both in terms of the return on equity and the interest coverage ratio. But, because of the magnitude of the outside financing needs, the debt-to-capitalization ratio cannot be reduced below 45 percent without issuing more stock. This increase in outside financing need is due to the assumption of increasing investment cost per net new phone.

Historical C Scenario (Tables B–10 and B–11)

The *Historical A Scenario* is further modified in the *Historical C Scenario* by letting local phone rates grow at the full inflation rate of 10 percent a year. Given an assumed volume growth of local phone conversations of 4 percent a year, local revenues increase 14.4 percent a year, instead of 9.5 percent as in the *Historical A* and *B* scenarios.[2]

As can be seen from Tables B–10 and B–11, financial results improve even more, return on equity shooting up to 29.6 percent, the interest coverage ratio increasing to 7.2 times, and the debt-to-capitalization ratio even falling below the 45-percent level.

These results would happen while local phone rates are kept constant in real terms relative to the 1975 level and while all other rates increase nominally at only 2.8 percent a year in a general inflation of 10 percent.

No-Inflation Scenario (Tables B–12 and B–13)

The *No-Inflation Scenario* suggests what AT&T's financial condition might be in an inflation-free environment. The growth rates for costs and revenues are taken directly from Chapter 2, where they were estimated as a measure of historical volume growth in the Bell System. One exception was made, in this and all following scenarios, for common costs, which are assumed to remain at a fixed proportion of total operating costs.

This scenario also assumes no increased inroads from competition in long distance transmission or station equipment. Finally, phone rates for all types of services remain at their 1975 level for the full 10-year period. It is further assumed that the investment cost per net new phone will remain at the same level over the projection years.

Given these assumptions of no inflation, relatively low growth, and generous profit margins, the financial condition of AT&T improves every year. As can be seen from Tables B–12 and B–13, return on equity increases regularly to 16.4 percent in 1985; the interest coverage ratio goes up to 5.2 and the debt-to-capitalization ratio falls below its allowed maximum in the final years of the projection period to 44.8 percent.

Nominal A Scenario (Tables B–14 and B–15)

The *No-Inflation Scenario* is modified in this and all following scenarios by factoring in a general inflation of 6 percent a year. Common costs are still limited to a fixed percentage of total operating costs.

In this and all following scenarios, joint costs grow with the number of phones (4.2 percent) and the general inflation (6 percent), while local costs grow with the number of conversations (4.5 percent) and the general inflation (6 percent). Finally, "other" costs are assumed to grow with volume only (10.2 percent), which means that unit costs will remain constant over the years because of the combined effect of productivity increases, economies of scale, and other factors.

Local phone revenues are assumed to grow with the number of conversations (4.5 percent a year) and the general inflation rate (6 percent). Other revenues grow with volume (10.2 percent as in the historical 1973–1975 period) and a rate of inflation of 2.2 percent (2.2 percent is in the same proportion to a 6-percent general inflation rate as the 2.6 percent 1971–1977 rate increase was to a general inflation of 7 percent). This effectively means that rates for "other" revenue services are allowed to lag behind the general price level. In real terms, the 1985 average price of these other services would be only 69 percent of what they were in 1975, allowing AT&T to meet the threat of increased competition.

Finally, in this and the following scenarios, the investment cost per net new phone is assumed to increase by 6 percent a year, even though no such trend was evident in the 1971–1977 period.

The results of this scenario, as they appear in Tables B–14 and B–15, still show a notable improvement over the years, both in terms of the return on equity and the interest coverage ratio. But, because of the magnitude of the outside financing needs, the debt-to-capitalization ratio cannot be reduced below 45 percent without issuing more stock, except in the last year.

Nominal B Scenario (Tables B–16 and B–17)

The *Nominal B Scenario* assumptions differ from the *Nominal A* only in that local phone rates are limited to growing at only half the inflation rate (as compared to the full inflation rate for the *A Scenario*). "Other" revenues and each cost category are assumed to grow at the same rate as in the *Nominal A Scenario*.

The results of this scenario appear in Tables B–16 and B–17. The return on equity, even though lower than in the *Nominal A Scenario*, improves every year over the 1975 levels, as does the interest coverage ratio. This clearly indicates that despite an environment of 6-percent general inflation, a more than adequate return will be generated by a slight increase in the "other" service rate structure and small (less than general inflation) increases in local rates.

Nominal C Scenario (Tables B–18 and B–19)

This scenario is generally based on the same hypotheses as the *Nominal A Scenario*, but differs from it in one major respect — the return on equity is constrained not to exceed 12 percent per year. Any potential surplus in net income is used to reduce proportionately the item called "other revenues." This reduction in other revenues is done by comparison with what other revenue would have been if it had continued to grow at the *Nominal A Scenario* rate, 12.6 percent.

As seen from Tables B–18 and B–19, the potential surplus, also called surplus available for diversion, starts at zero in 1976 and reaches $13.9 billion in 1985. As explained in Chapter 2, assuming that revenues vulnerable to competition amount to 65.9 percent of "other" revenues, this means that competitive service revenues could fall to 65.3 percent of what they would otherwise have been in 1985 and still allow AT&T to earn a 12-percent return on equity.[3] This revenue reduction could come through a number of combinations of rate, volume, and price elasticity variations. The surplus available for diversion in this scenario exceeds by a wide margin estimates made in Chapter 2 for revenues of competitors.

For consumers, the *Nominal C Scenario* means that local phone rates increase at the general inflation level (i.e., staying constant in real terms relative to the 1975 level). As for the three measures of financial integrity, two of them are constant over the years. Return on equity reaches its allowed maximum in 1977, while the debt-to-capitalization ratio reaches its allowed maximum in 1978. For its part, the interest coverage ratio increases in the early years and edges down slowly in the years following, while always remaining better than the initial 1975 ratio.

Nominal D Scenario (Tables B–20 and B–21)

This scenario is based on the same assumptions as *Nominal B Scenario*, with one modification — the return on equity is now constrained not to exceed 12 percent a year. Local phone rates still increase at only half the inflation rate, and any potential surplus in net income is again used to reduce "other" revenues from what they would otherwise have been. This reduction gives an indication of the competitive diversion that AT&T could tolerate without impairing achievement of its target return on equity.

The results appear in Tables B–20 and B–21. The three measures of financial integrity behave much as in the *Nominal C Scenario* because the only difference between the two scenarios is whether rate reductions occur only in competitive services or also in local service.

As for diversion to meet competition, if the market for competitive services were growing at 12.6 percent a year, AT&T could sustain a diversion of about $7.1 billion in 1985 without jeopardizing its financial position. For consumers, local phone rates fall in real relative terms over the 10-year period, to 75 percent of what they were in 1975, and rates for all other services fall to less than 61 percent of what they were in 1975.

Nominal E Scenario (Tables B–22, B–23, and B–24)

The object of this scenario is to evaluate the effect of using a higher depreciation rate for telco plant and equipment.

The principal assumptions of this scenario are: (1) a general inflation rate of 6 percent a year; (2) local phone rates increasing at this general inflation rate of 6 percent a year; (3) return on equity never exceeding 12 percent a year (any potential surplus in net income over and above 12 percent being used to reduce revenues coming from "all other" revenues); (4) the depreciation rate for shareholder reporting gradually doubling over a five-year period, thereby increasing allowed costs and the cash flow available to AT&T (by permitting smaller phone rate reductions for customers than would otherwise have been the case); and (5) no increase in AT&T investment outlays because of the accelerated depreciation.[4]

With the *E Scenario*, an analytical problem arises because AT&T uses a higher, but undisclosed, depreciation rate when reporting to the IRS than when reporting to the public. This factor, in conjunction with the investment tax credit, has the effect of reducing the taxes effectively paid to the IRS. This difference between tax charges (as reported to shareholders) and taxes paid to the IRS appears as deferred income taxes on AT&T's income statement and cumulatively as deferred credits on its balance sheet.

For *Nominal E* it is assumed that increasing the depreciation rate for shareholder reporting purposes (SRP) would not affect the depreciation rate used for IRS reporting purposes. It is further assumed that these higher SRP depreciation charges will be compensated by higher overall revenues, coming either from volume increases or higher rates than would otherwise be the case. Specifically, instead of using all growth-induced increases in revenues to reduce phone rates, it is assumed that part of this surplus is used to offset higher depreciation charges. This has the effect in the *Nominal E Scenario* of increasing the cash flow available to AT&T.[5] An increase of $1 in depreciation charges compensated by revenues being $1 higher will mean a decrease of $.40 in deferred income taxes and a net income stable at the same previous level, for an increase of $.60 in cash flow.

This increased cash flow for AT&T will mean smaller outside financial needs and a higher interest coverage ratio, even though the reported return on equity will be the same as in *C* and *D* scenarios. But, while AT&T is stronger financially, this causes phone rates, local or all other rates, to be higher than they would otherwise have been, or as in *C* and *D*.

In order to generate the *E* pro formas (see Tables B–22 and B–23), the combined effects of extra depreciation, higher revenues, lower outside financing needs, and lower interest charges must be taken into account. One especially important financial effect of this scenario is to reduce drastically the annual deferred income taxes, thus lowering the cumulative deferred credits that appear on the balance sheet. A special worksheet was used to compute these deferred credits, as shown in Table B–24.

On this worksheet, the first step is to obtain the amount effectively paid in taxes (lines 1 to 4) from a reference scenario. For this purpose *C Scenario* was selected since all basic elements of cost and revenue were grown at the same rate in *E Scenario* as in *C*. Lines 1 and 2 in Table B–24 are taken directly from *C Scenario*; line 3 is the annual difference between the amount appearing on line 2, and line 4 is the difference between line 1 and line 3.

The second step (lines 5 to 7) is to calculate the difference, if any, in revenues from *C* and *E* scenarios. Any such difference when positive would result in higher effective taxes paid under *E Scenario*. These extra taxes occur because, in the report given to the IRS, revenues are higher than in the *C Scenario*, but depreciation charges are the same.

The third step (lines 8 to 10) estimates the difference in annual interest charges between *C* and *E* scenarios. This difference arises because of a smaller outstanding debt in *E Scenario* due to lower outside financial needs. But interest savings also mean higher taxes paid.

The fourth step (lines 11 and 12) aggregates higher revenues and interest savings and then estimates taxes to be paid on this extra amount. The tax rate is assumed to be the same as what it has been historically, namely, 40 percent of profit before tax.

The fifth step (lines 13 to 15) estimates the annual deferred income taxes for *E Scenario*. This results from the difference between federal tax charges (as they appear on the *E* income statement) and the taxes effectively paid (line 4 plus line 12).

The final step (line 16) is to accumulate these annual deferred credits over the year. The result is then transcribed to the balance sheet.

As can be seen from line 15, the annual deferred credits are relatively small, and when accumulated, the total will not vary much over the years. However, the deferred credits for *C* and *E* scenarios will differ substantially. In 1985, *C Scenario* deferred credits equal $37.0 billion, while *E Scenario* deferred credits amount to only $14.6 billion. In essence, *E Scenario* deferred credits are smaller because the annual report published for shareholders parallels more closely the IRS report.

As can be seen from Table B–22, both the return on equity and the interest coverage ratio go down in the first five projection years and then gradually improve afterward, when the total depreciation rate stays at a stable level of 10 percent starting in 1981.

The seemingly lower performance in the initial years is essentially due to the fact that depreciation expense increases faster than earnings before interest and taxes. But these accounting conventions are clearly misleading, for the cash flow available to the company does increase year after year. As a rough indication, if in our definition of interest coverage ratio we included the extra depreciation expense (which does not require any cash outlay), the ratio would increase from 3.3 in 1975 to 5.5 in 1980 (instead of the 3.2 ratio appearing in Table B–22).

This transition period to a higher depreciation rate could potentially be painful both to AT&T and its shareholders if the extra depreciation expenses are not fully covered by higher revenues. In this scenario, for example, where return on equity decreases slightly for the first five years, total dividend payments would increase very little in dollar terms if they were kept at a fixed ratio of net income. This would effectively mean a slight decrease in dividends per share if equity is issued as assumed on the balance sheet. Of course, this need not happen. First, the dividend rate could be increased based on the fact that the cash flow is increased, and a different financial strategy could be used by issuing, for example, convertible debentures in the early years.

In any case, the measures of financial integrity improve in the last years, return on equity hitting its constrained maximum of 12 percent, the interest coverage ratio, as measured in the traditional way, nearing four times, and the debt-to-capitalization ratio falling to 44 percent.

As noted earlier, telephone rates will be higher in this scenario than they would otherwise have been, but this difference is not too large. As can be seen from Table 2–14 of Chapter 2, the *Nominal E Scenario* has the same local phone rates as in the *Nominal C Scenario,* and rates for "other" services are higher by 7 percentage points in real terms at the end of the 10-year period than in the *Nominal C Scenario.* Still, rates for other services would be only 60 percent in real terms as compared to their 1975 level.

- When the *Nominal E* rates are compared to those for *D,* the picture is as follows. Rates for other services are virtually the same, but local phone rates are higher in *E* than in *D* by 25 percent (but still equal to their 1975 level in real terms).

Overall, under *E Scenario,* AT&T could sustain a competitive diversion of $8.0 billion in 1985 while still improving its financial integrity. This scenario thus shows that growth in the telecommunications market could make possible both higher depreciation and lower service charges without threatening AT&T's financial position.

In the *E Scenario,* extra depreciation is assumed to happen gradually, as increases in revenues occur. Other scenarios could look at the effect of substantial capital write-offs that would not be offset by higher revenues. They, of course, would have a more negative effect on the financial status of AT&T than any gradual diversion of revenue induced by competition.

COMPARATIVE RESULTS: TESTING THE VALIDITY OF THE FINANCIAL RATIOS USED IN THE PROJECTIONS

To test the general validity of the structure of the models used in the pro forma projections, they can be compared with the actual results that were available for the first three years of the projection period, 1976, 1977, and 1978.[6]

For comparison with the actual results, we selected the three scenarios whose hypotheses best reproduced actual 1976, 1977, and 1978 events (see Tables B–25 through B–30).[7] For these three years, and for both the income statement and the balance sheet, the results are quite similar between the actual record and the three selected scenarios. Even though the differences between projected and actual figures may be greater for some individual items than others, the general validity of the model seems reasonably well established.

NOTES

1. Increased competition in customer-premise telephone equipment may result in smaller increases for AT&T-owned telephone instruments. This will not affect the projection materially, since the capital cost for the telephone instrument itself is only a very minor part of the overall required investment in plant and equipment (less than 3 percent on an historical cost basis).

2. When two or more annual growth rates are combined (e.g., volume and inflation), proper procedure is to multiply them, so that in the present case, we have: $1.10 \times 1.04 = 1.144$, where 14.4 is the combined growth rate. In the following scenarios, only the combined growth rates will appear in the tables, while the text will indicate what the components were. In addition, all growth rates are compounded on an annual basis.

3. $65.3 \text{ percent} = (1 - \dfrac{\text{diversion}}{\text{competitive services}}) \times 100$

$= (1 - \dfrac{\$13.9 \text{ billion}}{0.659 \times \$60.7 \text{ billion}}) \times 100$

4. This may not happen in reality, of course, since one purpose of using accelerated depreciation would be to encourage AT&T to modernize its equipment faster than it does now. This neutral investment assumption therefore exerts a downward bias on AT&T gross financing needs.

5. Cash flow is defined as being the sum of net income, depreciation charges, and deferred income taxes.

6. To replace the comparisons on a common basis, the historical revenues for 1976 and 1977 have been adjusted downward to 94.9 percent of what they are in the annual report to conform with the historical pattern of the EDC studies. The historical costs for 1976 and 1977 also do not have the same definitions as those in the Bell annual reports and have been reconstructed as being equal, by definition, to EDC revenues (interest charges plus federal taxes plus net income).

7. Following a California Public Utilities Commission Order prescribing a refund of revenues and a reduction of rates, and a related Internal Revenue Service ruling, AT&T modified its 1978 financial statements and restated those for prior years. AT&T reduced slightly the reported net income and equity account for 1978 and prior years; it also shifted $1 billion from the deferred credits account to current liabilities for the year 1978. Since this order has farreaching implications and given that AT&T is still trying to have it reversed, Tables B-29 and B-30 only take into account the slight reductions in net income and equity, and not the major reclassification between liability accounts.

Table B-6. HISTORICAL A Scenario[a,b] AT&T PRO FORMA INCOME STATEMENT, 1975–1985 (billions of $, unless otherwise noted)

Item	Hypotheses[c] (percent)	1975	1976	1977	1978	1979	1980	1981	1982	1983	1984	1985
I. Revenues												
Other	11.90	18.52	20.72	23.19	25.95	29.04	32.49	36.36	40.69	45.53	50.95	57.01
Local	9.50	8.95	9.80	10.73	11.75	12.87	14.09	15.43	16.89	18.50	20.26	22.18
Subtotal	—	27.47	30.52	33.92	37.70	41.91	46.58	51.79	57.58	64.03	71.21	79.19
II. Costs												
Other	10.10	8.58	9.45	10.40	11.45	12.61	13.88	15.28	16.83	18.53	20.40	22.46
Common	16.50	4.04	4.71	5.48	6.39	7.44	8.67	10.10	11.77	13.71	15.97	18.61
Joint	9.90	4.92	5.41	5.94	6.53	7.18	7.89	8.67	9.53	10.47	11.51	12.65
Local	10.00	2.31	2.54	2.80	3.07	3.38	3.72	4.09	4.50	4.95	5.45	5.99
Subtotal	—	19.85	22.11	24.62	27.44	30.61	34.16	38.14	42.63	47.66	53.33	59.71
III Earnings before interest and taxes (EBIT)	—	7.62	8.41	9.30	10.26	11.30	12.42	13.65	14.95	16.37	17.88	19.48
IV. Interest	10.00	2.30	2.46	2.54	2.75	3.02	3.32	3.67	4.07	4.53	5.06	5.67
V. Profit before taxes	—	5.32	5.95	6.76	7.51	8.28	9.10	9.98	10.88	11.84	12.82	13.81
VI. Federal tax	40.00	2.17	2.38	2.70	3.00	3.31	3.64	3.99	4.35	4.74	5.13	5.52
VII. Net income	—	3.15	3.57	4.06	4.51	4.97	5.46	5.99	6.53	7.10	7.69	8.29
VIII. Dividends	65.00	2.16	2.32	2.64	2.93	3.23	3.55	3.89	4.24	4.62	5.00	5.39
IX. Retained earnings	—	0.99	1.25	1.42	1.58	1.74	1.91	2.10	2.29	2.48	2.69	2.90
X. Net income/beginning-of-year equity (percent)	—	9.7	10.3	10.7	11.1	11.5	11.8	12.0	12.1	12.2	12.1	12.0
XI. EBIT/Interest (ratio)	—	3.3	3.4	3.7	3.7	3.7	3.7	3.7	3.7	3.6	3.5	3.4

Source: Calculations by Charles River Associates.

[a] Data for 1975 are calculated from AT&T's 1975 Annual Report and Embedded Direct Cost Study. Entries for subsequent years are projected according to the hypotheses explained in the text, using 1975 as the base year.

[b] Calculations are rounded to the nearest 100th.

[c] Hypotheses relate to assumed growth rates and financial ratios and are explained in the text.

Table B–7. Historical A Scenario[a,b] AT&T Pro Forma Balance Sheet, 1975–1985 (billions of $, unless otherwise noted)

Item	Hypotheses[c] (percent)	1975	1976	1977	1978	1979	1980	1981	1982	1983	1984	1985
I. Total phones (millions)	3.6	118.5	122.8	127.2	131.8	136.5	141.4	146.5	151.8	157.3	162.9	168.8
II. Change in phones (millions)	—	4.0	4.3	4.4	4.6	4.7	4.9	5.1	5.3	5.5	5.6	5.9
III. Change in plant and equipment (P+E)	10.0	7.0	6.6	7.5	8.6	9.6	11.0	12.6	14.5	16.5	18.5	21.4
IV. Gross plant in service (GPIS)	—	84.6	91.2	98.7	107.3	116.9	127.9	140.5	155.0	171.5	190.0	211.4
V. P+E net	83.0	70.4	75.7	81.9	89.1	97.0	106.2	116.6	128.7	142.3	157.7	175.5
VI. Investments	18.0	3.6	4.0	4.4	4.9	5.5	6.1	6.9	7.7	8.6	9.6	10.7
VII. Current assets and deferred charges	32.0	6.2	7.1	7.9	8.8	9.8	10.9	12.2	13.6	15.3	17.1	19.1
VIII. Total assets	—	80.2	86.8	94.2	102.8	112.3	123.2	135.7	150.0	166.2	184.4	205.3
IX. Current liabilities	37.0	6.9	8.2	9.1	10.2	11.3	12.6	14.1	15.8	17.6	19.7	22.1
X. Deferred credits	30.0	6.9	8.9	11.2	13.8	16.7	20.0	23.8	28.2	33.2	38.8	45.2
XI. Equity: beginning-of-year (BOY)	—	32.6	34.6	37.9	40.6	43.3	46.4	49.8	53.8	58.3	63.5	69.2
Retained earnings	—	1.0	1.3	1.4	1.6	1.7	1.9	2.1	2.3	2.5	2.7	2.9
Net issue	—	1.0	2.0	1.3	1.1	1.4	1.5	1.9	2.2	2.7	3.0	3.8
End-of-year (EOY)	—	34.6	37.9	40.6	43.3	46.4	49.8	53.8	58.3	63.5	69.2	75.9
XII. Debt: BOY	—	29.5	31.8	31.8	33.3	35.5	37.9	40.8	44.0	47.7	51.9	56.7
Net issue	—	2.3	0.0	1.5	2.2	2.4	2.9	3.2	3.7	4.2	4.8	5.4
EOY	—	31.8	31.8	33.3	35.5	37.9	40.8	44.0	47.7	51.9	56.7	62.1
XIII. Total liabilities	—	80.2	86.8	94.2	102.8	112.3	123.2	135.7	150.0	166.2	184.4	205.3
XIV. Debt/capitalization (EOY) (percent)	—	47.9	45.6	45.1	45.1	45.0	45.0	45.0	45.0	45.0	45.0	45.0

Source: Calculations by Charles River Associates.

[a]Data for 1975 are calculated from AT&T's 1975 Annual Report and Embedded Cost Study. Entries for subsequent years are projected according to the hypotheses explained in the text, using 1975 as the base year.

[b]Calculations are rounded to the nearest 10th.

[c]Hypotheses relate to assumed growth rates and financial ratios and are explained in the text.

Table B–8. Historical B Scenario[a,b] AT&T Pro Forma Income Statement, 1975–1985 (billions of $, unless otherwise noted)

Item	Hypotheses[c] (percent)	1975	1976	1977	1978	1979	1980	1981	1982	1983	1984	1985
I. Revenues												
Other	11.90	18.52	20.72	23.19	25.95	29.04	32.49	36.36	40.69	45.53	50.95	57.01
Local	9.50	8.95	9.80	10.73	11.75	12.87	14.09	15.43	16.89	18.50	20.26	22.18
Subtotal	—	27.47	30.52	33.92	37.70	41.91	46.58	51.79	57.58	64.03	71.21	79.19
II. Costs												
Other	10.10	8.58	9.45	10.40	11.45	12.61	13.80	15.28	16.83	18.53	20.40	22.46
Common	20.35	4.04	4.45	4.89	5.38	5.92	6.51	7.17	7.89	8.68	9.55	10.50
Joint	9.90	4.92	5.41	5.94	6.53	7.18	7.89	8.67	9.53	10.47	11.51	12.65
Local	10.00	2.31	2.54	2.80	3.07	3.38	3.72	4.09	4.50	4.95	5.45	5.99
Subtotal	—	19.85	21.85	24.03	26.43	29.09	32.00	35.21	38.75	42.63	46.91	51.60
III. Earnings before interest and taxes (EBIT)	—	7.62	8.67	9.89	11.27	12.82	14.58	16.58	18.83	21.40	24.30	27.59
IV. Interest	10.00	2.30	2.46	2.54	2.74	3.02	3.31	3.66	4.05	4.51	5.03	5.62
V. Profit before taxes	—	5.32	6.21	7.35	8.53	9.80	11.27	12.92	14.78	16.89	19.27	21.97
VI. Federal tax	40.00	2.17	2.48	2.94	3.41	3.92	4.51	5.17	5.91	6.76	7.71	8.79
VII. Net income	—	3.15	3.73	4.41	5.12	5.88	6.76	7.75	8.87	10.13	11.56	13.18
VIII. Dividends	65.00	2.16	2.42	2.87	3.33	3.82	4.39	5.04	5.77	6.58	7.51	8.57
IX. Retained earnings	—	0.99	1.31	1.54	1.79	2.06	2.37	2.71	3.10	3.55	4.05	4.61
X. Net income/beginning-of-year equity (percent)	—	9.7	10.8	11.7	12.6	13.6	14.6	15.6	16.5	17.4	18.3	19.2
XI. EBIT/interest (ratio)	—	3.3	3.5	3.9	4.1	4.2	4.4	4.5	4.6	4.7	4.8	4.9

Source: Calculations by Charles River Associates.

[a]Data for 1975 are calculated from AT&T's 1975 Annual Report and Embedded Direct Cost Study. Entries for subsequent years are projected according to the hypotheses explained in the text, using 1975 as the base year.

[b]Calculations are rounded to the nearest 100th.

[c]Hypotheses relate to assumed growth rates and financial ratios and are explained in the text.

Table B-9. Historical B Scenario[a,b] AT&T Pro Forma Balance Sheet, 1975–1985 (billions of $, unless otherwise noted)

Item	Hypotheses[c] (percent)	1975	1976	1977	1978	1979	1980	1981	1982	1983	1984	1985
I. Total phones (millions)	3.6	118.5	122.8	127.2	131.8	136.5	141.4	146.5	151.8	157.3	162.9	168.8
II. Change in phones (millions)	—	4.0	4.3	4.4	4.6	4.7	4.9	5.1	5.3	5.5	5.6	5.9
III. Change in plant and equipment (P+E)	10.0	7.0	6.6	7.5	8.6	9.6	11.0	12.6	14.5	16.5	18.5	21.4
IV. Gross plant in service (GPIS)	—	84.6	91.2	98.7	107.3	116.9	127.9	140.5	155.0	171.5	190.0	211.4
V. P+E net	83.0	70.4	75.7	81.9	89.1	97.0	106.2	116.6	128.7	142.3	157.7	175.5
VI. Investments	18.0	3.6	3.9	4.3	4.8	5.2	5.8	6.3	7.0	7.7	8.4	9.3
VII. Current assets and deferred charges	32.0	6.2	7.0	7.7	8.5	9.3	10.2	11.3	12.4	13.6	15.0	16.5
VIII. Total Assets	—	80.2	86.6	93.9	102.4	111.5	122.2	134.2	148.1	163.6	181.1	201.3
IX. Current liabilities	37.0	6.9	8.1	8.9	9.8	10.8	11.8	13.0	14.3	15.8	17.4	19.1
X. Deferred credits	30.0	6.9	8.9	11.2	13.8	16.7	20.0	23.8	28.2	33.2	38.8	45.2
XI. Equity: beginning-of-year (BOY)	—	32.6	34.6	37.8	40.6	43.3	46.2	49.7	53.6	58.1	63.0	68.7
Retained earnings	—	1.0	1.3	1.5	1.8	2.1	2.4	2.7	3.1	3.6	4.1	4.6
Net issue	—	1.0	1.9	1.3	0.9	0.8	1.1	1.2	1.4	1.3	1.6	2.0
End-of-year (EOY)	—	34.6	37.8	40.6	43.3	46.2	49.7	53.6	58.1	63.0	68.7	75.3
XII. Debt: BOY	—	29.5	31.8	31.8	33.2	35.5	37.8	40.7	43.8	47.5	51.6	56.2
net issue	—	2.3	0.0	1.4	2.3	2.3	2.9	3.1	3.7	4.1	4.6	5.5
EOY	—	31.8	31.8	33.2	35.5	37.8	40.7	43.8	47.5	51.6	56.2	61.7
XIII. Total liabilities	—	80.2	86.6	93.9	102.4	111.5	122.2	134.2	148.1	163.6	181.1	201.3
XIV. Debt/capitalization (EOY)(percent)	—	47.9	45.7	45.0	45.1	45.0	45.0	45.0	45.0	45.0	45.0	45.0

Source: Calculations by Charles River Associates.

[a]Data for 1975 are calculated from AT&T's 1975 Annual Report and Embedded Cost Study. Entries for subsequent years are projected according to the hypotheses explained in the text, using 1975 as the base year.

[b]Calculations are rounded to the nearest 10th.

[c]Hypotheses relate to assumed growth rates and financial ratios and are explained in the text.

Table B–10. Historical C Scenario[a,b] AT&T Pro Forma Income Statement, 1975–1985 (billions of $; unless otherwise noted)

Item	Hypotheses[c] (percent)	1975	1976	1977	1978	1979	1980	1981	1982	1983	1984	1985
I. Revenues												
Other	11.90	18.52	20.72	23.19	25.95	29.04	32.49	36.36	40.69	45.53	50.95	57.01
Local	14.40	8.95	10.24	11.71	13.40	15.33	17.54	20.06	22.95	26.26	30.04	34.36
Sub-total	—	27.47	30.96	34.90	39.35	44.37	50.03	56.42	63.64	71.79	80.99	91.37
II. Costs												
Other	10.10	8.58	9.45	10.40	11.45	12.61	13.88	15.28	16.83	18.53	20.40	22.46
Common	20.35	4.04	4.45	4.89	5.38	5.92	6.51	7.17	7.89	8.68	9.55	10.50
Joint	9.90	4.92	5.41	5.94	6.53	7.18	7.89	8.67	9.53	10.47	11.51	12.65
Local	10.00	2.31	2.54	2.80	3.07	3.38	3.72	4.09	4.50	4.95	5.45	5.99
Subtotal	—	19.85	21.85	24.03	26.43	29.09	32.00	35.21	38.75	42.63	46.91	51.60
III. Earnings before interest and taxes (EBIT)	—	7.62	9.11	10.87	12.92	15.28	18.03	21.21	24.89	29.16	34.08	39.77
IV. Interest	10.00	2.30	2.46	2.54	2.74	3.02	3.31	3.66	4.05	4.51	5.00	5.55
V. Profit before taxes	—	5.32	6.65	8.33	10.18	12.26	14.72	17.55	20.84	24.65	29.08	34.22
VI. Federal tax	40.00	2.17	2.66	3.33	4.07	4.90	5.89	7.02	8.34	9.86	11.63	13.69
VII. Net income	—	3.15	3.99	5.00	6.11	7.36	8.83	10.53	12.50	14.79	17.45	20.53
VIII. Dividends	65.00	2.16	2.59	3.25	3.97	4.78	5.74	6.84	8.13	9.61	11.34	13.34
IX. Retained earnings	—	0.99	1.40	1.75	2.14	2.58	3.09	3.69	4.37	5.18	6.11	7.19
X. Net income/ beginning-of-year equity (percent)	—	9.7	11.5	13.2	15.0	17.0	19.1	21.2	23.3	25.5	27.6	29.6
XI. Ebit/interest (ratio)	—	3.3	3.7	4.3	4.7	5.1	5.4	5.8	6.1	6.5	6.8	7.2

Source: Calculations by Charles River Associates.

[a]Data for 1975 are calculated from AT&T's 1975 Annual Report and Embedded Direct Cost Study. Entries for subsequent years are projected according to the hypotheses explained in the text, using 1975 as the base year.

[b]Calculations are rounded to the nearest 100th.

[c]Hypotheses relate to assumed growth rates and financial ratios and are explained in the text.

noted)

Item	Hypotheses[c] (percent)	1975	1976	1977	1978	1979	1980	1981	1982	1983	1984	1985
I. Total phones (millions)	3.6	118.5	122.8	127.2	131.8	136.5	141.4	146.5	151.8	157.3	162.9	168.8
II. Change in phones (millions)	—	4.0	4.3	4.4	4.6	4.7	4.9	5.1	5.3	5.5	5.6	5.9
III. Change in plant and equipment (P+E)	10.0	7.0	6.6	7.5	8.6	9.6	11.0	12.6	14.5	16.5	18.5	21.4
IV. Gross plant in service (GPIS)		84.6	91.2	98.7	107.3	116.9	127.9	140.5	155.0	171.5	190.0	211.4
V. P+E net	83.0	70.4	75.7	81.9	89.1	97.0	106.2	116.6	128.7	142.3	157.7	175.5
VI. Investments	18.0	3.6	3.9	4.3	4.8	5.2	5.8	6.3	7.0	7.7	8.4	9.3
VII. Current assets and deferred charges	32.0	6.2	7.0	7.7	8.5	9.3	10.2	11.3	12.4	13.6	15.0	16.5
VIII. Total assets	—	80.2	86.6	93.9	102.4	111.5	122.2	134.2	148.1	163.6	181.1	201.3
IX. Current liabilities	37.0	6.9	8.1	8.9	9.8	10.8	11.8	13.0	14.3	15.8	17.4	19.1
X. Deferred credits	30.0	6.9	8.9	11.2	13.8	16.7.	20.0	23.8	28.2	33.2	38.8	45.2
XI. Equity: Beginning-of-year (BOY)	—	32.6	34.6	37.8	40.6	43.3	46.2	49.7	53.6	58.1	63.3	69.4
Retained earnings	—	1.0	1.4	1.8	2.1	2.6	3.1	3.7	4.4	5.2	6.1	7.2
Net issue	—	1.0	1.8	1.0	0.6	0.3	0.4	0.2	0.1	0.0	0.0	0.0
End-of-year (EOY)	—	34.6	37.8	40.6	43.3	46.2	49.7	53.6	58.1	63.3	69.4	76.6
XII. Debt: BOY	—	29.5	31.8	31.8	33.2	35.5	37.8	40.7	43.8	47.5	51.3	55.5
Net issue	—	2.3	0.0	1.4	2.3	2.3	2.9	3.1	3.7	3.8	4.2	4.9
EOY	—	31.8	31.8	33.2	35.5	37.8	40.7	43.8	47.5	51.3	55.5	60.4
XIII. Total liabilities	—	80.2	86.6	93.9	102.4	111.5	122.2	134.2	148.1	163.6	181.1	201.3
XIV. Debt/capitalization (EOY) (percent)	—	47.9	45.7	45.0	45.1	45.0	45.0	45.0	45.0	44.8	44.4	44.1

Source: Calculations by Charles River Associates.

[a]Data for 1975 are calculated from AT&T's 1975 Annual Report and Embedded Cost Study. Entries for subsequent years are projected according to the hypotheses explained in the text, using 1975 as the base year.

[b]Calculations are rounded to the nearest 10th.

[c]Hypotheses relate to assumed growth rates and financial ratios and are explained in the text.

Table B–12. No-Inflation Scenario[a,b] AT&T Pro Forma Income Statement, 1975–1985 (billions of $, unless otherwise noted)

Item	Hypotheses[c]	1975	1976	1977	1978	1979	1980	1981	1982	1983	1984	1985
I. Revenues												
Other	10.20	18.52	20.41	22.49	24.78	27.31	30.10	33.17	36.55	40.28	44.39	48.92
Local	4.20	8.95	9.33	9.72	10.13	10.55	10.99	11.46	11.94	12.44	12.96	13.51
Subtotal	—	27.47	29.74	32.21	34.91	37.86	41.09	44.63	48.49	52.72	57.35	62.43
II. Costs												
Other	10.20	8.58	9.46	10.42	11.48	12.65	13.94	15.37	16.93	18.66	20.56	22.66
Common	20.35	4.04	4.34	4.67	5.03	5.42	5.84	6.31	6.80	7.36	7.95	8.60
Joint	4.20	4.92	5.13	5.34	5.57	5.80	6.04	6.30	6.56	6.84	7.12	7.42
Local	4.50	2.31	2.41	2.52	2.64	2.75	2.88	3.01	3.14	3.29	3.43	3.59
Subtotal	—	19.85	21.34	22.95	24.72	26.62	28.70	30.99	33.43	36.15	39.06	42.27
III. Earnings before interest and taxes (EBIT)	—	7.62	8.40	9.26	10.19	11.24	12.39	13.64	15.06	16.57	18.29	20.16
IV. Interest	8.00	2.30	2.40	2.42	2.54	2.70	2.88	3.07	3.25	3.44	3.66	3.87
V. Profit before taxes	—	5.32	6.00	6.84	7.65	8.54	9.51	10.57	11.81	13.13	14.63	16.29
VI. Federal tax	40.00	2.17	2.40	2.74	3.06	3.42	3.80	4.23	4.72	5.25	5.85	6.52
VII. Net income	—	3.15	3.60	4.10	4.59	5.12	5.71	6.34	7.09	7.88	8.78	9.77
VIII. Dividends	65.00	2.16	2.34	2.67	2.98	3.33	3.71	4.12	4.61	5.12	5.71	6.35
IX. Retained earnings	—	0.99	1.26	1.43	1.61	1.79	2.00	2.22	2.48	2.76	3.07	3.42
X. Net income/beginning-of-year equity (percent)	—	9.7	10.4	10.8	11.3	11.9	12.6	13.2	14.0	14.8	15.6	16.4
XI. EBIT/interest (ratio)	—	3.3	3.5	3.8	4.0	4.2	4.3	4.4	4.6	4.8	5.0	5.2

Source: Calculations by Charles River Associates.

[a]Data for 1975 are calculated from AT&T's 1975 Annual Report and Embedded Direct Cost Study. Entries for subsequent years are projected according to the hypotheses explained in the text, using 1975 as the base year.

[b]Calculations are rounded to the nearest 100th.

[c]Hypotheses relate to assumed growth rates and financial ratios and are explained in the text.

AT&T Pro forma balance sheet, 1975–1985 (billions of $, unless otherwise noted)

Item	Hypotheses[c] (percent)	1975	1976	1977	1978	1979	1980	1981	1982	1983	1984	1985
I. Total phones (millions)	4.2	118.5	123.5	128.7	134.1	139.7	145.6	151.7	158.0	164.7	171.6	178.8
II. Change in phones (millions)	—	4.0	5.0	5.2	5.4	5.6	5.9	6.1	6.3	6.7	6.9	7.2
III. Change in plant and equipment (P+E)	—	7.0	7.0	7.3	7.6	7.8	8.3	8.5	8.8	9.4	9.7	10.1
IV. Gross plant in service (GPIS)	—	84.6	91.6	98.9	106.5	114.3	122.6	131.1	139.9	149.3	159.0	169.1
V. P+E net	83.0	70.4	76.0	82.1	88.4	94.9	101.8	108.8	116.1	123.9	132.0	140.4
VI. Investments	18.0	3.6	3.8	4.1	4.4	4.8	5.2	5.6	6.0	6.5	7.0	7.6
VII. Current assets and deferred charges	32.0	6.2	6.8	7.3	7.9	8.5	9.2	9.9	10.7	11.6	12.5	13.5
VIII. Total assets	—	80.2	86.6	93.5	100.7	108.2	116.2	124.3	132.8	142.0	151.5	161.5
IX. Current liabilities	37.0	6.9	7.9	8.5	9.1	9.8	10.6	11.5	12.4	13.4	14.5	15.6
X. Deferred credits	30.0	6.9	9.0	11.2	13.5	15.8	18.3	20.9	23.5	26.3	29.2	32.2
XI. Equity: beginning-of-year (BOY)	—	32.6	34.6	37.9	40.6	43.0	45.4	48.0	50.5	53.3	56.3	59.4
Retained earnings	—	1.0	1.3	1.4	1.6	1.8	2.0	2.2	2.5	2.8	3.1	3.4
Net issue	—	1.0	2.0	1.3	0.8	0.6	0.6	0.3	0.3	0.2	0.0	0.0
End-of-year (EOY)	—	34.6	37.9	40.6	43.0	45.4	48.0	50.5	53.3	56.3	59.4	62.8
XII. Debt: BOY	—	29.5	31.8	31.8	33.2	35.1	37.2	39.3	41.4	43.6	46.0	48.4
Net issue	—	2.3	0.0	1.4	1.9	2.1	2.1	2.1	2.2	2.4	2.4	2.5
EOY	—	31.8	31.8	33.2	35.1	37.2	39.3	41.4	43.6	46.0	48.4	50.9
XIII. Total liabilities	—	80.2	86.6	93.5	100.7	108.2	116.2	124.3	132.8	142.0	151.5	161.5
Debt/capitalization (EOY) (percent)	—	47.9	45.6	45.0	44.9	45.0	45.0	45.0	45.0	45.0	44.9	44.8

Source: Calculations by Charles River Associates.

[a] Data for 1975 are calculated from AT&T's 1975 Annual Report and Embedded Cost Study. Entries for subsequent years are projected according to the hypotheses explained in the text, using 1975 as the base year.

[b] Calculations are rounded to the nearest 10th.

[c] Hypotheses relate to assumed growth rates and financial ratios and are explained in the text.

Table B–14. Nominal A Scenario[a,b] AT&T Pro Forma Income Statement, 1975–1985 (billions of $, unless otherwise noted)

Item	Hypotheses[c] (percent)	1975	1976	1977	1978	1979	1980	1981	1982	1983	1984	1985
I. Revenues												
Other	12.60	18.52	20.85	23.48	26.44	29.77	33.52	37.75	42.50	47.86	53.89	60.68
Local	10.80	8.95	9.92	10.99	12.17	13.49	14.95	16.56	18.35	20.33	22.53	24.96
Subtotal	—	27.47	30.77	34.47	38.61	43.26	48.47	54.31	60.85	68.19	76.42	85.64
II. Costs												
Other	10.20	6.58	9.46	10.42	11.48	12.65	13.94	15.37	16.93	18.66	20.56	22.66
Common	20.35	4.04	4.46	4.92	5.43	5.99	6.61	7.30	8.05	8.88	9.80	10.82
Joint	10.40	4.92	5.43	6.00	6.62	7.31	8.07	8.91	9.83	10.86	11.99	13.23
Local	10.80	2.31	2.56	2.84	3.14	3.48	3.86	4.27	4.74	5.25	5.81	6.44
Subtotal	—	19.85	21.91	24.18	26.67	29.43	32.48	35.85	39.55	43.65	48.16	53.15
III. Earnings before interest and taxes (EBIT)	—	7.62	8.86	10.29	11.94	13.83	15.99	18.46	21.30	24.54	28.26	32.49
IV. Interest	9.00	2.30	2.43	2.48	2.68	2.91	3.17	3.47	3.80	4.15	4.57	5.01
V. Profit before taxes	—	5.32	6.43	7.82	9.26	10.92	12.82	14.99	17.50	20.39	23.69	27.48
VI. Federal tax	40.00	2.17	2.57	3.12	3.70	4.37	5.13	6.00	7.00	8.16	9.48	10.99
VII. Net income	—	3.15	3.86	4.69	5.56	6.55	7.69	8.99	10.50	12.23	14.21	16.49
VIII. Dividends	65.00	2.16	2.51	3.05	3.61	4.26	5.00	5.84	6.83	7.95	9.24	10.72
IX. Retained earnings	—	0.99	1.35	1.64	1.95	2.29	2.69	3.15	3.67	4.28	4.97	5.77
X. Net income/beginning-of-year equity (percent)	—	9.7	11.2	12.2	13.5	15.0	16.4	17.8	19.4	20.9	22.6	24.2
XI. EBIT/interest (ratio)	—	3.3	3.6	4.1	4.5	4.8	5.0	5.3	5.6	5.9	6.2	6.5

Source: Calculations by Charles River Associates.

[a]Data for 1975 are calculated from AT&T's 1975 Annual Report and Embedded Direct Cost Study. Entries for subsequent years are projected according to the hypotheses explained in the text, using 1975 as the base year.

[b]Calculations are rounded to the nearest 100th.

[c]Hypotheses relate to assumed growth rates and financial ratios and are explained in the text.

Table B–15. Nominal A Scenario[a,b] AT&T Pro Forma Balance Sheet, 1975–1985 (billions of $, unless otherwise noted)

Item	Hypotheses[c] (percent)	1975	1976	1977	1978	1979	1980	1981	1982	1983	1984	1985
I. Total phones (millions)	4.2	118.5	123.5	128.7	134.1	139.7	145.6	151.7	158.0	164.7	171.6	178.8
II. Change in phones (millions)	—	4.0	5.0	5.2	5.4	5.6	5.9	6.1	6.3	6.7	6.9	7.2
III. Change in plant and equipment (P+E)	6.0	7.0	7.4	8.2	9.0	9.9	11.1	12.1	13.3	15.0	16.3	18.1
IV. Gross plant in service (GPIS)	—	84.6	92.0	100.2	109.2	119.1	130.2	142.3	155.6	170.6	186.9	205.0
V. P+E net	83.0	70.4	76.4	83.2	90.6	98.9	108.1	118.1	129.1	141.6	155.1	170.2
VI. Investments	18.0	3.6	3.9	4.4	4.8	5.3	5.8	6.5	7.1	7.9	8.7	9.6
VII. Current assets and deferred charges	32.0	6.2	7.0	7.7	8.5	9.4	10.4	11.5	12.7	14.0	15.4	17.0
VIII. Total assets	—	80.2	87.3	95.3	103.9	113.6	124.3	136.1	148.9	163.5	179.2	196.8
IX. Current liabilities	37.0	6.9	8.1	8.9	9.9	10.9	12.0	13.3	14.6	16.2	17.8	19.7
X. Deferred credits	30.0	6.9	9.1	11.6	14.3	17.3	20.6	24.2	28.2	32.7	37.6	43.0
XI. Equity: beginning-of-year (BOY)	—	32.6	34.6	38.3	41.1	43.8	47.0	50.4	54.2	58.4	63.0	68.1
Retained earnings	—	1.0	1.4	1.6	2.0	2.3	2.7	3.2	3.7	4.3	5.0	5.8
Net issue	—	1.0	2.3	1.2	0.7	0.9	0.7	0.6	0.5	0.3	0.1	0.0
End-of-year (EOY)	—	34.6	38.3	41.1	43.8	47.0	50.4	54.2	58.4	63.0	68.1	73.9
XII. Debt: BOY	—	29.5	31.8	31.8	33.7	35.9	38.4	41.3	44.4	47.7	51.6	55.7
Net issue	—	2.3	0.0	1.9	2.2	2.5	2.9	3.1	3.3	3.9	4.1	4.5
EOY	—	31.8	31.8	33.7	35.9	38.4	41.3	44.4	47.7	51.6	55.7	60.2
XIII. Total liabilities	—	80.2	87.3	95.3	103.9	113.6	124.3	136.1	148.9	163.5	179.2	196.8
XIV. Debt/capitalization (EOY) (percent)	—	47.9	45.4	45.1	45.0	45.0	45.0	45.0	45.0	45.0	45.0	44.9

Source: Calculations by Charles River Associates.

[a]Data for 1975 are calculated from AT&T's 1975 Annual Report and Embedded Cost Study. Entries for subsequent years are projected according to the hypotheses explained in the text, using 1975 as the base year.

[b]Calculations are rounded to the nearest 10th.

[c]Hypotheses relate to assumed growth rates and financial ratios and are explained in the text.

Table B–16. Nominal B Scenario[a,b] AT&T Pro Forma Income Statement, 1975–1985 (billions of $, unless otherwise noted)

Item	Hypotheses[c] (percent)	1975	1976	1977	1978	1979	1980	1981	1982	1983	1984	1985
I. Revenues												
Other	12.60	18.52	20.85	23.48	26.44	29.77	33.52	37.75	42.50	47.86	53.89	60.68
Local	7.30	8.95	9.60	10.30	11.06	11.86	12.73	13.66	14.66	15.73	16.87	18.11
Subtotal	—	27.47	30.45	33.78	37.50	41.63	46.25	51.41	57.16	63.59	70.76	78.79
II. Costs												
Other	10.20	8.58	9.46	10.42	11.48	12.65	13.94	15.37	16.93	18.66	20.56	22.66
Common	20.35	4.04	4.46	4.92	5.43	5.99	6.61	7.30	8.05	8.88	9.80	10.82
Joint	10.40	4.92	5.43	6.00	6.62	7.31	8.07	8.91	9.83	10.86	11.99	13.23
Local	10.80	2.31	2.56	2.84	3.14	3.48	3.86	4.27	4.74	5.25	5.81	6.44
Subtotal	—	19.85	21.91	24.18	26.67	29.43	32.48	35.85	39.55	43.65	48.16	53.15
III. Earnings before interest and taxes (EBIT)	—	7.62	8.54	9.60	10.83	12.20	13.77	15.56	17.61	19.94	22.60	25.64
IV. Interest	9.00	2.30	2.43	2.48	2.68	2.91	3.17	3.47	3.80	4.15	4.57	5.01
V. Profit before taxes	—	5.32	6.11	7.12	8.15	9.29	10.60	12.09	13.81	15.79	18.03	20.63
VI. Federal tax	40.00	2.17	2.44	2.85	3.26	3.72	4.24	4.84	5.52	6.32	7.21	8.25
VII. Net Income	—	3.15	3.67	4.27	4.89	5.57	6.36	7.25	8.29	9.47	10.82	12.38
VIII. Dividends	65.00	2.16	2.39	2.78	3.18	3.62	4.13	4.71	5.39	6.16	7.03	8.05
IX. Retained earnings	—	0.99	1.28	1.49	1.71	1.95	2.23	2.54	2.90	3.31	3.79	4.33
X. Net income/beginning-of-year equity (percent)	—	9.7	10.6	11.1	11.9	12.7	13.5	14.4	15.3	16.2	17.2	18.2
XI. EBIT/interest (ratio)	—	3.3	3.5	3.9	4.0	4.2	4.3	4.5	4.6	4.8	4.9	5.1

Source: Calculations by Charles River Associates.

[a]Data for 1975 are calculated from AT&T's 1975 Annual Report and Embedded Direct Cost Study. Entries for subsequent years are projected according to the hypotheses explained in the text, using 1975 as the base year.

[b]Calculations are rounded to the nearest 100th.

[c]Hypotheses relate to assumed growth rates and financial ratios and are explained in the text.

Table ... Nominal ... Scenario ... AT&T Pro Forma Balance Sheet, 1975–1985 (millions of $, unless otherwise noted)

Item	Hypotheses[c] (percent)	1975	1976	1977	1978	1979	1980	1981	1982	1983	1984	1985
I. Total phones (millions)	4.2	118.5	123.5	128.7	134.1	139.7	145.6	151.7	158.0	164.7	171.6	178.8
II. Change in phones (millions)	—	4.0	5.0	5.2	5.4	5.6	5.9	6.1	6.3	6.7	6.9	7.2
III. Change in plant and equipment (P+E)	6.0	7.0	7.4	8.2	9.0	9.9	11.1	12.1	13.3	15.0	16.3	18.1
IV. Gross plant in service (GPIS)	—	84.6	92.0	100.2	109.2	119.1	130.2	142.3	155.6	170.6	186.9	205.0
V. P+E net	83.0	70.4	76.4	83.2	90.6	98.9	108.1	118.1	129.1	141.6	155.1	170.2
VI. Investments	18.0	3.6	3.9	4.4	4.8	5.3	5.8	6.5	7.1	7.9	8.7	9.6
VII. Current assets and deferred charges	32.0	6.2	7.0	7.7	8.5	9.4	10.4	11.5	12.7	14.0	15.4	17.0
VIII. Total assets	—	80.2	87.3	95.3	103.9	113.6	124.3	136.1	148.9	163.5	179.2	196.8
IX. Current liabilities	37.0	6.9	8.1	8.9	9.9	10.9	12.0	13.3	14.6	16.2	17.8	19.7
X. Deferred credits	30.0	6.9	9.1	11.6	14.3	17.3	20.6	24.2	28.2	32.7	37.6	43.0
XI. Equity: beginning-of-year (BOY)	—	32.6	34.6	38.3	41.1	43.8	47.0	50.4	54.2	58.4	63.0	68.1
Retained earnings	—	1.0	1.3	1.5	1.7	2.0	2.2	2.5	2.9	3.3	3.8	4.3
Net issue	—	1.0	2.4	1.3	1.0	1.2	1.2	1.3	1.3	1.3	1.3	1.4
End-of-year (EOY)	—	34.6	38.3	41.1	43.8	47.0	50.4	54.2	58.4	63.0	68.1	73.8
XII. Debit: BOY	—	29.5	31.8	31.8	33.7	35.9	38.4	41.3	44.4	47.7	51.6	55.7
Net issue	—	2.3	0.0	1.9	2.2	2.5	2.9	3.1	3.3	3.9	4.1	4.6
EOY	—	31.8	31.8	33.7	35.9	38.4	41.3	44.4	47.7	51.6	55.7	60.3
XIII. Total liabilities	—	80.2	87.3	95.3	103.9	113.6	124.3	136.1	148.9	163.5	179.2	196.8
XIV. Debt/capitalization (EOY) (percent)	—	47.9	45.4	45.1	45.0	45.0	45.0	45.0	45.0	45.0	45.0	45.0

Source: Calculations by Charles River Associates.

[a] Data for 1975 are calculated from AT&T's 1975 Annual Report and Embedded Cost Study. Entries for subsequent years are projected according to the hypotheses explained in the text, using 1975 as the base year.

[b] Calculations are rounded to the nearest 10th.

[c] Hypotheses relate to assumed growth rates and financial ratios and are explained in the text.

Table B–18. Nominal C Scenario[a,b] AT&T Pro Forma Income Statement, 1975–1985 (billions of $, unless otherwise noted)

Item	Hypotheses[c] (percent)	1975	1976	1977	1978	1979	1980	1981	1982	1983	1984	1985
I. Revenues												
Other	12.60	18.52	20.85	23.48	26.44	29.77	33.52	37.75	42.50	47.86	53.89	60.68
Surplus for diversion	—	0.00	0.00	(0.17)	(1.08)	(2.19)	(3.46)	(4.95)	(6.71)	(8.75)	(11.14)	(13.92)
Other net	—	18.52	20.85	23.31	25.36	27.58	30.06	32.80	35.79	39.11	42.75	46.76
Local	10.80	8.95	9.92	10.99	12.17	13.49	14.95	16.56	18.35	20.33	22.53	24.96
Subtotal	—	27.47	30.77	34.30	37.53	41.07	45.01	49.36	54.14	59.44	65.28	71.72
II. Costs												
Other	10.20	8.58	9.46	10.42	11.48	12.65	13.94	15.37	16.93	18.66	20.56	22.66
Common	20.35	4.04	4.46	4.92	5.43	5.99	6.61	7.30	8.05	8.88	9.80	10.82
Joint	10.40	4.92	5.43	6.00	6.62	7.31	8.07	8.91	9.83	10.86	11.99	13.23
Local	10.80	2.31	2.56	2.84	3.14	3.48	3.86	4.27	4.74	5.25	5.81	6.44
Subtotal	—	19.85	21.91	24.18	26.67	29.43	32.48	35.85	39.55	43.65	48.16	53.15
III. Earnings before interest and taxes (EBIT)	—	7.62	8.86	10.12	10.86	11.64	12.53	13.51	14.59	15.79	17.12	18.57
IV. Interest	9.00	2.30	2.43	2.48	2.68	2.91	3.17	3.47	3.80	4.15	4.57	5.01
V. Profit before taxes	—	5.32	6.43	7.64	8.18	8.73	9.36	10.04	10.79	11.64	12.55	13.56
VI. Federal tax	40.00	2.17	2.57	3.04	3.25	3.47	3.72	3.99	4.29	4.63	4.99	5.39
VII. Net Income	—	3.15	3.86	4.60	4.93	5.26	5.64	6.05	6.50	7.01	7.56	8.17
VIII. Dividends	65.00	2.16	2.51	2.99	3.20	3.42	3.67	3.93	4.23	4.56	4.91	5.31
IX Retained earnings	—	0.99	1.35	1.61	1.73	1.84	1.97	2.12	2.27	2.45	2.65	2.86
X. Net income/beginning-of-year equity (percent)	12.0	9.7	11.2	12.0	12.0	12.0	12.0	12.0	12.0	12.0	12.0	12.0
XI. EBIT/interest (ratio)	—	3.3	3.6	4.1	4.1	4.0	4.0	3.9	3.8	3.8	3.7	3.7

Source: Calculations by Charles River Associates.

[a]Data for 1975 are calculated from AT&T's 1975 Annual Report and Embedded Direct Cost Study. Entries for subsequent years are projected according to the hypotheses explained in the text, using 1975 as the base year.
[b]Calculations are rounded to the nearest 100th.

Item	Hypotheses[c] (percent)	1975	1976	1977	1978	1979	1980	1981	1982	1983	1984	1985
I. Total phones (millions)	4.2	118.5	123.5	128.7	134.1	139.7	145.6	151.7	158.0	164.7	171.6	178.8
II. Change in phones (millions)	—	4.0	5.0	5.2	5.4	5.6	5.9	6.1	6.3	6.7	6.9	7.2
III. Change in plant and equipment (P+E)	6.0	7.0	7.4	8.2	9.0	9.9	11.1	12.1	13.3	15.0	16.3	18.1
IV. Gross plant in service (GPIS)	—	84.6	92.0	100.2	109.2	119.1	130.2	142.3	155.6	170.6	186.9	205.0
V. P+E net	83.0	70.4	76.4	83.2	90.6	98.9	108.1	118.1	129.1	141.6	155.1	170.2
VI. Investments	18.0	3.6	3.9	4.4	4.8	5.3	5.8	6.5	7.1	7.9	8.7	9.6
VII. Current assets and deferred charges	32.0	6.2	7.0	7.7	8.5	9.4	10.4	11.5	12.7	14.0	15.4	17.0
VIII. Total assets	—	80.2	87.3	95.3	103.9	113.6	124.3	136.1	148.9	163.5	179.2	196.8
IX. Current liabilities	37.0	6.9	8.1	8.9	9.9	10.9	12.0	13.3	14.6	16.2	17.8	19.7
X. Deferred credits	30.0	6.9	9.1	11.6	14.3	17.3	20.6	24.2	28.2	32.7	37.6	43.0
XI. Equity: beginning-of-year (BOY)	—	32.6	34.6	38.3	41.1	43.8	47.0	50.4	54.2	58.4	63.0	68.1
Retained earnings	—	1.0	1.4	1.6	1.7	1.8	2.0	2.1	2.3	2.5	2.7	2.9
Net issue	—	1.0	2.3	1.2	1.0	1.4	1.4	1.7	1.9	2.1	2.4	2.8
End-of-year (EOY)	—	34.6	38.3	41.1	43.8	47.0	50.4	54.2	58.4	63.0	68.1	73.8
XII. Debt: BOY	—	29.5	31.8	31.8	33.7	35.9	38.4	41.3	44.4	47.7	51.6	55.7
Net issue	—	2.3	0.0	1.9	2.2	2.5	2.9	3.1	3.3	3.9	4.1	4.6
EOY	—	31.8	31.8	33.7	35.9	38.4	41.3	44.4	47.7	51.6	55.7	60.3
XIII. Total liabilities	—	80.2	87.3	95.3	103.9	113.6	124.3	136.1	148.9	163.5	179.2	196.8
XIV. Debt/capitalization (EOY) (percent)	—	47.9	45.4	45.1	45.0	45.0	45.0	45.0	45.0	45.0	45.0	45.0

Source: Calculations by Charles River Associates.

[a]Data for 1975 are calculated from AT&T's 1975 Annual Report and Embedded Direct Cost Study. Entries for subsequent years are projected according to the hypotheses explained in the text, using 1975 as the base year.

[b]Calculations are rounded to the nearest 10th.

[c]Hypotheses relate to assumed growth rates and financial ratios and are explained in the text.

Table B–20 Nominal D Scenario [a,b] AT&T Pro Forma Income Statement, 1975–1985 (Billions of $, unless otherwise noted)

Item	Hypotheses[c] (percent)	1975	1976	1977	1978	1979	1980	1981	1982	1983	1984	1985
I. Revenues												
Other	12.60	18.52	20.85	23.48	26.44	29.77	33.52	37.75	42.50	47.86	53.89	60.68
Surplus for diversion	—	0.00	0.00	0.00	0.00	(0.56)	(1.24)	(2.05)	(3.02)	(4.15)	(5.48)	(7.07)
Other net	—	18.52	20.85	23.48	26.44	29.21	32.28	35.70	39.48	43.71	48.41	53.61
Local	7.30	8.95	9.60	10.30	11.06	11.86	12.73	13.66	14.66	15.73	16.87	18.11
Subtotal	—	27.47	30.45	33.78	37.50	41.07	45.01	49.36	54.14	59.44	65.28	71.72
II. Costs												
Other	10.20	8.58	9.46	10.42	11.48	12.65	13.94	15.37	16.93	18.66	20.56	22.66
Common	20.35	4.04	4.46	4.92	5.43	5.99	6.61	7.30	8.05	8.88	9.80	10.82
Joint	10.40	4.92	5.43	6.00	6.62	7.31	8.07	8.91	9.83	10.86	11.99	13.23
Local	10.80	2.31	2.56	2.84	3.14	3.48	3.86	4.27	4.74	5.25	5.81	6.44
Subtotal	—	19.85	21.91	24.18	26.67	29.43	32.48	35.85	39.55	43.65	48.16	53.15
III. Earnings before interest and taxes (EBIT)	—	7.62	8.54	9.60	10.83	11.64	12.53	13.51	14.59	15.79	17.12	18.57
IV. Interest	9.00	2.30	2.43	2.48	2.68	2.91	3.17	3.47	3.80	4.15	4.57	5.01
V. Profit before taxes	—	5.32	6.11	7.12	8.15	8.73	9.36	10.04	10.79	11.64	12.55	13.56
VI. Federal tax	40.00	2.17	2.44	2.85	3.26	3.47	3.72	3.99	4.29	4.63	4.99	5.39
VII. Net income	—	3.15	3.67	4.27	4.89	5.26	5.69	6.05	6.50	7.01	7.56	8.17
VIII. Dividends	65.00	2.16	2.39	2.78	3.18	3.42	3.67	3.93	4.23	4.56	4.91	5.31
IX. Retained earnings	—	0.99	1.28	1.49	1.71	1.84	1.97	2.12	2.27	2.45	2.65	2.86
X. Net income/beginning-of-year equity (percent)	12.0	9.7	10.6	11.1	11.9	12.0	12.0	12.0	12.0	12.0	12.0	12.0
XI. EBIT/interest (ratio)	—	3.3	3.5	3.9	4.0	4.0	4.0	3.9	3.8	3.8	3.7	3.7

Source: Calculations by Charles River Associates.

[a]Data for 1975 are calculated from AT&T's 1975 Annual Report and Embedded Direct Cost Study. Entries for subsequent years are projected according to the hypotheses explained in the text, using 1975 as the base year.

[b]Calculations are rounded to the nearest 100th.

AT&T Pro Forma Balance Sheet, 1975–1985 (billions of $, unless otherwise noted)

Item	Hypotheses[c] (percent)	1975	1976	1977	1978	1979	1980	1981	1982	1983	1984	1985
I. Total phones (millions)	4.2	118.5	123.5	128.7	134.1	139.7	145.6	151.7	158.0	164.7	171.6	178.8
II. Change in phones (millions)	—	4.0	5.0	5.2	5.4	5.6	5.9	6.1	6.3	6.7	6.9	7.2
III. Change in plant and equipment (P+E)	6.0	7.0	7.4	8.2	9.0	9.9	11.1	12.1	13.3	15.0	16.3	18.1
IV. Gross plant in service (GPIS)	—	84.6	92.0	100.2	109.2	119.1	130.2	142.3	155.6	170.6	186.9	205.0
V. P+E net	83.0	70.4	76.4	83.2	90.6	98.9	108.1	118.1	129.1	141.6	155.1	170.2
VI. Investments	18.0	3.6	3.9	4.4	4.8	5.3	5.8	6.5	7.1	7.9	8.7	9.6
VII. Current assets and deferred charges	32.0	6.2	7.0	7.7	8.5	9.4	10.4	11.5	12.7	14.0	15.4	17.0
VIII. Total assets	—	80.2	87.3	95.3	103.9	113.6	124.3	136.1	148.9	163.5	179.2	196.8
IX. Current liabilities	37.0	6.9	8.1	8.9	9.9	10.9	12.0	13.3	14.6	16.2	17.8	19.7
X. Deferred credits	30.0	6.9	9.1	11.6	14.3	17.3	20.6	24.2	28.2	32.7	37.6	43.0
XI. Equity: beginning-of-year (BOY)	—	32.6	34.6	38.3	41.1	43.8	47.0	50.4	54.2	58.4	63.0	68.1
Retained earnings	—	1.0	1.3	1.5	1.7	1.8	2.0	2.1	2.3	2.5	2.7	2.9
Net issue	—	1.0	2.4	1.3	1.0	1.4	1.4	1.7	1.9	2.1	2.4	2.8
End-of-year (EOY)	—	34.6	38.3	41.1	43.8	47.0	50.4	54.2	58.4	63.0	68.1	73.8
XII. Debt: BOY	—	29.5	31.8	31.8	33.7	35.9	38.4	41.3	44.4	47.7	51.6	55.7
Net issue	—	2.3	0.0	1.9	2.2	2.5	2.9	3.1	3.3	3.9	4.1	4.6
EOY	—	31.8	31.8	33.7	35.9	38.4	41.3	44.4	47.7	51.6	55.7	60.3
XIII. Total liabilities	—	80.2	87.3	95.3	103.6	113.6	124.3	136.1	148.9	163.5	179.2	196.8
XIV. Debt/capitalization (EOY) (percent)	—	47.9	45.4	45.1	45.0	45.0	45.0	45.0	45.0	45.0	45.0	45.0

Source: Calculations by Charles River Associates.

[a] Data for 1975 are calculated from AT&T's 1975 Annual Report and Embedded Cost Study. Entries for subsequent years are projected according to the hypotheses explained in the text, using 1975 as the base year.

[b] Calculations are rounded to the nearest 10th.

[c] Hypotheses relate to assumed growth rates and financial ratios and are explained in the text.

Table B–22. Nominal E Scenario[a,b] AT&T Pro Forma Income Statement, 1975–1985 (billions of $, unless otherwise noted)

Item	Hypotheses[c] (percent)	1975	1976	1977	1978	1979	1980	1981	1982	1983	1984	1985
I. Revenues												
Other	12.60	18.52	20.85	23.48	26.44	29.77	33.52	37.75	42.50	47.86	53.89	60.68
Surplus for diversion	—	0.00	0.00	0.00	0.00	0.00	0.00	0.00	(1.06)	(3.08)	(5.42)	(8.03)
Other net	—	18.52	20.85	23.48	26.44	29.77	33.52	37.75	41.44	44.78	48.47	52.65
Local	10.80	8.95	9.92	10.99	12.17	13.49	14.95	16.56	18.35	20.33	22.53	24.96
Subtotal	—	27.47	30.77	34.47	38.61	43.26	48.47	54.31	59.79	65.11	71.00	77.61
II. Costs												
Other	10.20	8.58	9.46	10.42	11.48	12.65	13.94	15.37	16.93	18.66	20.56	22.66
Common	20.35	4.04	4.46	4.92	5.43	5.99	6.61	7.30	8.05	8.88	9.80	10.82
Joint	10.40	4.92	5.43	6.00	6.62	7.31	8.07	8.91	9.83	10.86	11.99	13.23
Local	10.80	2.31	2.56	2.84	3.14	3.48	3.86	4.27	4.74	5.25	5.81	6.44
Subtotal	—	19.85	21.91	24.18	26.67	29.43	32.48	35.85	39.55	43.65	48.16	53.15
III. Earnings before interest and taxes (EBIT)	—	7.62	8.86	10.29	11.94	13.83	15.99	18.46	20.24	21.46	22.84	24.46
Extra depreciation	—	0.00	0.90	2.00	3.30	4.80	6.50	7.10	7.80	8.50	9.30	10.30
EBIT (revised)	—	7.62	7.96	8.29	8.64	9.03	9.49	11.36	12.44	12.96	13.54	14.16
IV. Interest	9.00	2.30	2.43	2.48	2.62	2.77	2.93	3.07	3.19	3.32	3.50	3.70
V. Profit before taxes	—	5.32	5.53	5.81	6.02	6.26	6.56	8.29	9.25	9.64	10.04	10.46
VI. Federal tax	40.00	2.17	2.21	2.32	2.41	2.50	2.62	3.32	3.68	3.83	3.99	4.16
VII. Net income	—	3.15	3.32	3.49	3.61	3.76	3.94	4.97	5.57	5.81	6.05	6.30
VIII. Dividends	65.00	2.16	2.16	2.27	2.35	2.44	2.56	3.23	3.62	3.78	3.93	4.10
IX. Retained earnings	—	0.99	1.16	1.22	1.26	1.32	1.38	1.74	1.95	2.03	2.12	2.20
X. Net income/beginning-of-year equity (percent)	12.0	9.7	9.6	9.2	9.0	9.0	9.1	11.1	12.0	12.0	12.0	12.0
XI. EBIT/interest (ratio)	—	3.3	3.3	3.3	3.3	3.3	3.2	3.7	3.9	3.9	3.9	3.8

Source: Calculations by Charles River Associates.

[a]Data for 1975 are calculated from AT&T's 1975 Annual Report and Embedded Direct Cost Study. Entries for subsequent years are projected according to the hypotheses explained in the text, using 1975 as the base year.

Item	Hypotheses[c] (percent)	1975	1976	1977	1978	1979	1980	1981	1982	1983	1984	1985
I. Total phones (millions)	4.2	118.5	123.5	128.7	134.1	139.7	145.6	151.7	158.0	164.7	171.6	178.8
II. Change in phones (millions)	—	4.0	5.0	5.2	5.4	5.6	5.9	6.1	6.3	6.7	6.9	7.2
III. Change in plant and equipment (P+E)	6.0	7.0	7.4	8.2	9.0	9.9	11.1	12.1	13.3	15.0	16.3	18.1
IV. Gross plant in service (GPIS)	—	84.6	92.0	100.2	109.2	119.1	130.2	142.3	155.6	170.6	186.9	205.0
V. P+E net	83.0	70.4	76.4	83.2	90.6	98.9	108.1	118.1	129.1	141.6	155.1	170.2
Extra depreciation Annual (percent)	10.0	0.0	1.0	2.0	3.0	4.0	5.0	5.0	5.0	5.0	5.0	5.0
(billions)	—	0.0	0.9	2.0	3.3	4.8	6.5	7.1	7.8	8.5	9.3	10.3
Cumulative (billions)	—	0.0	0.9	2.9	6.2	11.0	17.5	24.6	32.4	40.9	50.2	60.5
Revised P+E net	—	70.4	75.5	80.3	84.4	87.9	90.6	93.5	96.7	100.7	104.9	109.7
VI. Investments	18.0	3.6	3.9	4.4	4.8	5.3	5.8	6.5	7.1	7.9	8.7	9.6
VII. Current assets and deferred charges	32.0	6.2	7.0	7.7	8.5	9.4	10.4	11.5	12.7	14.0	15.4	17.0
VIII. Total assets	—	80.2	86.4	92.4	97.7	102.6	106.8	111.5	116.5	122.6	129.0	136.3
IX. Current liabilities	37.0	6.9	8.1	8.9	9.9	10.9	12.0	13.3	14.6	16.2	17.8	19.7
X. Deferred credits	—	6.9	8.7	10.4	11.8	12.9	13.6	14.5	15.3	16.4	17.6	18.9
XI. Equity: Beginning-of-year (BOY)	—	32.6	34.6	37.8	40.2	41.8	43.3	44.7	46.4	48.4	50.4	52.5
Retained earnings	—	1.0	1.2	1.2	1.3	1.3	1.4	1.7	2.0	2.0	2.1	2.2
Net issue	—	1.0	2.0	1.2	0.3	0.2	0.0	0.0	0.0	0.0	0.0	0.0
End-of-year (EOY)	—	34.6	37.8	40.2	41.8	43.3	44.7	46.4	48.4	50.4	52.5	54.7
XII. Debt: BOY	—	29.5	31.8	31.8	32.9	34.2	35.5	36.5	37.3	38.2	39.6	41.1
Net issue	—	2.3	0.0	1.1	1.3	1.3	1.0	0.8	0.9	1.4	1.5	1.9
EOY	—	31.8	31.8	32.9	34.2	35.5	36.5	37.3	38.2	39.6	41.1	43.0
XIII. Total liabilities	—	80.2	86.4	92.4	97.7	102.6	106.8	111.5	116.5	122.6	129.0	136.3
XIV. Debt/capitalization (EOY) (percent)	—	47.9	45.7	45.0	45.0	45.1	45.0	44.6	44.1	44.0	43.9	44.0

Source: Calculations by Charles River Associates.

[a]Data for 1975 are calculated from AT&T's 1975 Annual Report and Embedded Cost Study. Entries for subsequent years are projected according to the hypotheses explained in the text, using 1975 as the base year.

[b]Calculations are rounded to the nearest 10th.

[c]Hypotheses relate to assumed growth rates and financial ratios and are explained in the text.

Table B–24. Nominal E Scenario[a,b] Deferred Credits Worksheet, 1975–1985 (billions of $)

Item	Scenario	1975	1976	1977	1978	1979	1980	1981	1982	1983	1984	1985
1. Federal taxes	C	2.17	2.57	3.04	3.25	3.47	3.72	3.99	4.29	4.63	4.99	5.39
2. Deferred credits	C	6.88	9.10	11.56	14.26	17.23	20.56	24.19	28.18	32.68	37.57	43.00
3. Change in deferred credits	C	2.04	2.22	2.46	2.70	2.97	3.33	3.63	3.99	4.50	4.89	5.43
4. Tax effectively paid	C	0.13	0.35	0.58	0.55	0.50	0.39	0.36	0.30	0.13	0.10	(0.04)
5. Subtotal revenues	E	27.47	30.77	34.47	38.61	43.26	48.47	54.31	59.79	65.11	71.00	77.61
6. Subtotal revenues	C	27.47	30.77	34.30	37.53	41.07	45.01	49.36	54.14	59.44	65.28	71.72
7. Extra revenues	E	0.00	0.00	0.17	1.08	2.19	3.46	4.95	5.65	5.67	5.72	5.89
8. Interest charges	C	2.30	2.43	2.48	2.68	2.91	3.17	3.47	3.80	4.15	4.57	5.01
9. Interest charges	E	2.30	2.43	2.48	2.62	2.77	2.93	3.07	3.19	3.32	3.50	3.70
10. Interest savings	E	0.00	0.00	0.00	0.06	0.14	0.24	0.40	0.61	0.83	1.07	1.31
11. Extra revenue + interest	E	0.00	0.00	0.17	1.14	2.33	3.70	5.35	6.26	6.50	6.79	7.20
12. Extra taxes paid	E	0.00	0.00	0.07	0.46	0.93	1.48	2.14	2.50	2.60	2.72	2.88
13. Federal tax charges	E	2.17	2.21	2.32	2.41	2.50	2.62	3.32	3.68	3.83	3.99	4.16
14. Taxes paid	E	0.13	0.35	0.65	1.01	1.43	1.87	2.50	2.80	2.73	2.82	2.84
15. Change in deferred credits	E	2.04	1.86	1.67	1.40	1.07	0.75	0.82	0.88	1.10	1.17	1.32
16. Cumulative deferred credits	E	6.88	8.74	10.41	11.81	12.88	13.63	14.45	15.33	16.43	17.60	18.92

Source: Calculations by Charles River Associates.
[a]Data for 1975 are calculated from AT&T's 1975 Annual Report and Embedded Direct Cost Study. Entries for subsequent years are projected according to the hypotheses explained in the text, using 1975 as the base year.
[b]Calculations are rounded to the nearest 100th.

Table B–25. Comparison of "Actual" and "Projected" AT&T Income Statement for 1976

Item	Units	Actual	Scenario Projections		
			Historical B	Historical C	Nominal A
I. Revenues subtotal	Billions of $	31.14	30.52	30.96	30.77
Change from previous year	Percentage	13.4	11.1	12.7	12.0
II. Costs subtotal	Billions of $	22.22	21.85	21.85	21.91
Change from previous year	Percentage	12.5	10.1	10.1	10.4
III. Earnings before interest and taxes (EBIT)	Billions of $	8.92	8.57	9.11	8.86
IV. Interest charges	Billions of $	2.40	2.46	2.46	2.43
Percent of beginning-of-year (BOY) debt	Percentage	7.5	7.7	7.7	7.6
V. Profit before taxes (PBT)	Billions of $	6.52	6.21	6.65	6.43
VI. Federal taxes	Billions of $	2.69	2.48	2.66	2.57
Percent of PBT	Percentage	41.2	40.0	40.0	40.0
VII. Net income (NI)	Billions of $	3.83	3.73	3.99	3.86
Change from previous year	Percentage	21.6	18.4	26.7	22.5
VIII. Dividends	Billions of $	2.49	2.42	2.59	2.51
Percent of NI	Percentage	65.0	65.0	65.0	65.0
IX. Retained earnings	Billions of $	1.34	1.31	1.40	1.35
X. NI/BOY equity	Percentage	11.1	10.8	11.5	11.2
XI. EBIT/interest	Ratio	3.7	3.5	3.7	3.6

Source: Calculations by Charles River Associates, based on information provided in AT&T's annual reports.

Table B–26. Comparison of "Actual" and "Projected" AT&T Balance Sheet for 1976

Item	Units	Actual	Scenario Projections		
			Historical B	Historical C	Nominal A
I. Number of phones	Millions	123.1	122.8	122.8	123.5
II. Change from previous year	Millions	4.6	4.3	4.3	5.0
	Percentage	3.9	3.6	3.6	4.2
III. Plan and equipment (P+E) Change from previous year	Billions of $	6.7	6.6	6.6	7.4
IV. Gross plant in service (GPIS)	Billions of $	91.3	91.2	91.2	92.0
V. P+E Net	Billions of $	75.9	75.7	75.7	76.4
VI. Investments	Billions of $	3.66	3.9	3.9	3.9
Percent of operating costs	Percentage	16.5	18.0	18.0	18.0
VII. Current assets and deferred charges (CADC)	Billions of $	7.13	7.0	7.0	7.0
Percent of operating costs	Percentage	32.1	32.0	32.0	32.0
VIII. Total assets	Billions of $	86.72	86.6	86.6	87.3
IX. Current liabilities	Billions of $	7.86	8.1	8.1	8.1
Percent of operating costs	Percentage	35.4	37.0	37.0	37.0
X. Deferred credits	Billions of $	9.11	8.9	8.9	9.1
Percent of GPIS	Percentage	33.0	30.0	30.0	30.0

Table B–26 (continued)

XI.	Equity				
	Beginning of year	Billions of $	34.6	34.6	34.6	34.6
	Retained earnings	Billions of $	1.34	1.3	1.4	1.4
	Net issue	Billions of $	1.2	1.9	1.8	2.3
	End of year	Billions of $	37.2	37.8	37.8	38.3
XII.	Debt				
	Beginning of year	Billions of $	31.8	31.8	31.8	31.8
	Net issue	Billions of $	0.73	0.0	0.0	0.0
	End of year	Billions of $	32.53	31.8	31.8	31.8
XIII.	Total liabilities	Billions of $	86.72	86.6	86.6	87.3
XIV.	Debt/capitalization	Percentage	46.7	45.7	45.7	45.4

Source: Calculations by Charles River Associates, based on information provided in AT&T's annual reports.

Table B–27. Comparison of "Actual" and "Projected" AT&T Income Statement for 1977

			Scenario Projections		
Item	Units	Actual	Historical B	Historical C	Nominal A
I. Revenues subtotal	Billions of $	34.63	33.92	34.90	34.47
Change from previous year	Percentage	11.2	11.1	12.7	12.0
II. Costs subtotal	Billions of $	24.65	24.03	24.03	24.18
Change from previous year	Percentage	10.9	10.0	10.0	10.4
III. Earnings before interest and taxes (EBIT)	Billions of $	9.98	9.89	10.87	10.29
IV. Interest charges	Billions of $	2.44	2.54	2.54	2.48
Percent of beginning of year (BOY) debt	Percentage	7.5	8.0	8.0	7.8
V. Profit before taxes (PBT)	Billions of $	7.54	7.35	8.33	7.81
VI. Federal taxes	Billions of $	3.00	2.94	3.33	3.12
Percent of PBT	Percentage	39.8	40.0	40.0	40.0
VII. Net income (NI)	Billions of $	4.54	4.41	5.00	4.69
Change from previous year	Percentage	18.5	18.2	25.3	21.5
VIII. Dividends	Billions of $	2.82	2.87	3.25	3.05
Percent of NI	Percentage	62.1	65.0	65.0	65.0
IX. Retained earnings	Billions of $	1.72	1.54	1.75	1.64
X. NI/BOY equity	Percentage	12.2	11.7	13.2	12.2
XI. EBIT/interest	Ratio	4.1	3.9	4.3	4.1

Source: Calculations by Charles River Associates, based on information provided in AT&T's annual reports.

Table B-28. Comparison of "Actual" and "Projected" AT&T Balance Sheet for 1977

Item	Units	Actual	Scenario Projections		
			Historical B	Historical C	Nominal A
I. Number of phones	Millions	128.5	127.2	127.2	128.7
II. Change from previous year	Millions	5.4	4.4	4.4	5.2
	Percentage	4.4	3.6	3.6	4.2
III. Plant and equipment (P+E) Change from previous year	Billions of $	7.4	7.5	7.5	8.2
IV. Gross plant in service (GPIS)	Billions of $	98.72	98.7	98.7	100.2
V. P+E net	Billions of $	82.40	81.9	81.9	83.2
VI. Investments (I)	Billions of $	3.80	4.3	4.3	4.4
Percent of operating costs	Percentage	15.4	18.0	18.0	18.0
VII. Current assets & Deferred Charges (CADC)	Billions of $	7.77	7.7	7.7	7.7
Percent of operating costs	Percentage	31.5	32.0	32.0	32.0
VIII. Total assets	Billions of $	93.97	93.9	93.9	95.3
IX. Current liabilities (CL)	Billions of $	9.06	8.9	8.9	8.9
Percent of operating costs	Percentage	36.8	37.0	37.0	37.0

Table continued on following page.

Table B–28 (continued)

Item	Units	Actual	Scenario Projections		
			Historical B	Historical C	Nominal A
X. Deferred Credits	Billions of $	11.51	11.2	11.2	11.6
(DC) Percent of GPIS	Percentage	32.0	30.0	30.0	30.0
XI. Equity					
Beginning of year	Billions of $	37.2	37.8	37.8	38.3
Retained earnings	Billions of $	1.7	1.5	1.8	1.6
Net issue	Billions of $	2.0	1.3	1.0	1.2
End of year	Billions of $	40.9	40.6	40.6	41.1
XII. Debt					
Beginning of year	Billions of $	32.5	31.8	31.8	31.8
Net issue	Billions of $	0.0	1.4	1.4	1.9
End of year	Billions of $	32.5	33.2	33.2	33.7
XIII. Total liabilities	Billions of $	93.97	93.9	93.9	95.3
XIV. Debt/capitalization	Percentage	44.3	45.0	45.0	45.1

Source: Calculations by Charles River Associates, based on information provided in AT&T's annual reports.

Table B–29. Comparison of "Actual" and "Projected" AT&T Income Statement for 1978

Item	Units	Actual	Scenario Projections		
			Historical B	Historical C	Nominal A
I. Revenues subtotal	Billions of $	38.90	37.70	39.35	38.61
Change from previous year	Percentage	12.3	11.1	12.8	12.0
II. Costs subtotal	Billions of $	27.44	26.43	26.43	26.67
Change from previous year	Percentage	11.3	10.0	10.0	10.3
III. Earnings before interest and taxes (EBIT)	Billions of $	11.46	11.27	12.92	11.94
IV. Interest charges	Billions of $	2.69	2.74	2.74	2.68
Percent of beginning-of-year (BOY) debt	Percentage	8.3	8.3	8.3	8.0
V. Profit before taxes (PBT)	Billions of $	8.77	8.53	10.18	9.26
VI. Federal taxes	Billions of $	3.49	3.41	4.07	3.70
Percent of PBT	Percentage	39.8	40.0	40.0	40.0
VII. Net income (NI)	Billions of $	5.27	5.12	6.11	5.56
Change from previous year	Percentage	17.6	16.1	22.2	18.6
VIII. Dividends	Billions of $	3.20	3.33	3.97	3.61
Percent of NI	Percentage	60.7	65.0	65.0	65.0
IX. Retained earnings	Billions of $	2.07	1.79	2.14	1.95
X. NI/BOY equity	Percentage	12.9	12.6	15.0	13.5
XI. EBIT/interest	Ratio	4.3	4.1	4.7	4.5

Source: Calculations by Charles River Associates, based on information provided in AT&T's annual reports.

Table B–30. Comparison of "Actual" and "Projected" AT&T Balance Sheet for 1978

Item	Units	Actual	Scenario Projections		
			Historical B	Historical C	Nominal A
I. Number of phones	Millions	133.4	131.8	131.8	134.1
II. Change from previous year	Millions	5.0	4.6	4.6	5.4
	Percentage	3.9	3.6	3.6	4.2
III. Plant and equipment (P+E) Change from previous year	Billions of $	8.75	8.6	8.6	9.0
IV. Gross plant in service (GPIS)	Billions of $	107.5	107.3	107.3	109.2
V. P+E net	Billions of $	90.4	89.1	89.1	90.6
VI. Investments (I)	Billions of $	4.0	4.8	4.8	4.8
Percent of operating costs	Percentage	14.7	18.0	18.0	18.0
VII. Current assets & deferred charges (CADC)	Billions of $	8.9	8.5	8.5	8.5
Percent of operating costs	Percentage	32.6	32.0	32.0	32.0
VIII. Total assets	Billions of $	103.3	102.4	102.4	103.9
IX. Current liabilities (CI)	Billions of $	10.6	9.8	9.8	9.9
Percent of operating costs	Percentage	38.8	37.0	37.0	37.0

Table B–30 *(continued)*

X.	Deferred Credits	Billions of $	14.1	13.8	13.8	14.3
	(DC) Percent of GPIS	Percentage	29.2	30.0	30.0	30.0
XI.	Equity					
	Beginning of year	Billions of $	40.7	40.6	40.6	41.1
	Retained earnings	Billions of $	2.1	1.8	2.1	2.0
	Net issue	Billions of $	1.2	0.9	0.6	0.7
	End of year	Billions of $	44.1	43.3	43.3	43.8
XII.	Debt					
	Beginning of year	Billions of $	32.5	33.2	33.2	33.7
	Net issue	Billions of $	2.0	2.3	2.3	2.2
	End of year	Billions of $	34.5	35.5	35.5	35.9
XIII.	Total liabilities	Billions of $	103.3	102.4	102.4	103.9
XIV.	Debt/capitalization	Percentage	43.9	45.1	45.1	45.0

Source: Calculations by Charles River Associates, based on information provided in AT&T's annual reports.

Alternative Rate Structures

In Chapter 3, alternative rate structures were presented that seemed better for meeting efficiency and welfare goals previously listed. This appendix contains a more detailed explanation of these rate structures. The first step in estimating the effect of rate restructuring is to obtain information on price elasticities. Table C–1 presents a list of telephone service categories and the range of elasticity estimates for each category drawn from six previous studies.

The rate structure approach adopted in Chapter 3 above involves (a) distance-insensitive and generally lower message toll charges; (b) greater emphasis on peak-load pricing for both local and toll calls; (c) usage-sensitive pricing for local calls; and (d) maintenance of the initial level of total revenue. Table C–2 shows data on revenues and telephones for 1975. The companies include both Bell and the independents and account for approximately 99 percent of the relevant total industry revenues. The exercise here concerns only the first two revenue items shown in Table C–2 — local subscriber revenue (excluding PBX and KTS revenue) and domestic message toll revenue. The other revenue categories are assumed to be affected only slightly by the rate revisions as presented and are therefore excluded.[1] The initial level of local and message toll revenue to be maintained under rate revision is thus approximately $26.6 billion, equal to the total for

Bell plus the independent companies for local and message toll service.

Table C-1. Estimates of Price Elasticities for Telephone Service Categories

Service Category	Range of Price Elasticity Estimates
Local service (residential)	
Calls	−0.1 to −0.4
Connections	−0.2 to −0.7
Toll service	
Overall	−0.4 to −1.0
Day	−0.2 to −0.9
Evening/Night	−0.6 to −1.7
Residential	−0.5 to −1.9
Business	−0.2 to ?
Toll cross-elasticity between day and evening/night periods	0.1 to 0.4

Sources: S. C. Littlechild, "Peak Load Pricing of Telephone Calls," *The Bell Journal of Economics and Management Science* 1 (Autumn 1970): 191–210; S. C. Littlechild and J. J. Rousseau, "Pricing Policy of a U.S. Telephone Company," *Journal of Public Economics* 4 (1975): 35–56; James H. Alleman, "The Pricing of Local Telephone Service," U.S. Department of Commerce, Office of Telecommunications Special Publication 77–14, April 1977; A. Rodney Dobell, L. D. Taylor, L. Waverman, T. H. Liu, and M. Copeland, "Telephone Communications in Canada: Demand, Production and Investment Decisions," *The Bell Journal of Economics and Management Science* 3 (Spring 1972): 175–219; B. E. Davis, G. J. Caccappolo, and M. A. Chaudry, "An Econometric Planning Model for American Telephone and Telegraph Company," *The Bell Journal of Economics and Management Science* 4 (Spring 1973): 29–56; L. Waverman, "The Pricing of Telephone Services in Great Britain: Quasi-Optimality Considered," *I. O. Review,* 5 (1977): 1–0.

Table C–2.   Telephone Statistics, 1975

Service	Revenues (millions of dollars)		
	Independents	Bell System	Total
Local subscriber revenue	2,182	10,989	13,171
Domestic message toll	2,342	11,018	13,360
Overseas message toll[a]	NA	499	576
Total message toll	2,419	11,517	13,936
Local private line	26	200	226
PBX and KTS revenue[b]	227	2,803	3,030
Miscellaneous and directory advertising (less uncollectibles)	159	989	1,148
Pay phone, service station, other local	85	376	461
Toll private line	107	1,280	1,387
WATS	80	1,427	1,507
Other toll	9		9
Total operating revenue	5,294	29,582	34,876

Type of Telephone	Telephones (millions)
	Industry Total
Residence main stations[c]	66.7
Business main and Centrex main[d]	9.4
PBX[e]	11.5
Key[f]	10.5
Extensions[g] (excluding PBX, Key)	48.3
Total[g]	146.4

Sources: Data for Independent Telephone Companies taken from United States Independent Telephone Association, *Independent Telephone Statistics,* Vol. 1, 1976 (for year 1975). Other sources as listed in footnotes; Total PBX and KTS revenue from Federal Communications Commission, *First Report,* FCC Docket No. 20003, Table III–1, September 23, 1976. Bell PBX and KTS revenue figure from "Historical Data Regarding Bell System and Non-Telephone Company Provided Terminal Equipment," Second Supplemental Response, FCC Docket No. 20003, Bell Exhibit 20–A, 1976.

[a]Figures for overseas message toll service from Federal Communications Commission, *Statistics of Communications Common Carriers.* F.C.C., Washington, D.C., for year ended December 31, 1975, Table 14, pp. 23.

[b]Total of $3.06 billion reduced by 1 percent to make coverage comparable to FCC plus USITA totals for other services. Independents figure obtained by subtracting Bell figure from total.

[c]This figure of 66.7 million represents the sum of 13.4 million independent main stations (United States Independent Telephone Association, *Independent Telephone Statistics,* Vol. 1, 1976) plus 53.3 million Bell System residential main stations. The number of Bell System residential main stations is 44.28 percent of the total, which is the figure for the Bell System from American Telephone and Telegraph, *Bell System Statistical Manual,* 1950–1975, p. 508.

[d]Total industry business phones figure of 37.7 million was estimated using data from the Federal Communications Commission, *Statistics of Communications Common Car-*

riers, 1975, and the USITA, *Independent Telephone Statistics*, op. cit. This figure of 37.7 million, less both 10.5 million Key telephones (see note below) and 11.5 million PBX phones, yields 15.7 million business mains, extensions, and Centrex. Sixty percent of this figure was used as an estimate of total industry business mains and Centrex mains. Sixty percent is approximately the proportion of business mains to total business phones, and Centrex mains to total Centrex phones.

eBell System PBX estimated using 15.8 million PBX from the FCC, *Common Carrier Statistics* less 5.4 million Centrex phones (AT&T, Bell System *Statistical Manual*, 1950-1975, May 1976).

fData for total industry Key telephones from Stanford Research Institute, "Market Forecasts for Terminal Equipment Areas," Bell Exhibit 19, FCC Docket No. 20003, Table 2.

gTotal industry company owned telephones excluding coin phones, from the FCC *Common Carrier Statistics* and the USITA, *Independent Telephone Statistics*. Extension phones were then estimated to be the remainder after all previous categories of telephones were subtracted from the industry total. The estimate for extensions is consistent with numbers of extensions from the FCC and USITA sources less an adjustment for key extensions and plus an adjustment for centrex extensions.

Illustrative local telephone rate revisions are shown in Table C–3. All of the structures shown include a charge for all calls and higher charges for peak period calls than for shoulder period calls and for off-peak calls. Structures A and C maintain a monthly business charge at least twice as large as the residential charge. There are two reasons for this, the first being historical continuity. The second reason is that businesses appear to receive roughly twice as many calls as residences.[2] It could be argued that usage-sensitive pricing should apply to both parties to a call. For simplicity this has not been done in the examples, but the higher monthly business charges have been used to reflect the fact that businesses on balance receive more calls.

Tables C–4 and C–5 show estimated initial prices and volumes of message toll telephone minutes. These initial prices and volumes are used with the elasticities and prices shown in Tables C–6a through C–6e to calculate volume and revenue under rate revision. As these tables show, the worst case in terms of total revenue occurs when elasticities take the lower of two alternative values; capacity is assumed to be just over 30 percent above initial average peak usage, and price is cut to the capacity constrained level.

Table C-3. Local Charges, Volume, and Revenue for Telephone Services (M = millions; B = billions)

Rate Structure	Item	Residential Access Charge[a]	Business Access Charge[b]	Phones[c]	Peak Calls[a]	Shoulder Calls[a]	Off-Peak Calls[a]	Total Revenue
A	Charge	$3/month	$10/month	$1/month	$0.06	$0.04	$0.02	
	volume/year	70.7M	10.7M	128.4M	96.7B	40.2B	28.3B	
	Revenue/year	$2.55B	$1.28B	$1.54B	$5.80B	$1.61E	$0.57B	$13.4B
B	Charge	$3/month	$6/month	$1/month	$0.08	$0.05	$0.02	
	volume/year	69.7M	10.7M	127.4M	90.4B	37.7B	26.6B	
	Revenue/year	$2.51B	$.77B	$1.53B	$7.23B	$1.89E	$.53B	$14.5B
C	Charge	$3/month	$12/month	$1/month	$0.10	$0.06	$0.04	
	volume/year	68.1M	10.7M	125.8M	83.8B	34.9B	24.6B	
	Revenue/year	$2.45B	$1.54B	$1.51	$8.38B	$2.09B	$.98	$17.0B

Source: Calculations by Charles River Associates, based on sources cited in footnotes below.

[a]For base volume of residential main stations see Table C–2. Residential demand is assumed to depend on the sum of access charge and for the charge per phone, as well as the call charges. Residential demand for connections is assumed to follow the relationship in B. M. Mitchell, "Optimal Pricing of Local Telephone Service," *American Economic Review* 68. (September 1978) p. 527, with prices adjusted upward by a factor of 1.324 to reflect increases from 1970 to 1975 (see American Telephone and Telegraph, *Statistical Report 1976*, pp. 2–3), The adjusted Mitchell relationship indicates increases over the initial level of connections of approximately 6 percent for structure A, 4 percent for structure B, and 2 percent for structure C.

[b]For base volumes see Table C–2. Business access charge levied on main stations and PBX trunks. The number of PBX phones per trunk in the USITA statistics is 7.7. A figure of 10 PBX phones per trunk is used to estimate the number of Bell System trunks, however, on the assumption that the Bell companies have more very large PBX systems, with a higher volume of intraoffice relative to outgoing calls. This yields approximately 1.3 billion PBX trunks based on 10.4 million Bell PBX phones and .24 million independent PBX trunks. Since this is a lower number of PBX trunks than would result from applying the USITA ratio the assumption is a conservative one from the standpoint of generating access charge revenue in the analysis.

Footnotes continued on following page.

Table C-3 (continued)

cVolume of phones represents the sum of residential main stations, business main stations (9.4 million), business main stations (9.4 million), and extension phones (48.3 million — see Table C-2), but excludes Key telephones and PBX phones.

dPeak usage is from 9 A.M.–5 P.M.; shoulder usage from 5 P.M.–11 P.M.; and off-peak usage from 11 P.M.–9 A.M. After time-of-day pricing is introduced, the following distribution of calls is assumed: 35 percent of residential and 80 percent of business calls pay a peak charge; and 40 percent of residential and 10 percent of business calls pay the shoulder charge. This call distribution is based roughly on the data in John K. Hopley, "The Response of Local Telephone Usage to Peak Pricing," in *Assessing New Pricing Concepts in Public Utilities*, Proceedings of the Institute of Public Utilities Ninth Annual Conference, Michigan State University, 1978.

The demand for residential calls is assumed to depend on the shoulder call charge and follows the Mitchell (1978, p. 527) relationship with prices adjusted as in note a. This yields 93.0 residential calls per month in structure A. 88.4 calls per month in structure B, and 83.7 calls per month in structure C. The demand for business calls is assumed to be 12 calls per day in rate structure A at $0.06 peak charge. Assuming a business call elasticity of −0.2 implies a business demand of 11.2 calls per day in structure B at $.08 peak charge, and 10.4 calls per day in structure C. The total number of business calls, from which the peak, shoulder, and off-peak calls are calculated, is estimated using 28.8 million total business mains, PBX, and Centrex, which is the total from FCC and USITA, op. cit., and assuming 250 business days per year. American Telephone and Telegraph, *Bell System Statistical Manual, 1950–1975* indicates around 190 billion local calls annually for all companies. (Page 802 of AT&T's Statistical Manual gives 206,561 million Bell and non-Bell conversations. Thus, using the proportion of total Bell System conversations that are local from page 803, yields (144,178 ÷ 154,903) × 206,561 = 192.3.) If residence phones currently average 120 calls per month (B.M. Mitchell, "Optimal Pricing of Local Telephone Service," Rand Corporation R-1962–MF, November 1976, p. 28) this total implies approximately 13 business calls per business day. It should be noted that the Bell figure for total calls is lower than figures indicated in Federal Communications Commission, *Common Carrier Statistics* 1975, and United States Independent Telephone Association, *Independent Telephone Statistics*, 1976. Again, using the lower figure is conservative with respect to the revenue analysis.

Table C–4. Initial Prices and Volumes of Message Toll Minutes

Time Period	Toll Minute Distribution[a]	Interstate		Intrastate	
		Price per Minute[b]	Volume (billion minutes)	Price per Minute[c]	Volume[d] (billion minutes)
8 A.M.–12 noon	24.2%	$0.33	7.8	$0.22	7.4
12 noon–5 P.M.	27.5	0.33	8.9	0.22	8.4
5 P.M.–11 P.M.	28.3	0.21	9.1	0.14	8.7
11 P.M.–8 A.M.	3.7	0.12	1.2	0.08	1.1
Weekend[e]	16.3	0.12	5.2	0.08	5.0
Total[f]	100%	0.25	32.2		30.6

Sources: Calculations based on sources cited in footnotes below.

[a]Calculated from 1975 New York Public Service study of New York Telephone Company toll minute distribution, reproduced in P. J. Berman and A. G. Oettinger, "The Medium and the Telephone: The Politics of Information Resources," Harvard University Program on Information Technologies and Public Policy, June 1976.

[b]Average price per minute of call in 197- to 292-mile range. See Federal-State Joint Board in Telephone Separations (1977), J.B.-49. p. 4. American Telephone and Telegraph, *Long Lines Statistics 1950–75*, p. 115, indicates that 43 percent of interstate calls fall within a 0–200-mile band and an additional 22 percent are in the 200–500-mile range. Thus the average length of haul for interstate calls appears to be in the 197–292-mile range.

[c]Assumed to be two-thirds of interstate average price as follows (see Table C–5). For the states surveyed the average day rates are $0.29 (first minute) and $0.19 (additional minutes) for the average length of haul. These rates are, respectively, 63 percent and 61 percent of the interstate rates used in note a. Note that these average price relationships are *not* for interstate and intrastate calls of equal distance but rather for the average length of haul within each category.

[d]Using $5.18 billion as intrastate revenue and the time-of-day volume and price distributions shown yields estimate of total intrastate minutes of 30.6 billion. Intrastate volumes calculated as follows:

$$\text{(Intrastate volume) } 5.18 = 0.22T(0.242 + 0.275) + 0.14(0.283T) + 0.08(.2T)$$
$$\text{where } T = \text{Total intrastate minutes}$$

[e]Includes Sunday evening.

[f]Total interstate minutes calculated from American Telephone and Telegraph *Long Line Statistics,* 1950–1975, pp. 107, 118. Using the time of day distribution and estimated average prices yields total estimated interstate revenue of $8.18 billion which is 61.2 percent of total message toll revenue shown in Table 2, "Telephone Statistics–1975." This estimate agrees quite well with Bell System data showing 61.0 percent of message toll revenue to be interstate and 39.0 percent intrastate. See AT&T "Monthly Report No. 4" (December 1975, Filed with the F.C.C.). Total volume of both interstate and intrastate minutes represents the estimated annual volume of

paid minutes. Other sources cite a figure of 75.7 billion interstate minutes for the year 1976 (see, for example, "Part 6 of Exchange Network Facilities for Interstate Access (ENFIA)," Cost Study Methodology and Summary of Results, Bell Tariff Filing, 1978, p. 131) which at a 12 percent growth rate would imply 67.6 billion in 1975— still considerably higher than our estimate of 32.2 billion minutes. However, 67.6 billion represents holding time minutes estimated for separations purposes. In the separations process the number of minutes are counted at both ends in order to divide toll revenue among the specific telephone companies involved. Thus this measure of the number of interstate or intrastate minutes is at least twice as high as the number of paid minutes (holding time minutes also include busies and incomplete calls, which paid minutes do not).

Table C–5. Daytime Intrastate Toll Rates for Selected U.S. Cities and States

State	Average Length of Haul[a] (miles)	Call to:	Miles[b]	Average First-Minute Price[c]	Average Additional Minute Price[d]
New York/Buffalo	58	Rochester	54	0.26	0.25
Georgia/Atlanta	61	Rome	60	0.22	0.22
Illinois/Chicago	36	Rockford	82	0.30	0.20
Maryland/Baltimore	36	Havre de Grace	30	0.32	0.18
Michigan/Detroit	40	Ann Arbor	40	0.22	0.22
Washington/Tacoma	54	Aberdeen	64	0.41	0.31
California/Los Angeles	65	Riverside	65	0.31	0.17
Iowa/Des Moines	40	Ames	33	0.31	0.16
Minnesota/St. Paul	60	St. Cloud	60	0.24	0.24
Oregon/Portland	55	Salem	50	0.24	0.23
Average Price Weighted by Number of Intrastate Messages[a]			—	0.29	0.19

Source: Calculations by Charles River Associates, based on sources cited below.

[a]Average intrastate toll length of haul, by state. Taken from FCC Federal-State Joint Board on Separations, Docket 20981: Data Request. Exhibit J–44, by state.

[b]Measured straight-line distance between the cities using Rand-McNally, *Road Atlas: United States, Canada and Mexico,* 1975, by state.

[c]Average first-minute day (8 A.M.–5 P.M.) rate calculated from white pages phone book of each city in column 1.

[d]Marginal minute (each additional minute), day (8 A.M.–5 P.M.) intrastate rate taken from white pages phone book of each city in column 1.

Table C–6a. Estimated Prices and Volumes of Message Toll Minutes Assuming Low Elasticity and Low Capacity

Time Period	Elas-ticity[a]	Price per Minute[b]	Volume (billion minutes)			Assumed Capac-ity[c]	Total Revenue (billions of dollars)
			Intra-state	Inter-state	Total		
8 A.M.–12 noon	−0.3	0.25	7.1	8.5	15.6	16	$ 3.9
12 noon–5 P.M.	−0.3	0.19	8.7	10.5	19.2	20	3.6
5 P.M.–11 P.M.	−1.1	0.14	8.7	13.9	22.6	24	3.2
11 P.M.–8 A.M.	−1.1	0.02	5.2	8.9	14.1	28	0.3
Weekend	−1.1	0.06	6.8	11.0	17.8	20	1.1
Total		0.135[d]	36.5	52.8	89.3	108	$12.1

Source: Calculated by Charles River Associates.

[a]Constant elasticity demand function: $q = \beta p^{-\epsilon}$ where β is a constant and $-\epsilon$ the assumed elasticity. Beta's were calculated using initial prices and volumes given in Table C–4.

[b]Prices estimated such that resulting volumes not exceed assumed capacity.

[c]Capacity assumed to be 4.0 billion minutes (annually) per available hour of time period. Initial peak (morning) volume average is 3.8 per hour. Weekend assumption is 12.5 available hours per day times 0.4 as many days as other periods equals 5.0 hours. Night capacity reduced to 7 available hours to allow buffer period between night and higher priced periods.

[d]A weighted average of the total revenue divided by the total number of minutes.

Table C–6b. Estimated Prices and Volumes of Message Toll Minutes Assuming Low Elasticity and Moderate Capacity

Time Period	Elas-ticity[a]	Price per Minute[b]	Volume (billion minutes) Intra-state	Inter-state	Total	Assumed Capac-ity[c]	Total Revenue (billions of dollars)
8 A.M.–12 noon	−0.3	0.13	8.7	10.3	19.0	20	$ 2.5
12 noon–5 P.M.	−0.3	0.11	10.3	12.4	22.7	25	2.5
5 P.M.–11 P.M.	−1.1	0.11	11.3	18.1	29.4	30	3.2
11 P.M.–8 A.M.	−1.1	0.01	11.1	19.0	30.1	35	0.3
Weekend	−1.1	0.05	8.4	13.5	21.9	25	1.1
Total		0.078[d]	49.8	73.3	123.1	135	$ 9.6

Source: Calculated by Charles River Associates.

[a]Constant elasticity demand function: $q = \beta p^{-\epsilon}$ where β is a constant and $-\epsilon$ the assumed elasticity. Beta's were calculated using initial prices and volumes given in Table C–4.

[b]Prices estimated such that resulting volumes not exceed assumed capacity.

[c]Capacity assumed to be 5.0 billion minutes (annually) per available hour of time period. Initial peak (morning) volume average is 3.8 per hour. Weekend assumption is 12.5 available hours per day times 0.4 as many days as other periods equals 5.0 hours. Night capacity reduced to 7 available hours to allow buffer period between night and higher priced periods.

[d]Weighted average of total revenue divided by the total number of minutes.

Table C–6c. Estimated Prices and Volumes of Message Toll Minutes Assuming Low Elasticity and Low Capacity, and Maintaining Initial Revenue Level

Time Period	Elas-ticity[a]	Price per Minute[b]	Volume (billion minutes)			Assumed Capac-ity[c]	Total Revenue (billions of dollars)
			Intra-state	Inter-state	Total		
8 A.M.–12 noon	−0.3	0.27	7.0	8.3	15.3	16	$ 4.1
12 noon–5 P.M.	−0.3	0.27	7.8	9.5	17.3	20	4.7
5 P.M.–11 P.M.	−1.1	0.14	8.7	13.9	22.6	24	3.2
11 P.M.–8 A.M.	−1.1	0.02	5.2	8.9	14.1	28	0.3
Weekend	−1.1	0.06	6.8	11.0	17.8	20	1.1
Total		0.154[d]	35.5	51.6	87.1	108	$13.4

Source: Calculated by Charles River Associates.

[a]Constant elasticity demand function: $q = \beta p^{-\epsilon}$ where β is a constant and $-\epsilon$ the assumed elasticity. Beta's were calculated using initial prices and volumes given in Table C–4.

[b]Prices estimated such that the resulting elastic period volumes not exceed assumed capacity, and total revenue not fall below $13.4 billion.

[c]Capacity assumed to be 4.0 billion minutes (annually) per available hour of time period. Initial peak (morning) volume average is 3.8 per hour. Weekend assumption is 12.5 available hours per day times 0.4 as many days as other periods equal 5.0 hours. Night capacity reduced to 7 available hours to allow buffer period between night and higher priced periods.

[d]Weighted average of total revenue divided by the total number of minutes.

Table C–6d. Estimated Prices and Volumes of Message Toll Minutes Assuming High Elasticity and Low Capacity

Time Period	Elas-ticity[a]	Price per Minute[b]	Volume (billion minutes)			Assumed Capac-ity[e]	Total Revenue (billions of dollars)
			Intra-state	Inter-state	Total		
8 A.M.–12 noon	−0.7	0.26	6.7	9.2	15.9	16	$ 4.1
12 noon–5 P.M.	−0.7	0.23	8.1	11.5	19.6	20	4.5
5 P.M.–11 P.M.	−1.3	0.15	8.2	14.1	22.3	24	3.3
11 P.M.–8 A.M.	−1.3	0.02	6.5	16.2	22.7	28	0.5
Weekend	−1.3	0.06	7.4	11.6	19.0	20	1.1
Total		0.136[d]	36.9	62.6	99.5	108	$13.5

Source: Calculated by Charles River Associates.

[a]Constant elasticity demand function: $q = \beta p^{-\epsilon}$ where β is a constant and $-\epsilon$ the assumed elasticity. Beta's were calculated using initial prices and volumes given in Table C–4.

[b]Prices estimated such that the resulting volumes not exceed assumed capacity.

[e]Capacity assumed to be 4.0 billion minutes (annually) per available hour of time period. Initial peak (morning) volume average is 3.8 per hour. Weekend assumption is 12.5 available hours per day times 0.4 as many days as other periods equals 5.0 hours. Night capacity reduced to 7 available hours to allow buffer period between night and higher priced periods.

[d]Weighted average of total revenue divided by the total number of minutes.

Table C–6e. Estimated Prices and Volumes of Message Toll Minutes Assuming High Elasticity and Moderate Capacity

Time Period	Elas-ticity[a]	Price per Minute[b]	Volume (billion minutes) Intra-state	Inter-state	Total	Assumed Capac-ity[c]	Total Revenue (billions of dollars)
8 A.M.–12 noon	−0.7	0.19	8.3	11.5	19.8	20	$ 3.8
12 noon–5 P.M.	−0.7	0.17	10.0	14.2	24.2	25	4.1
5 P.M.–11 P.M.	−1.3	0.12	11.0	18.9	29.9	30	3.6
11 P.M.–8 A.M.	−1.3	0.02	6.5	16.2	22.7	35	0.5
Weekend	−1.3	0.05	9.3	14.7	24.0	25	1.2
Total		0.109[d]	45.1	75.5	120.6	135	$13.2

Source: Calculated by Charles River Associates.

[a]Constant elasticity demand function: $q = \beta p^{-\epsilon}$ where β is a constant and $-\epsilon$ the assumed elasticity. Beta's were calculated using initial prices and volumes given in Table C–4.

[b]Prices estimated such that resulting volumes not exceed assumed capacity.

[c]Capacity assumed to be 5.0 billion minutes (annually) per available hour of time period. Initial peak (morning) volume average is 3.8 per hour. Weekend assumption is 12.5 available hours per day times 0.4 as many days as other periods equals 5.0 hours. Night capacity reduced to 7 available hours to allow buffer period between night and higher priced periods.

[d]Weighted average of total revenue divided by total number of minutes.

Table C–7 summarizes the rate structures and revenues derived from the local and toll tables. The local and toll structures can be paired to yield total revenue of approximately $26.6 billion. For instance, local price structure C would compensate for the adverse decline in revenues in toll version 5. Local version B matches the next to worst toll case, version 1, where elasticity and capacity are both low. Local version A, with a $0.04 per call shoulder charge, when matched with any of toll structures 3 through 5 would result in approximately the initial level of revenue.

Table C–7. Alternative Rate Structure Parameters and Expected Revenue for Local and Long Distance Telephone Service

Rate Structure[a]	Monthly Charges			Local Call Charges			Total Revenue (in billions)
	Residence Access[b]	Business Access[c]	Phones[d]	Peak	Shoulder	Off-Peak	
Initial	$7.43	$21.00					$13.2
A	3.00	10.00	$1.00	$0.06	$0.04	$0.02	13.4
B	3.00	6.00	1.00	0.08	0.05	0.02	14.5
C	3.00	12.00	1.00	0.10	0.06	0.04	17.0

Rate Structure[e]	Elas-ticity[f]	Capac-ity[g]	Message Toll Charge Per Minute					Total Revenue (in billions)
			Morn-ing	After-noon	Eve-ning	Night	Week-end	
Initial	—	—	0.28	0.28	0.18	0.10	0.10	$13.4
1	Low	Low	0.25	0.19	0.14	0.02	0.06	12.1
2	Low	Moderate	0.13	0.11	0.11	0.01	0.05	.9.6
3	Low	Low	0.27	0.27	0.14	0.02	0.06	13.4
4	High	Low	0.26	0.23	0.15	0.02	0.06	13.5
5	High	Moderate	0.19	0.17	0.12	0.02	0.05	13.2

Sources: Calculations based on sources cited in footnotes below.

[a]Initial Structure represents estimated average charges. Initial residence monthly charge is the figure from B. M. Mitchell, "Optimal Pricing of Local Telephone Service," *American Economic Review* 68:4 (September 1978): Table 3, adjusted upward by Bell Local service price index; see American Telephone and Telegraph, *Statistical Report 1976*, pp. 2–3. The figure for initial business monthly charge is a rough estimate for the charge per main station line that results after deducting from total revenue the estimated residential revenue at the initial monthly charge shown above, $2.0 billion in message charge revenue, $1.50 per month for extension phones, $31.50 per month for PBX trunks, and an assumed figure of $1.4 billion for installation charges and surcharges for items such as touch-tone service and luxury extensions.

[b]Structures A through C represent average charge per residence main station.

[c]Structures A through C represent average charge per business main station and PBX trunk.

[d]Phone category includes residential main stations, business main stations, and extension phones, but excludes key telephones and PBX phones.

[e]Initial structure is the estimated average price weighted by estimated volumes of interstate and intrastate minutes (see Table C–4). Structures 1–5 are taken from Table C–6.

[f]Low elasticity: −0.3 day, −1.1 evening, night, weekend. High elasticity: −0.7 day, −1.3 evening, night, weekend.

[g]Low capacity approximately 5 percent above present peak period average hourly minutes. Moderate capacity approximately 32 percent above present peak average.

In contrast to the Littlechild studies the rate levels shown here are much higher. This occurs because Littlechild seeks to cover total cost (1967 prices) of the best available technology, whereas the rate structures shown here seek to maintain initial total revenue (1975 prices). Nevertheless, there are similarities, as shown in Table C–8, where the average local call is assumed to last three minutes.[3] The relative prices of interstate and local usage charges are in the same range as Littlechild's results for all price structures except C–2. A similar result occurs in comparison with the Mitchell results for local residential prices;[4] structure C–2 is furthest out of line.[5]

Table C–8. Comparison of Price Structures with Littlechild and Mitchell Studies

Price Structure	Residence Monthly Rate[a]	Average Local Call Charge[b]	Average Interstate Price[c]	Ratio of Interstate to Average Local Price	Ratio of Monthly to Average Local Call Charge
Initial[c]	$7.43		$0.75		
A–3	4.00	$0.05	0.46	9.2	80
A–4	4.00	0.05	0.41	8.2	80
A–5	4.00	0.05	0.33	6.6	80
B–1	4.00	0.06	0.23	3.8	67
C–2	4.00	0.08	0.23	2.9	50
Littlechild (1970)		0.02	0.12	6.0	
Littlechild and Rousseau (1975)[d]		0.02	0.17	8.5	
Mitchell (1976)[e]	2.19	0.03			73

Source: Calculations based on sources cited below.

[a]For structures A–C this represents the sum of the access charge plus the equipment charge of one dollar.

[b]Per three-minute call. All local calls are assumed to be three minutes long. Average local call charges are a weighted average of peak, shoulder, and off-peak charges from Table C–3. Average interstate prices taken from Table C–6a through Table C–6e.

[c]Initial Residence Charge taken from Table C–7. Initial average interstate price is the weighted average of the interstate prices from Table C–4.

[d]Long-run Optimal prices resulting from surplus maximization subject to break-even profit constraint.

[e]Average of optimal prices for nine variations of costs and for demand conditions I.

Conclusion

The rate structures presented in Tables C–3, C–6, and C–7 show that it is possible to achieve the telephone system's current level of revenue with a simple nationwide pricing scheme that incorporates peak load and usage-sensitive pricing and toll rates that are lower than present levels and distance insensitive. Although they are not radical departures from historical practices, the structures shown here improve on present pricing in terms of simplicity, relationship to cost, and accessibility of telephone service to the public.

Qualifications

Several qualifying remarks to the above rate structure discussion should be made. They involve metering costs, peak pricing of local service, reliability of data, and distance-insensitive toll pricing.

Metering Current estimates of the total capital investment per subscriber line required for metering range from $30 in step-by-step offices, to $10 in crossbar offices, to negligible amounts in electronic offices. Currently around 10 percent of residential phones and 50 percent of business phones are metered.[6] Thus, universal metering would incur additional costs for around 75 million lines.

The revenue requirement used above does not reflect cost reductions due to decreased local calling and thus allows some room for increased cost due to metering. If the initial number of peak period calls were 39.0 billion[7] and the long-run marginal cost of such calls were $0.03,[8] the peak pricing structures shown in Table C–3 would yield cost savings of around $0.04 billion. This would cover annual metering costs of $5 per phone, which is the highest figure used by Mitchell in his analysis.

Local Peak Pricing Garfinkel[9] presents data showing an early evening peak for residence phone calls. In addition, Cohen[10] indicates an evening peak for local calls on some weekdays and a morning peak on others. Garfinkel cautions that peak pricing schemes must take into account the business versus residential mix in a particular jurisdiction to avoid the possibility of accentuating an evening peak and creating capacity and investment problems.

A nationwide peak surcharge on morning calls may thus not be desirable. One alternative that still maintains a simple structure would be no surcharge in areas with an evening peak. Since such areas are probably rural the impact on revenue calculations shown previously would most likely be minimal.

Data Reliability Although the data used above represent much effort, there are still inconsistencies between different sources. For instance, the FCC[11] and the U.S. Independent Telephone Association[12] indicate a total of over 250 billion local calls in 1975 for the entire telephone industry. The *Bell System Statistical Manual*,[13] however, indicates only around 190 billion local calls for all companies and a Bell System figure nearly 30 percent below the FCC's figure for the Bell System. Where possible, therefore, the attempt has been to choose the more conservative of conflicting data, in this example the lower local call figure.

Toll Pricing Distance-insensitive toll prices achieve rate simplicity and a move in the direction of what appears to be the future cost relationship. It should be noted, however, that most intrastate calls are over relatively short distances; the average length of haul of the states shown in Table C–5 is in the range of 35 to 65 miles. Uniform toll prices result in price increases for intrastate calls in many cases in Table C–4 and distance insensitivity contrasts both with present practices and Littlechild's results using 1967 cost data. Further investigation of the cost of relatively short toll calls would be helpful.

NOTES

1. Some shift of revenue, such as from private line and WATS to message toll, might be expected. Given the approximate nature of many of the assumptions used below, however, excluding the final categories in Table C–2 seems appropriate.
2. L. Garfinkel, "Usage Sensitive Pricing: Studies of a New Trend," *Telephony* 192 (February 10, 1975).
3. See Garfinkel, "Usage Sensitive Pricing," op. cit.
4. Since Mitchell optimizes only with respect to residential monthly and local call charges, his monthly charge may be lower than a broader scope optimization would produce.
5. An additional study by Jeffrey Rohlfs ("Economically-Efficient Bell-System Pricing," Bell Laboratories Economic Discussion Paper No. 138, January 1979), became available after the present report was substantially completed. The Rohlfs study uses estimates of marginal costs as well as elasticity estimates to calculate its optimal prices. Its major conclusions are:

(1) Economic efficiency can be increased by having separate charges for access and local usage. This subjects consumers' decisions whether to subscribe, as well as their usage decisions, to the discipline of appropriate prices. (2) Economically

efficient pricing would also have the following results: A subscriber who made no long-distance calls and rented no optional terminal equipment would pay approximately 80 percent more than at present. However, long-distance rates would be reduced to approximately half of current rates.

The conclusion that long distance rates would be reduced by approximately one-half agrees with the average results in our Table 3–6, for instance, but differs according to what time period during the day is considered, as seen by referring to Table 3–5. (Rohlfs does not address the possibility of changing the time-of-day pricing structure.) However, the Rohlfs conclusion about the increase in payment by customers that made no long distance calls does not agree with present results. For instance, in rate structure A an average residential customer making no long distance calls would pay $7.91 ($4.00 for access and phone plus 93 calls per month distributed among peak, shoulder, and off-peak prices) and in structure C, $9.78. These payments are, respectively, 6 percent and 32 percent above the initial monthly charge of $7.43, in contrast to the Rohlfs conclusion of 80 percent. (Rohlfs also does not consider the possibility of pricing business and residential access differently.)

Two features of the Rohlfs study appear to underlie the difference in conclusions. These are Rohlfs' use of lower (closer to zero) elasticity estimates and the inclusion of marginal cost data. Rohlfs concludes that his results are *not* very sensitive to changes in elasticity assumptions. It is not clear, however, over what range of elasticities this sensitivity is examined, and thus the difference in conclusions due to elasticity assumptions is not known. The inclusion of marginal cost, however, appears to require a substantial increase in local charges since the marginal cost of access alone in the Rohlfs study is $130 per year (p. 66). The prices in the Rohlfs study must cover marginal cost to be economically efficient.

6. Garfinkel, "Usage Sensitive Pricing: Studies of a New Trend," pp. 28–29.
7. Toll calls of 190 billion (see Data Reliability discussion below) and peak patterns indicated in Garfinkel (1975).
8. Littlechild, "Peak-Load Pricing of Telephone Calls" (1970) obtains roughly $.03 for day calls based on 1967 data.
9. Garfinkel, "Usage Sensitive Pricing: Studies of a New Trend," p. 28.
10. Gerald Cohen, "Measured Rates versus Flat Rates: A Pricing Experiment," General Telephone and Electronics Service Corporation, Rates and Tariff Department. Presented at the Fifth Annual Telecommunications Policy Research Conference. Airlie House, Va.: March 1977.
11. Federal Communications Commission, *Statistics of Communications Common Carriers,* December 31, 1975.
12. United States Independent Telephone Association, *Independent Telephone Statistics for the Year 1975: Vol. II,* July 1976.
13. American Telephone and Telegraph Company, *Bell System Statistical Manual 1950–1975,* May 1976, pp. 802–803.

Econometric Studies of Economies of Scale in Telephone Systems

Appendix D contains supplementary technical information on the econometric studies of economies of scale in telephone systems discussed and summarized in Chapter 4.[1] The major empirical results of eight such studies of economies of scale are presented in Table D–1 along with descriptions of the functional forms estimated, the estimation methods and data bases used, and the definition of the variable, if any, used to measure technological change. Parameters of aggregate production functions are estimated in seven of the studies. The parameters of a cost function for multiple outputs are estimated in the eighth study. Most of the studies deal with the U.S. Bell System, although three of the studies consider Bell Canada alone or in addition to the U.S. system.

The ordinary least squares (OLS) estimation method is used in most of the studies. An innovative estimation method, the canonical ridge, is used by Vinod in his 1975 and 1976 studies and in Bell Exhibit 60 in FCC Docket 20003. The canonical ridge estimation method is at a developmental stage. Its superiority to ordinary least squares has not been proved, and expressions for the variance of the canonical ridge estimator are not known.[2]

Table D–1. Summary of Econometric Estimates of Scale Elasticities for Telephone Systems as a Whole, Estimated from Aggregate Production or Cost Functions

Author, Year (Exhibit Number in FCC Docket 20003)	Reported Scale Elasticity (Confidence Limits)	Functional Form	Estimation Method	Data Base	Definition of Technological Progress Variable	Scale Elasticities from Estimated Equations of Similar Statistical Fit
Dobell, et al., 1972 (Exhibit 44)[a]	1.11 (significantly greater than 1.0 at 97.6 percent level)	Cobb-Douglas production function: $\ln Y = a + b\ln K + c\ln L + dT$	Ordinary least squares applied to logarithms of the variables	Bell Canada, annual data, 1952–1967	T = percent of station-to-station direct-dialed toll calls	
Vinod, 1972 (Exhibit 41)[b]	Increasing from 0.24 to 2.46 for individual years over the sample period	Multiplicative nonhomogeneous production function: $\ln Y = a + b\ln L + c\ln K + d\ln L\ln K$	Ordinary least squares applied to logarithms of the variables	U.S. Bell System, annual data, 1947–1970	Not applicable (comments that factor neutral technological change may be "inappropriate")	
Sudit, 1973 (Exhibit 43)[c]	From 1.13 to 2.00 for individual years in the study period, generally increasing in size over time	Additive nonhomogeneous production function: $Y = a + bL + cK + d\ln K + eK\ln L$	Ordinary least squares	U.S. Bell System, annual data, 1947–1971	Not applicable	
	From 1.09 to 1.76 for individual years, generally increasing over time	"	"	Bell Canada, annual data, 1951–1967	"	
	Increasing from 0.25 to 2.15 for individual years over the sample period	Multiplicative nonhomogeneous (see above, Vinod, 1972)	"	"	"	

Table D–1. *(continued)*

Study	Estimate	Functional form	Estimation method	Data	Technical progress proxy	
Mantell, 1974 (Exhibit 40)[a]	1.16 (1.03-1.29 at 95-percent level)	Cobb-Douglas production function with labor, plant, and factor-neutral disembodied technological progress proxy variable	Ordinary least squares applied to logarithms of the variables	U.S. Bell System, annual data, 1946–1970	Percent of long distance calls annually dialed direct	1.04 (uses a different labor variable)
	1.04-1.24	"	"	"	Combinations of percent of phones served by crossbar exchanges and average age of plant	
	1.17	Vinod's (1972) multiplicative nonhomogeneous production function with an exogenous technology variable	"	"	Percent of long distance calls annually dialed direct	
	1.00 and .98, depending on the labor variable used	"	"	"	Percent of phones served by crossbar exchanges	
	1.15	Linear form with output deflated by technology coefficient from Cobb-Douglas result: $Y = a + bL + cK$	Ordinary least squares	"	Not applicable	
	1.13	Sudit's (1973) additive nonhomogeneous production function with output deflated by technology coefficient from Cobb-Douglas result	Ordinary least squares	U.S. Bell System, annual data, 1946–1970	Not applicable	

Table D-1. (continued)

Author, Year (Exhibit Number in FCC Docket 20003)	Reported Scale Elasticity (Confidence Limits)	Functional Form	Estimation Method	Data Base	Definition of Technological Progress Variable	Scale Elasticities from Estimated Equations of Similar Statistical Fit
Vinod, 1975 (Exhibit 42)[e]	1.04 (0.91-1.17 at 95 percent level)	Reestimation of one of Mantell's regression equations that yielded a scale elasticity estimate of 1.04	Ordinary least squares applied to logarithms of the variables	U.S. Bell System, annual data, 1947–1971	Percent of long distance calls annually dialed direct	
	1.20	"	Weighted least squares assuming first order serial correlation (Cochrane-Orcutt)	"	"	
	1.17	Cobb-Douglas production function	"Canonical ridge" estimation (to "adjust" for multicollinearity among the independent variables)	"	"	
Vinod, 1976 (Exhibit 59)[f]	1.19-1.23	Additive separable joint translog type production function with two outputs (toll and local service), labor and capital inputs, and factor-neutral disembodied technical change	Canonical ridge estimation	U.S. Bell System, annual data, 1947–1973	A proxy constructed from R&D expenses deflated by the CPI. The construction allows deflated R&D to have a maximum impact after 6 years	

Table D–1. *(continued)*

No author, 1976 (Exhibit 60)[g]	For all results of this study, first value is when there is neutral technological change alone; second value when there is also factor augmenting change. Confidence limits are not reported: 0.78, 1.08	Cobb-Douglas production function	Ordinary least squares (OLS results generally include at least one insignificant coefficient)	U.S. Bell System, annual data, 1947–1973	Indexes of embodied and disembodied technological progress built up from constructed indexes for innovation in transmission, switching, and computer use
	0.74, 1.27	Multiplicative nonhomogeneous production function (see Vinod, 1972, above)	"	"	"
	0.98, 1.18	Additive nonhomogeneous production function (see Sudit, above)	"	"	"
	0.88, 1.31	Translog production function function: $lnY = a + blnL + clnK + dlnLlnK + e(lnL)^2 + f(lnK)^2 + glnT$	"	"	"

Table D–1. (continued)

Author, Year (Exhibit Number in FCC Docket 20003)	Reported Scale Elasticity (Confidence Limits)	Functional Form	Estimation Method	Data Base	Definition of -Technological Progress Variable	Scale Elasticities from Estimated Equations of Similar Statistical Fit
	0.79	"XP-3": $lnY = a + blnL + clnK + dlnT + elnLlnK + flnLlnT + glnKlnT$	Ordinary least squares (OLS results generally include at least one insignificant coefficient)	U.S. Bell System, annual data, 1947–1973	Indexes of embodied and disembodied technological progress built up from constructed indexes for innovation in transmission, switching, and computer use	
	1.64, 1.09	Cobb-Douglas	Canonical ridge estimation	"	"	
	1.64, 1.05	Multiplicative nonhomogeneous	"	"	"	
	2.08, 1.32	Additive nonhomogeneous	"	"	"	
	1.82, 1.14	Translog	"	"	"	
	1.34	"XP-3" (see above)	"	"	"	
Fuss and Waverman, 1977 (Not an exhibit)[h]	1952: 0.85 (0.64 − 1.06 at 95-percent level) 1964: 0.90 (0.62 − 1.17) 1975: 0.92 (0.59 − 1.25)	Translog joint cost function in a simultaneous demand and production system. Three outputs (local, toll, and the remaining services) and three inputs (labor, capital, materials) are used.	Iterative three stage least squares	Bell Canada, annual data, 1952–1975	Time trend to represent factor-neutral disembodied technological progress	

[a]Dobell, A. Rodney; Lester D. Taylor; Leonard Waverman; Tsuang-Hua Liu; and Michael D. G. Copeland. "Telephone Communications in Canada: Demand, Production, and Investment Decisions." *Bell Journal of Economics and Management Science* 3, 1 (Spring 1972): 175–219. (Also appears as Bell Exhibit 44 in FCC Docket 20003.)

[b]Vinod, H. D. "Nonhomogeneous Production Functions and Applications to Telecommunications." *Bell Journal of Economics and Management Science* 3, 2 (Autumn 1972): 531–543. (Also appears as Bell Exhibit 41 in FCC Docket 20003.)

[c]Sudit, Ephrain F. "Additive Nonhomogeneous Production Functions in Telecommunications." *Bell Journal of Economics and Management Science* 4, 2 (Autumn 1973): 499–514. (Also appears as Bell Exhibit 43 in FCC Docket 20003.)

[d]Mantell, Leroy H. "An Econometric Study of Returns to Scale in the Bell System." Staff Research Paper, Office of Telecommunications Policy, Executive Office of the President, Washington, D.C., February 1974. (Also appears as Bell Exhibit 40 in FCC Docket 20003.)

[e]Vinod, H. D. "Application of New Ridge Regression Methods to a Study of Bell System Scale Economies." Bell Exhibit 42, FCC Docket 20003, April 21, 1975.

[f]_____. "Bell System Scale Economies and Estimation of Joint Production Functions." Bell Exhibit 59, FCC Docket 20003 (Fifth Supplemental Response), August 20, 1976.

[g]American Telephone and Telegraph Company. "An Econometric Study of Returns to Scale in the Bell System." Bell Exhibit 60, FCC Docket 20003 (Fifth Supplemental Response), August 20, 1976.

[h]Fuss, Melvyn, and Leonard Waverman. "Multi-product Multi-input Cost Functions for a Regulated Utility: The Case of Telecommunications in Canada." Paper presented at the National Bureau of Economic Research Conference on Public Regulation, Washington, D.C., December 15–17, 1977.

The most common treatment of technological change in the studies is use of the percent of station-to-station direct-dialed toll calls as a proxy for technological progress in the telephone system. Other fairly simple variables used as proxies for technological change are the percent of phones served by crossbar exchanges and a simple time trend. Additionally, there are two attempts to develop and use more complicated proxies for technological change. Vinod, in his 1976 study, constructs a technological change variable from R&D expenses, and the authors of Bell Exhibit 60 in FCC Docket 20003 utilize new, disaggregate indexes of technological progress by which they attempt to allow for both embodied and disembodied technological change.

The results presented in this latter study are suspect on several technical grounds. Results are derived using ordinary least squares and canonical ridge estimation methods. The ordinary least squares results are suspect because the scale elasticities are calculated from parameters of capital and labor variables at least one of which is generally negative or insignificant. Second, the results from the canonical ridge estimations for the same functional forms are difficult to evaluate because, as previously noted, the statistical significance of parameters estimated by this method is not known. Finally, the way technological change is handled — whether it is entered only in a capital and labor augmenting form or whether a factor neutral disembodied form is also used — makes a substantial difference in the estimated scale elasticity, and the direction of the difference is not consistent across the two estimation techniques.

Particular definitions of the variables other than the technological progress variable are not reported in Table D–1. In general, the studies estimating aggregate production functions utilize a measure of constant dollar aggregate revenue less the cost of purchased inputs (i.e., value added) as the output variable. Capital input is generally measured as the constant dollar depreciated capital stock. The exception to this treatment occurs in Mantell's study. He uses a gross capital stock variable defined as the sum of annual capital increments, each of which is deflated by a price index appropriate to the year of purchase.[3] Labor input is defined as total skill-weighted man-hours except in Bell Exhibit 60 in which various augmentation procedures are performed on the value of labor hours of various components of the labor force. The augmentation procedures are supposed to allow for embodied technological progress. The authors of Exhibit 60 also perform augmentation procedures on some components of constant dollar capital accounts, but they do not make it clear whether these accounts are gross or net of depreciation.

Procedures for deriving the variables used in estimating aggregate production functions are not always clearly explained. This is especially the case in the studies from Bell Labs; confidentiality problems are sometimes noted. Where more than one way of deriving a data series is possible — the one reported example is Mantell's two ways of deriving his labor input variable — there are generally no clear selection criteria. None of the studies presents the data series used in the estimations.

Fuss and Waverman's joint cost function formulation uses indexes of aggregate output for toll, local, and "competitive"[4] services that are derived from constant dollar revenues for subaggregates of the three categories. For inputs the authors utilize skill-weighted man-hours, net value of capital in constant dollars, and a materials input variable derived from the cost of materials, services, rent and supplies, and indirect taxes.[5]

The point estimates of scale elasticity reported in Table D-1 (excluding the suspect results of Bell Exhibit 60) vary within roughly the same range when there is a change in any of the following: (1) the definition of the input variables other than the technological progress variable; (2) the estimation method; or (3) the definition of the technological progress proxy variable.

Because ordinary least squares estimates are sensitive to the large degree of multicollinearity present in the time-series data bases, changes in the way a particular variable is derived can have a large impact on the point estimate of the scale elasticity derived from aggregate production functions estimated by this method.

For example, Mantell reports the results of two regressions that differ only in that the labor input variable is calculated in two different ways. One regression implies that the scale elasticity is 1.16 and that this value is significantly different from unity at the 99-percent confidence level. The confidence interval at the 95-percent level is (1.03, 1.29). The second regression implies that the scale elasticity is 1.04. This estimate of the scale elasticity, however, has a confidence interval of (0.91, 1.17) at the 95-percent level.[6]

Differences in point estimates of scale elasticity of a size similar to those occuring when variable definitions change are also reported when alternative estimation techniques are employed. For example, Vinod in his 1975 paper reestimates the second of the two equations discussed above by his canonical ridge estimation technique and gets a scale elasticity estimate of 1.17. However, since an expression for the variance of his estimator is unknown, there is no way to test this estimate to determine if it is significantly different from unity.

Differences in point estimates of a similar size also result when different proxies for the technological progress variable are employed. For example, Mantell estimated a multiplicative nonhomogeneous production function by ordinary least squares that yielded a point estimate for the scale elasticity of 1.17 when percent of direct-dialed toll calls was used as the technological progess variable. When percent of phones served by crossbar exchanges was substituted as the technological progress proxy, the estimate of scale elasticity dropped to 1.00 and 0.98, depending on the derivation of the labor input variable included in the regression.

These variations in estimated scale elasticity make precise estimation of the degree of economies of scale impossible. There are in addition two potential misspecifications of equations in the various studies. Either serial correlation of the disturbance term or simultaneous equation bias would bias the estimates of scale elasticity.

Many of the results from ordinary least squares regressions of aggregate production functions including technological progress variables reported in the reviewed studies exhibit evidence of possible positive serial correlation of the disturbance term. Although serial correlation of the disturbance term does not bias coefficients estimated by ordinary least squares, it may be an indication that an important variable has been left out of the equation or that an included variable does not adequately represent the variable of theoretical interest. In particular, many of the ordinary least squares regressions in Bell Exhibit 60 have Durbin-Watson statistics that clearly indicate positive serial correlation of the disturbance term. Most other ordinary least squares regressions presented in Table D-1 have Durbin-Watson statistics in the indeterminant range. The exception is the regression for Bell Canada reported by Dobell and co-authors, which does not exhibit serial correlation by the Durbin-Watson test.

Results from OLS estimation of equations in which technological change variables are excluded have not been discussed in this appendix because such regressions consistently exhibit serial correlation. Indeed, proxies for technological change were first introduced to reduce serial correlation. Their inclusion has generally reduced, but not eliminated, serial correlation.

It should be noted that the two commonly used proxies for technological progress — percent of direct-dialed toll calls and percent of phones served by crossbar exchanges — will be of limited use in studies covering future years. They have tracked the major technological innovations of the 1950s and 1960s, but introduction of these innovations has been completed. A more general description of technological progress is required. Alternatively, additional "dummy" variables to pick up the effects of specific new innovations will be required if serial correlation is to be minimized.

The serial correlation remaining after a technological variable has been included could result from the joint effects of construction cycles and the econometrician's use of a capital stock variable in place of a theoretically superior but unavailable capital services variable. Specifically, if plant was built in anticipation of demand levels, as is suggested by the discussion in Bell Exhibit 58,[7] a capital stock variable would tend to overstate capital service inputs more than average during a period of heavy construction and output would tend to be overestimated during this period as well.

This proposition has not been tested empirically. To do so would require the econometrician to make a careful analysis of the timing of construction programs. The limited evidence is suggestive, however. Figure D–1 is a plot of residuals from an ordinary least squares regression of an aggregate Bell System production function with percent direct-dialed toll calls as the technological progress variable. The positive residuals in the figure tend to occur during the 1950s when direct distance dialing was introduced and during the late 1960s when solid state circuitry was introduced.[8]

Evidence with respect to simultaneous equation bias in the econometric studies of economies of scale in telephone systems is also limited but suggestive. The evidence is limited because only one study, that of Fuss and Waverman, utilizes a simultaneous estimation technique. It is therefore difficult to generalize from the fact that they find no conclusive evidence of economies of scale. However, Fuss and Waverman do address the important topic of modeling joint production in a simultaneous equations system that is a promising direction for empirical research on the question of economies of scale. And, unlike Vinod, they impose a structure on their equations derived from economic theory. When Vinod, in contrast, estimates a joint production function in his 1976 paper, he allows canonical correlation analysis to pick out the relationships between inputs and outputs.

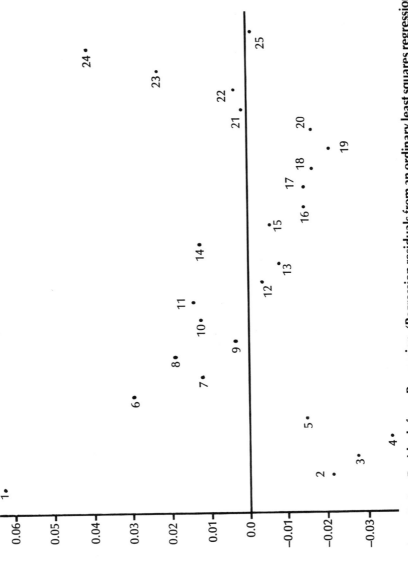

Figure D-1. Residuals from Regression. (Regression residuals from an ordinary least squares regression of an aggregate Bell System Production function with percent of direct-dialed toll calls as the technological progress variable. Data are from the 1946—1976 period.) *Source:* H. D. Vinod, "Application of New Ridge Regression Methods to a Study of Bell System Scale Economies," FCC Docket 20003, Bell Exhibit 42, April 21, 1975.

NOTES

1. The value of evidence from econometric studies of economies of scale in comparison to evidence from engineering and simulation studies and the practical problems in conducting econometric studies and interpreting their results are discussed in Chapter 4.
2. H. D. Vinod, "Canonical Ridge and Econometrics of Joint Production," *Journal of Econometrics* 4 (May 1976): 147–166. The possible superiority of canonical ridge estimates to ordinary least squares estimates in a mean square error sense and the author's inability to prove this superiority because expressions for the variance of the canonical ridge estimator are not known are discussed at p. 150.
3. He argues that depreciation does not impair the productivity of telephone system equipment, particularly when maintained at Bell's high standards.
4. The three subcategories contained in the "competitive" services category are other toll (private line, data communications, broadband, TWX), miscellaneous (consulting and other services), and directory advertising.
5. The source of their data is a Bell Canada submission to the Canadian Radio-Television Commission, Response to Interrogatories of the Province of Ontario, Item 101, 12 February 1977 (BCS). The net capital stock submitted is generated according to a process described in a 1970 memo by R. E. Olley, "Productivity Gains in a Public Utility — Bell Canada 1952 to 1967," paper presented at the Annual Meeting of the Canadian Economics Association, Winnipeg.
6. Mantell did not test for a significant difference of the scale elasticity from unity for the second regression. However, Vinod in his 1975 study reestimated the equation using Mantell's data, got the same scale elasticity estimate, and calculated the confidence interval.
7. "Principles of Network Engineering," FCC Docket 20003, Exhibit 58 (Fifth Supplemental Response), August 20, 1976, pp. 12–22.
8. Again, a complete analysis would require identification of the specific timing of the introduction of new technologies and the associated construction programs. It is noted in Bell Exhibit 60 that construction associated with the introduction of solid state circuitry was heavy after 1966 (AT&T, "An Econometric Study of Returns to Scale in the Bell System," Bell Exhibit 60, FCC Docket 20003 (Fifth Supplemental Response), August 20, 1976). Gideon Fishelson identifies 1956–1958 as the installation period for direct distance dialing in the United States ("Telecommunications, CES Production Function," *Applied Econometrics* 9 [March 1977]: 13). The first full-scale installation of direct distance dialing in Canada occurred in 1958, the same year in which heavy investment to upgrade long distance service was completed in that country (A. Rodney Dobell et al., "Telephone Communications in Canada: Demand, Production and Investment Decisions," *The Bell Journal of Economics and Management Science* 3 [Spring 1972]: 175–219).

Index

A

Accountability, political, and
  decreased regulation, 221n
ACCT. *See* Ad Hoc Committee for
  Competitive Telecommunications
Adams, W. J., 165
Ad Hoc Committee for Competitive
  Telecommunications (ACCT),
  45, 216
Administration, simplifying, 81, 88
Airline industry, regulation in,
  169-170
Allocative efficiency, static, 76-77
  83, 84
American Satellite Corporation (ASC),
  185
American Telephone & Telegraph
  (AT&T)
  creation of, 2
  cost calculations of, 11
  cost findings of, 10
  experienced growth rates for, 52
  financial projection scenarios for,
    50-67 *(see also* Financial
    scenarios)
  net income of, 53
  1975 costs, 5
  1975 revenues, 5
  projected financial performance for,
    64, 65, 67-68
Antitrust laws, and predatory pricing,
  213
Appropriability, question of, 161
Arrow, Kenneth J., 161
Arthur D. Little (ADL) study, 167
ASC. *See* American Satellite
  Corporation
Assets, in sheet ratios, 252
Average variable cost, 212
"Averch-Johnson effect," 20n, 163,
  213, 222n

B

Balance sheet ratios, 248-252

Balance sheets, 257-260
    actual vs. projected, 290-291, 293,
        294, 296-297
    financial ratios used in, 250
"Basic transmission" costs, 135, 138
    average, 142
    development possibilities in, 137
Baumol, William J., 103n, 163-164
Beginning-of-year (BOY) equity
    account, 259-260
Bell Canada operations, 127
Bell Exhibit, 60, 151n
Bell Labs, 158
Bell System. *See also* American
    Telephone and Telegraph
    cross-section cost estimates for, 180
    developing of, 1-2
    expenses of, 34, 35
    growth experience of, 39
    and incipient competition, 28
    and Kingsbury commitment, 27
    open competition with, 26 *(see also*
        Competition)
    operating expenses for, 178, 183
    operating revenues, 4
    revenues of, 34, 35, 93, 176, 301
Bell System Statistical Manual, 315
Berliner patent, 26
"Big John" cars, 169
Board-to-board principle, 27
Business telephone service
    distributed expenditures for,
        237-238
    pricing policy for, 77
    and rate setting, 13
    rates and revenues for, 227-229, 230
Business Week, 44

C

Cable television, 189-192
    coverage, 192
    growth experience of, 39
Canadian hearings, 28
Canadian National/Canadian Pacific
    Telecommunications, 108n

Capacity
    and alternative rate structure, 99
    and demand, 10
    in long distance analysis, 98
Capital, econometric data on, 132
Capped rate regulation, 209
Carrier-of-last-resort issue, 219n-220n
Cartelization, danger of, 123
Carterfone decision, 28, 36, 37, 42-43
    71n
"Cellular system," 187
Changes in financial position (CFP)
    statements, 251
Chase Manhattan Bank study, 146
Civil Aeronautics Board (CAB),
    169-170
Coal industry, regulation of, 168
Cobb-Douglas production function,
    318-322
Cohen, Gerald, 314
Comanor, William, 165, 166
Common costs
    classification as, 211
    in projected financial scenarios, 58
Communication policy. *See* Public
    policy
Competition
    aspects of, 218
    and AT&T's financial performance,
        66
    and Bell System, 68
    and business telephones, 44-45
    cream-skimming behavior of, 149n
    "fair," 105n
    forecasting impact of, 49-68
    and freedom of entry, 202
    and growth potential, 42
    and growth rate, 37
    historical, 23-37
    impact on costs, 143-146
    incipient, 28
    and local services, 157
    open, 26
    and price reductions, 69-70n
    in projected scenarios, 54
    and regulation, 8
    services vulnerable to, 72n

Competitors, AT&T, revenues of, 47.
    See also Independents
Concentration
    and measures of innovativeness, 167
    and research effort, 165
Consumer Expenditure Survey, 253
Consumers
    average telephone expenditure of,
        231-233, 234
    of business telephone service, 237
    and predatory pricing, 212
Continuity, historical, 81, 88-89
"Contribution," 3-5, 18n
    and local service, 15
    of service categories, 13
Cooperatives, and growth rates, 31
Cost analysis, benefits of, 13
Costs
    basic transmission, 135, 137, 138,
        142
    classification of, 211
    of FCC commissioners, 145
    impact of competition on, 218n
    "institutional," 14
    for long distance transmission, 135
    of regulation, 143-146
    traffic-sensitive vs. nontraffic-
        sensitive, 27
Cost trends, in late 19th century, 34
Cream skimming, 204, 205
Cross-section data, in econometric
    analysis, 127
"Cross-subsidization," 3-5, 18n, 21n
Customer premise equipment, 46

D

Datran, 45-46, 80
Debt-to-capitalization ratio, in
    projected financial scenarios, 57,
    65
Debt-to-debt-plus-equity ratio, 67
deButts, John, 12
Deferred charges, in balance sheet
    ratios, 252
Deferred credits, in balance sheet

ratios, 252
DeForest, Lee, 160
Demand
    and alternative rate structure, 99
    and capacity needs, 10
    cross-elasticities of, 149n
    local residential, 96
    and pricing policy, 118, 119
    residential, 109n
Demsetz, Harold, 162
Density, economies of, 199n
Depreciation policies, 251
Depreciation rate, in projected
    financial scenarios, 53
Deregulation, argument against, 156
Dial service, introduction of, 160
Dividends, and income statement
    ratios, 248
Dobell, W., 96, 318
Dupuit, Jules, 79
Durbin-Watson test, 326

E

Earnings before interest and taxes
    (EBIT), 253, 254, 256
Econometric estimates, of scale
    elasticities for telephone systems,
    318-323
Econometrics, of economies of scale,
    125-134
Economies of density, 199n
Economies of fill, 152n
Economies of joint production, 220n
Economies of scale, 125-134
    costs of regulation, 115-119
    defined, 148
    econometric studies of, 317
    vs. "economies of fill", 152n
    in long distance telecommunications,
        134-143
    in multiproduct case, 179
    in natural monopoly, 112
    and regulation, 114, 116, 118
    slight, 115
    and technological change, 148, 150n

in telecommunications industry, 18
Economy, U. S., growth experience of, 39
Elasticities, price, estimates of, 300
Elasticity
  scale, 128-129, 318-322
  scale vs. average, 137
Electronic communication, conversion of paper media to, 70n
Electronic households, 189
Electronic media, growth rate for, 38, 39
Embedded direct cost (EDC) studies, 14, 49, 53
  and implicit return of capital, 253
  modifications to, 254
  scenario results of, 61, 62
End-of-year (EOY) equity account, 259-260
ENFIA proceedings, 28
Entrants, new
  restriction on, 172
  and technological change, 164
  and telephone system costs, 144
Entry
  as alternative to regulation, 124
  as alternative technology, 122
  argument for controls on, 204
  freedom of, 202, 219n
  prevention of uneconomic, 79-80, 87-88
Equipment investment, and balance sheet ratios, 248-252
Equity
  and alternative rate structure, 91
  in pricing policy, 77-78, 85-86
Execunet service, 46
  competition with, 72n
  proceedings on, 28
Externalities, call related, 103n
Expenditure, telephone, average annual, 231-233

F

Families. *See also* Residential service

telephone service expenditures of, 238
total annual telephone expenditures by, 240
welfare change for, 241n-242n
Farmer cooperatives, on rural areas, 173-174
Federal Communications Commission (FCC), 27
  bandwidth expansion allowed by, 187
  costs of, 145
  data base of, 199n
  Docket 20003, 166-168, 196
  and ratemaking policy, 10
  regulation expenses of, 146
Federal tax, and income statement ratios, 248
Financial ratio
  historical, 245-253
  testing validity of, 268
Financial scenarios, 55, 261
  competition in, 54
  debt-to-capitalization ratio in, 57, 65
  of Embedded Direct Cost Study, 61, 62
  historical A, 270-271
  historical B, 262, 272-273
  historical C, 262, 274-275
  interest coverage ratio in, 56, 64
  no-inflation, 63, 64, 65, 67-68, 262-263, 276-277
  nominal, 63, 64, 65, 67-68
  nominal A, 263, 278-279
  nominal B, 264, 280-281
  nominal C, 264, 282-283
  nominal D, 265, 284-285
  nominal E, 265-268, 286-288
  projected, 50-67
  return on equity, 63, 64, 65, 67-68
  telephone rate indexes for, 60
Financial statements, 253-260, 269
Fisher Franklin M., 161
Florida, pricing policy in, 85-86
Forbearance regulatory principle, 208
Fuss, Melvyn, 325, 327

G

Gabel, Richard, 160
Garfinkel, L., 314
General Telephone and Electronics,
    phone production of, 71n
Georgia, pricing policy in, 85-86
Globerman, Steven, 166
Government regulation, in natural
    monopoly, 113-114. *See also*
    Regulation
Great Depression
  Bell System during, 27
  impact on telecommunications
    industry, 174
Gross national product, and growth
    rates, 39
Growth
  forecasting impact of, 49-68
  historical patterns of, 23-27, 29
  potential, 38-42
  relative rates for, 30-38
"Growth dividend," 68
Growth rates
  and Carterfone decision, 37
  combining, 269n
  and incipient competition, 36

H

High-technology industry, innovation
    in, 166-167
Holding time, costs of, 108n
Horizontal equity principle, 209
Household effects, of alternative rate
    structure, 225
Households, telephone usage by, 41.
    See also Residential service

I

Illinois, pricing policy in, 85-86
Income
  and annual telephone expenditures,
    240
  and average telephone expenditures,
    234
    and telephone expenditure, 232-233
Income statement ratios, 246-248
Income statements, 254-257
  actual vs. projected, 289, 292, 295
  financial ratios used in, 247
Incremental costs, 8
Independents
  acquisition of, 26
  cross-section cost estimates for, 180
  and growth rates, 31
  1975 revenues, 93
  operating expenses for, 178, 183
  revenues for, 176, 301
  in rural areas, 173-174
Independent Telephone Association,
    U.S., 315
Indiana, pricing policy in, 85-86
Industry. *See also* Telecommunica-
    tion industry
  cost structure of, 116
  predatory pricing, 214
Inflation
  and financial self-sufficiency, 87
  in projected financial scenarios, 52
  and regulation, 8
  and technological innovation, 16-17,
    163-164
Interest coverage ratio, in projected
    financial scenarios, 64
Interstate Commerce Commission
    (ICC), 6
  accounts system instituted by, 26
  impact ratesetting by, 168-169
  and railroad industry, 169
Interstate rates, 99. *See also* Rate
    structures
Interest coverage ratio, projected in
    financial scenarios, 56
Installation, local loop and instrument,
    107n
Investments, in balance sheet ratios,
    252

J

Joint costs, 211. *See also* Costs

K

Kamien, Morton I., 162-163
Kingsbury agreement, 27, 31
Klevorick, Alvin K., 164

L

Labor, econometric data on, 132
Laspeyre index, 226
Lease policy, of AT&T, 44
Liabilities, in balance sheet ratios, 252
"Lifeline" rates, 76
Littlechild, S. C., 101, 109n
Littlechild study, 313
Loading coil, 160
Loans, telephone, 198n
Local exchange rates, sample of, 85
Local exchanges, and technological advance, 160
Local loop, bypassing, 185-187
Local peak pricing, 314-315. *See also* Pricing
Local telephone service
  alternative structure for, 95, 187-192
  alternative rate structure parameters for, 312
  and competition, 157
  costs of, 5
  and elasticity of demand, 15
  and Kingsbury commitment, 27
  rate regulation in, 208, 209
  rate structure for, 11-12
Local telephone service rates
  alternatives to, 93-96
  forecasting impact of competition on, 49-68
  and market growth, 49-68
  revenues from, 303-304
  volume and, 303-304
Long distance market, competition in, 167
Long distance rates, 2
  alternatives to, 96-101
  and pricing policy, 91 *(see also* Pricing policy)

regulation of, 3
Long distance telephone service, 186
  alternative rate structure for, 97, 312
  competition in, 14, 45-46, 68, 157
  contribution of, 12
  economies of scale in, 134-143
  growth in, 40
  growth rate for, 147
  high markups on, 157
  and Kingsbury commitment, 27
  and option value, 18n-19n
  through satellites, 140
  and technological innovation, 192-193
Long lines plant, book cost of, 139
Long Lines System, diversion of traffic from, 140

M

MacAvoy, Paul W., 168
Mantell, L., 130
Mantell's scale, statistical significance of, 130
Mantell's study, 318, 324-325
Marginal costs, 212
Market entry, and telephone system costs, 145
Market growth. *See also* Growth
  and economies of scale, 18
  during monopoly, 38
  and technological change, 17
Market power, in natural monopoly, 119-122
Market structures
  preferred, 184-193
  and technological innovation, 158
  and theoretical analysis, 161-164
Market test, 12
MCI, 46
Media, growth rate for, 38, 39
Message toll minutes
  estimated prices for, 307, 308, 309, 310, 311
  estimated volumes for, 307, 308, 309, 310, 311

initial prices and volumes for, 305
Message toll service (MTS), 48. *See
  also* Toll service
  competition with, 72n
  and entry controls, 205
  revenues generated by, 187
Metering, 314
Microwave systems, costs of, 151n
Mitchell, Bridger, 101, 109n
Mitchell study, 313
Modular jacks, 44
Monopoly, natural, 111, 112-125
  earnings per telephone during, 34, 36
  and economies of scale, 115-119
  and effectiveness of regulation,
    122-123
  and growth rate, 37
  history of, 25-26, 148n
  impact of registration on, 222n
  market power in, 119
  and technological innovation, 159
  of telecommunications industry,
    17-18
  and technological innovation, 162
Monopoly extortion, 213
Monopoly market, erosion of, 147-148
Morgan, J. P., 26, 69n
MSN. *See* Multiple Supplier Network
  study
MTS. *See* Message toll service
Multiple Supplier Network Study
  (MSN), AT&T's, 10, 105n,
  140-143
Multiplexing, scale economies in, 135
Multiplexing equipment
  costs of, 152n
  investment costs for, 139

N

Newspaper industry, growth
  experience of, 39
Northern Telecom, Inc., 71n

O

Operating costs

Bell vs. independent, 178
  in income statement ratios, 246
Optic fiber technology
  attraction of, 201n
  costs of, 108n
  potential of, 188
Option value, 18n-19n
Ordinary least squares (OLS)
  estimates, 317, 324, 326
Other common carriers (OCC), 21n
  cost comparisons for, 141
  and lower-density routes, 142
Overhead costs, in projected financial
  scenarios, 58
Overinvestment, wasteful, 68

P

Paasche index, 226
Packet communications technology,
  152n-153n
Parsimony, principle of, 202
P&E categories, 253
Patents
  monopoly on, 25-26
  and technological innovation, 164,
    165
Peak load pricing, 90, 92
Penetration rate, and market growth,
  30, 31
Petroleum industry
  inefficiency in, 171
  production controls in, 170
Phillips, Almarin, 164, 165
Plant investment, and balance sheet
  ratios, 248-252
Policy instrument, regulatory lag
  as, 163
Political accountability, and decreased
  regulation, 221n
Price index, and welfare change,
  238-240
Pricing
  constrained by alternative
    technology, 121
  and duration of calls, 108n
  local peak, 314-315

monopoly, 118, 119
peak load, 90
predatory, 210-217
and telephone service rates, 3
usage-sensitive, 92, 108n
value-of-service, 6
Pricing policy
 and administrative simplicity, 81, 88
 alternative rate structures, 89-101
  102-103
 and elasticity of demand, 118, 119
 equity in, 78, 85-86
 and financial self-sufficiency, 86-87
 goals of, 75-81, 101-102
 and historical continuity, 81, 88-89
 Littlechild study, 101, 109n
 Mitchell study, 101, 109n
 and prevention of uneconomic entry,
  79-80, 87-88
 restructuring of, 17
 and static allocative efficiency,
  76-77, 83, 84
 and technological change, 81, 88
 and universal service, 75, 82-83
Private line service, 92, 105n
 cost of, 141
 and entry controls, 205
 interstate, 11
Probe Research Inc., 253
Production process, in econometrics,
 126
Productivity change, and technological
 innovation, 164
Profit before tax (PBT), 256
Profitability, and increased
 competition, 49
Profits, and rate-of-return regulation, 8
Profit seeking, and predatory pricing,
 214
Public policy
 and alternative rate structure, 91
 domestic telecommunications policy,
  217-218
 and effectiveness of regulation, 123
 and entry controls, 204-206
 ·and freedom of entry, 203
 and predatory pricing, 210-217

and rate regulation, 207-210
and uneconomic entry, 210-217
and universal service, 76, 103n

R

Railroad industry
 and pricing policy, 79
 regulation in, 156, 157, 169
Ramsey, F. D., 79
Ratemaking
 and elasticity of demand, 14
 goals of, 75-81
 and market growth, 33
 policy issues for, 10-18
 value-of-service, 6-7, 79, 104n
Rate-of-return regulation, 8, 9
Rate structures, 172-183
 A-5, 227-229, 230, 234, 236, 239
 alternative, 89-101, 102-103,
  225-243, 299-316
 and Littlechild study, 313
 for local service, 93-96
 for long distance service, 96-101
 and Mitchell study, 313
 and technological change, 7-8
 and welfare change, 238-241
Recession, and market growth, 36
Regulation, rate, 207-210
 alternatives to, 124-125
 vs. competition, 143-146, 147
 and conventional wisdom, 201-202
 cost of, 117, 149n
 costs and benefits of, 113
 and cost minimization, 20n
 direct costs of, 147
 effectiveness of, 122-123
 indirect costs of, 153
 inertia of, 117-118
 justification of, 146, 172
 of monopoly service, 222n
 in natural monopoly, 113-114
 in petroleum industry, 170
 and preferred market structure, 184
 problem of, 3-10
 rate-of-return, 8, 9, 171

social benefits of, 155
and technological innovation,
    155-156, 158-159, 168-172,
    193-194
Regulatory commissions, 168
Research and development
and competition, 28
impact on market structure, 165
Residence main stations, 106n
Residential service
average annual expenditure on,
    231-233, 234
average toll expenditures, 235-237
extension phones, 242n
local expenditure for, 231-235
rates and revenues for, 227-229, 230
Revenues
and alternative rate structures, 100
Bell System vs. independents, 301
for local service, 303-304
operating, 177
Revenues per station, and growth
    rate, 31, 33-34
Rohlfs, Jeffrey, 315n
Rohlfs study, 315n-316n
Rural areas
independents in, 174 *(see also*
    Independents)
provision for service in, 193
rate structure for, 172-183
and revenues per telephone, 176
Rural Electrification Act of 1936,
    174, 193
Rural Electrification Administration
    (REA), 173
loans of, 198n
and small companies, 193
on toll revenues, 175
Rural Telephone Bank, 174

S

Satellite systems, 140
switching facilities for, 152n
Satellite Data Exchange Service
    (SDX), 185, 186

Scale economies. *See* Economies of
    scale
SCCs. *See* Specialized common
    carriers
Scenarios. *See* Financial scenarios
Scherer, F. M., 165, 166
Schwartz, Nancy L., 162-163
Schumpeter, Joseph, 161, 163
SDX. *See* Satellite Data Exchange
    Service
Service. *See also* Local telephone
    service; Long distance telephone
    service; Private line service
and pricing policy, 78
Service categories
contribution of, 13
cost analysis for, 14
Sloss, James, 168
Small Business Administration, 215
Small Business Investment
    Corporations, 215
Smith v. *Illinois*, 27
Solicitation calls, 83
SPCC, 46
Specialized common carriers (SCCs),
    28
competition with, 72n
hearings on, 28
landline segment of, 72n
potential growth for, 46
types of, 45
Stability, market, and regulation, 157
Stanford Research Institute (SRI),
    13, 166-167
Station-to-station concept, adoption
    of, 28
Statistical methods, 133
Steady market growth, 27-28
Stigler, George, 164
Strowger Switch, 160
Sudit, Ephrain F., 318
Surcharge, on touch-tone dialing, 83
Sustainability argument, 104n-105n,
    149n, 186, 206, 219n
Switching equipment
costs of, 135
investment costs, 138-140

T

Tandem dial, 198n
Tariffs, telephone
  and administrative simplicity, 88
  equity in, 85-86
  and financial self-sufficiency, 86-87
  and historical continuity, 88-89
  and prevention of uneconomic entry,
    87-88
  and public policy, 91-93
  and static allocative efficiency,
    83, 84
  structure of, 81-89
  and technological change, 88
  and universal service, 82-83
Technological innovation, 81, 88
  as alternative to regulation, 124
  econometric data on, 132-133
  and economies of scale, 148, 150n
  in electronic communication, 38
  and freedom of entry, 220n
  historical record of, 159-160
  impact of, 164-168
  optic fiber technology, 188
  in long distance service, 192-193
  market structure and, 158-172
  and regulation, 9, 155-156, 193-194
  suppression of, 171-172
  theoretical analysis of, 161-164
Telecommunications industry
  annual growth of, 70n
  developing of, 1-2
  econometric evidence of economies
    of scale in, 125-134
  economies of scale, 111, 319
  entry controls, 204
  freedom of entry in, 20
  growth rates for, 51
  historical perspective on, 23-37
    acquisition of independents, 26
    growth period, 37
    incipient competition, 28
    Kingsbury commitment, 27-28
    monopoly period, 25-26
    open competition, 26
    steady growth, 27-28

increased competition in, 66
  as national monopoly, 17-18
  point estimates of scale elasticity in,
    129
  potential growth for, 38-42
  universal service achieved by, 82-83
Telecommunications Industry Task
    Force, 105n
Telenet, 46, 142
Telephone rate indexes, in projected
    financial scenarios, 60
Telephones
  business 44-45 *(see also* Business
    service)
  customer-owned, 44
  earnings per, 34-37
  growth in numbers of, 32, 73n
  and operating revenues, 177
Telephone service. *See also* Service
  average annual expenditure on,
    231-233
  price elasticities for, 300
Television industry
  cable, 189-192
  growth experience of, 39
Telex rates, 87
Telex services, 192
Temin, Peter, 161
Time-phased maximum rate
    regulation, 209
Time-series estimates, 150n
  and cross-section cost results, 182
  in econometric analysis, 127
Toll pricing, 315
Toll rates. *See also* Long distance rates
  categories dedicated to, 198n
  daytime intrastate, 306
  distance-insensitive, 90, 92
Toll service (MTS). *See also* Message
    toll service
  rate regulation in, 209
  residential expenditures on, 235-237
  revenue from, 175
Touch-tone dialing, surcharge on, 83
Transmission systems
  investment cost, 136
  volume growth for, 41

Trucking industry, regulation in, 156, 157, 184
TWX, 192
Tymnet, cost efficiency of, 142

U

Uncertainty, in market structure, 163
Universal service, 75-76, 83-83, 103n
Usage-sensitive pricing, 90, 92, 108n
Utilities, and pricing policy, 77

V

Vacuum tube, 160
Value-of-service rates, 6-7, 79, 104n
Vail, Theodore, 26-27, 30
Video transmission, 40, 70n
Vinod, H. D., 317, 318, 320, 324, 327
Volume, for local service, 303-304

W

Wages, and telephone service rates, 3. *See also* Income
Wall Street Journal, The, 46
WATS. *See* Wide area telephone service

Waverman, Leonard, 96, 325, 327
Welfare analysis, 231, 242, 243
Western Electric
  competition of, 45
  phone production of, 44
Western Union
  local loop network of, 192
  patent challenge of, 25
Wide area telephone service (WATS), 92
  competition with, 72n
  and entry controls, 205
  and pricing policy, 87
  rate regulation in, 209
  revenues generated by, 187
Widener, Peter, 26, 69n
Wilson, R. W., 166
Windfall profits, tax on, 171
Wire, ratio of aerial to underground, 179
World War I, Bell System during, 70n
World War II, impact on Bell System, 27, 36

X

Xerox Telecommunications Network, 186
Xerox telecommunications service, proposed, 187

About the Authors

John R. Meyer, 1907 Professor in Transportation, Logistics and Distribution at the Harvard Graduate School of Business Administration, has been a faculty member at Harvard University since 1955. He also teaches in the Department of Economics and at the John F. Kennedy School of Government. Additionally, he is on the Board of Directors of Charles River Associates. Professor Meyer is a Fellow of the American Academy of Arts and Sciences and the Econometric Society, and a member of the Council on Foreign Relations, the American Economic Association, the American Statistical Association, and the Economic History Association. Professor Meyer, whose research interests include the future of the New England economy and industrial location decisions, is preparing a new book on urban transportation problems and solutions.

Robert W. Wilson, a Senior Research Associate at Charles River Associates, received a Ph.D. in economics from Yale University and a bachelor's degree in physics from the Massachusetts Institute of Technology. He has published articles on the economics of technological change in professional journals and is coauthor of a book on innovation, competition, and government policy in the semiconductor industry. He has also coauthored government studies dealing with the price discrimination statutes and with regulation of the petroleum industry. Dr. Wilson has conducted economic research in support of antitrust litigation and served for two years with the Antitrust Division of the U.S. Department of Justice.

M. Alan Baughcum received his Ph.D. in economics from the University of North Carolina at Chapel Hill and his undergraduate degree in mathematics from Emory University. The Director of Telecommunications at Charles River Associates, he has had principal responsibility for the conduct of contract research and litigation support for antitrust and regulatory proceedings in telecommunications. He has additional experience in the economics of industrial organization, agriculture, transportation, and occupational and professional licensure. His primary interest is analyzing the ways in which government policy influences and is influenced by the private sector. Prior to joining the CRA staff, Dr. Baughcum was a member of the Department of Economics of the University of North Carolina at Chapel Hill and North Carolina State University. In addition, he was an economist with the North Carolina Department of Justice and served as an expert witness in regulatory hearings on the cost of equity capital to telephone utilities.

Ellen Burton, Senior Research Associate at Charles River Associates, has analyzed the impacts of government policy on employment, operations, investment, and innovation in telecommunications, energy, and transportation industries. The coauthor of a book on the impacts of trade policy on the U.S. automobile industry, she has research and management experience that includes designing models of components of coal, petroleum, and uranium supply and of the market penetration process for new energy-using technologies. Ms. Burton is a candidate for the Ph.D. in economics at the Massachusetts Institute of Technology.

Louis Caouette is a Senior Research Associate at Charles River Associates. He holds an M.B.A. from the Harvard Graduate School of Business Administration, and an M.A. in economics and a bachelor's degree in electrical engineering from Laval University in Quebec. Mr. Caouette has performed financial, technical, and economic analyses for the semiconductor, broadcasting, newspaper publishing, chemical, mineral, and financial service industries, among others.